Current Topics in Microbiology 100 and Immunology

Editors

T Cell Hybridomas

A Workshop at the Basel Institute for Immunology

Organized and Edited by H. v. Boehmer, W. Haas,
G. Köhler, F. Melchers and J. Zeuthen
With the Collaboration of S. Buser-Boyd

With 52 Figures

Springer-Verlag
Berlin Heidelberg New York 1982

Editors

H.V. Boehmer, W. Haas
G. Köhler, F. Melchers
J. Zeuthen

Basel Institute for Immunology
Grenzacherstrasse 487
CH-4005 Basel
Switzerland

ISBN-13: 978-3-642-68588-0 e-ISBN-13: 978-3-642-68586-6
DOI: 10.1007/978-3-642-68586-6

Library of Congress Catalog Card Number 15-12910
Softcover reprint of the hardcover 1st edition 1982

Typesetting: FotoSatz Pfeifer, Germering
Printing and binding: Universitätsdruckerei H. Stürtz AG. Würzburg
2121/3321-543210

Table of Contents

Indexed in Current Contents

Preface

For more than ten years cell fusion techniques have been applied in studies on various lymphocyte functions. Ig expression was first studied in hybrids obtained by fusing myeloma cells with fibroblasts (1) or lymphomas (2), both of which do not produce Ig, and with Ig-producing myelomas (3) or human blood lymphocytes (4). Köhler and Milstein (5) fused a myeloma with spleen cells from immunized mice. Up to 10% of the hybrids obtained secreted antibodies specific for the immunizing antigen. This suggested that plasma cells preferentially fused with the myeloma cells, a finding which was of enormous practical value. It was found that both B and T lymphocytes could be fused with the T cell tumor BW5147, which is however not permissive for Ig synthesis (6). A very large number of T cell hybridomas were generated by fusing BW5147 with cell populations containing in vivo or in vitro activated cells (7). The hybrids showed no specific T cell functions and binding assays for T cell receptors were not available. In particular, no hybrids were obtained which expressed specific cytolytic activity that could be tested in short-term ^{51}Cr-release assays (8). However, the frustrations expressed about these failures, published in January, 1978 (9), were relieved by Taniguchi and Miller's publication a few months later of T cell hybridomas producing antigen-specific suppressor factors (10). Unfortunately, their hybrids rapidly lost factor production. Subsequently, many laboratories generated murine and human T cell hybridomas which produced antigen-specific molecules that suppressed or induced various lymphocyte functions (this volume; 8,11). Many T cell hybridomas could be induced to produce various lymphokines (this volume; 12). Some T cell hybrids expressed antigen receptors which could be identified by binding antigen or antiidiotypic antibodies or antigen-induced lymphokine production (this volume; 13). Indeed, T cell hybridomas expressing specific lytic activity could also be generated (this volume; 14). It is remarkable that BW5147 is permissive for expression of all these T cell functions. Many more fusion experiments will be required - including interspecies fusions - to determine the optimal conditions for fusion and expression of particular T cell functions in hybrid cells. Although the problem of stable expression of particular T cell functions in hybridomas has not yet been completely solved, several laboratories have generated sufficient numbers of hybrid cells to allow purification and biochemical analysis of antigen-specific T cell factors. Most advanced is the analysis of suppressor factors (this volume).

T cell hybridomas are not the only source of monoclonal T cell products. First, several murine and human T cell tumors can be induced to express normal T cell functions such as lymphokine production (15). Second, transformed T cell lines expressing specific functions can be obtained by infection of mice or cells in vitro with radiation leukemia virus (16). Third, clones of all major T cell classes can now be grown continuously in tissue culture (17).

To help evaluate the potential of T cell hybridomas for understanding the functioning of the immune system as well as for practical purposes, the workshop on "T Cell Hybridomas: Sources of Specific Mediators in the Immune System" was held January 27-29, 1982, at the Basel Institute for Immunology. The techniques used, the difficulties encountered and the present state of art were discussed, and most of this is presented in the following papers.

References

1. Periman (1970) Nature 228: 1086; Coffino et al. (1971) Nature New Biol 231: 87; Bevan et al. (1972) Prog Biophys Mol Biol 25: 131
2. Mohit (1971) Proc Natl Acad Sci USA 68: 3045
3. Cotton and Milstein (1973) Nature 244: 42
4. Schwaber and Cohen (1974) Proc Natl Acad Sci USA 71: 2203
5. Köhler and Milstein (1975) Nature 256: 495
6. Iverson et al. (1978) Curr Top Microbiol Immunol 81: 192
7. Hämmerling (1977) Eur J Immunol 7: 743; Goldsby et al. (1977) Nature 267: 707; Köhler et al. (1977) Eur J Immunol 7: 758
8. Curr Top Microbiol Immunol 81: 195-231 (1978)
9. Melchers (1978) Nature 271: 9
10. Taniguchi and Miller (1978) J Exp Med 148: 373
11. Research Monographs in Immunology, Vol. 3: Monoclonal Antibodies and T Cell Hybridomas: Perspectives and Technical Advances, G.J. Hämmerling, U. Hämmerling, J.F. Kearney, eds., Elsevier/North Holland, 1982; Watanabe et al. (1978) J Immunol 121: 2113; Hewitt and Liew (1979) Eur J Immunol 9: 572; Okada et al. (1981) Proc Natl Acad Sci USA 78: 7717; Hiramatsu et al. (1981) J Immunol 127: 1118; Irigoyen et al. (1981) J Exp Med 154: 1827
12. Harwell et al. (1981) J Exp Med 152: 893; Kappler et al. (1981) J Exp Med 153: 1198
13. Kappler et al. (1981) J Exp Med 153: 1198
14. Nabholz et al. (1981) Nature 287: 437
15. Behring Institute Research Communications No. 57, Interleukin 2, F.R. Seiler and H.G. Schwick, eds.; Lymphokines, Vol. 5, Monoclonal T Cells and Their Products, M. Feldmann and M. Schreier, eds., Academic Press, New York, 1982
16. Ricciardi Castagnoli et al., this volume; Finn et al. (1979) Proc Natl Acad Sci USA 76: 4033
17. Immunol Rev 54 (1981); Nabel et al. (1981) Cell 23: 19

Acknowledgments

The Basel Institute for Immunology, founded and supported by F. Hoffmann-La Roche and Co., Ltd., and Springer-Verlag jointly sponsored this workshop. Our sincere thanks are due for their support.

The work of B. Badovinac, J. Bron, M. Dekany, R. Rodrigues, J.-P. Rudloff, G. Scialdone, and A. Wyss of the Basel Institute for Immunology in managing the technical details of this workshop is much appreciated.

We also thank Marga Botsch and Rick Mills of Springer-Verlag for their assistance in the publication of the workshop proceedings.

We are grateful to all contributors to this book for meeting the deadline for submission of their manuscripts, which greatly facilitated rapid publication. The articles appear essentially as submitted by the authors; no editing for style or content has been done.

The Organizers

Lectin-Dependent Cytolytic and Cytolymic T Helper Clones and Hybridomas

K. Fischer Lindahl, A.A. Nordin, M.H. Schreier

A. Introduction

Any cytotoxic T lymphocyte (CTL) will kill P815 or EL4 target cells on addition of a lectin such as phytohemagglutinin (PHA) or concanavalin A (Con A) (1,2), and lectin-dependent killing has been considered a prerogative of CTL (3). However, lectin-dependent killers could be found under conditions where no specific CTL were induced (4-6), and nearly every T cell clone stimulated by Con A and T cell growth factor (TCGF) caused ^{51}Cr-release in the presence of PHA (7). It therefore seemed possible that any T cell can become a lectin-dependent killer. To find out how common this activity might be, we tested a set of antigen-specific, I-A-restricted T helper cell lines, clones and hybridomas (8-10).

B. Materials and Methods

Helper Cell Lines and Clones. All T cells were from C57BL/6 mice. The methods used to establish and clone these cells in vitro have been published together with a functional characterization (8-11). All were I-A^b-restricted. They had been cultured serum-free for more than a year by continued restimulation with antigen and irradiated C57BL/6 *nu/nu* spleen cells or in medium supplemented with TCGF. Before testing, the lines were expanded for at least two weeks in medium with TCGF in the absence of adherent cells and antigen.

Hybridomas. All hybridomas originated from two fusions of a chicken egg albumin- (EA) specific clone, EA-14, with BW5147 and are described elsewhere in this volume (12). They were cultured in RPMI 1640 supplemented with 5% heat-inactivated fetal calf serum, glutamine, pyruvate, 2-mercaptoethanol, and penicillin-streptomycin.

^{51}Cr-Release Assay. A standard method (13) was used with 10^4 ^{51}Cr-labeled target cells and effector-to-target ratios of 10:1, 3:1 and 1:1. Triplicates were set up in round-bottom microtiter plates in 200 µl of RPMI 1640 with 5% heat-inactivated fetal calf serum. PHA-P (Difco) was added to a final concentration of 10 µg/ml, Con A (Sigma) to 5 µg/ml (3), and the cultures were incubated for 3.5 hours. The experimental ^{51}Cr-release is expressed as a percentage of the total (measured in the presence of a detergent), and the spontaneous release has been subtracted in the tables. Spleen cells stimulated with Con A for three days or in allogeneic mixed lymphocyte cultures

(MLC) for five days (13) were used as control effector cells.

Binding Tests. Monoclonal antibodies were labeled biosynthetically with ^{3}H-leucine (14). IgGs were purified on protein A columns, IgMs by repeated precipitations with 50% ammonium sulfate and 0.005 M phosphate. One or two million cells were incubated at room temperature for 90 min with 25-50,000 cpm in 100 µl PBS with 1.5% BSA and 0.2% NaN_3, and rapidly washed in the same buffer. The radioactivity bound was measured by liquid scintillation counting and compared to that of control cells with known surface antigens.

Survival Test. Unlabeled P815 cells were incubated in round-bottom sterile tubes with lectins and irradiated (3000 rads) effector cells at the same concentrations and under the same conditions as in the ^{51}Cr-release assay. After 3.5 to 4 hours' incubation the cells were resuspended and diluted (4-fold in Table 7, 10-fold in Fig. 2) with fresh medium without lectin. The cell suspensions were distributed in triplicate in flat-bottom microtiter plates, 250 µl/well, and incubated at 37°C in a humidified atmosphere of 7% CO_2 in air. Uptake of ^{3}H-thymidine was determined after a 4-hour pulse with 1 µCi/well on the third day. The lectin remaining in the experimental wells after dilution was known not to affect the growth of P815.

Table 1. Effect of lectin on ^{51}Cr-release by T helper cells

Effector cells	Net % of total ^{51}Cr released		
	No lectin	PHA	Con A
S26-14	4.4	40.5	44.1
S26-14-37	11.8	52.1	50.8
EA Line (uncloned)	1.8	50.9	51.4
OII-68	0.9	9.1	22.7
OII-16	-1.3	35.6	12.4
OII-74	-0.1	48.2	4.8
OII-96	-0.5	43.3	4.9

Results are given for ten effector cells per P815 target cell. The spontaneous release ranged from 11.4% to 14.4% of total. Clones with the designation OII are derived from the uncloned EA-specific T cell line after 10 months of in vitro propagation.

C. Results

Helper Cell Lines and Clones. Of two clones specific for sheep erythrocytes (SRBC) one (S26-14) gave significant ^{51}Cr-release from P815 in the presence of either PHA or Con A, and 25 of 31 subclones from it caused more than 20% ^{51}Cr-release at ten effectors per target cell. S26-14 mediated $I\text{-}A^{b}$-restricted, SRBC-specific delayed hypersensitivity in vivo (10) and was a potent helper in vitro, as were many of its subclones. Helper and killer activity was not significantly correlated among them (15).

An uncloned T helper line specific for chicken ovalbumin (EA) was a very active lectin-dependent effector cell. Of seven early clones from it three were active; of 66 established more recently 35 were found to be active with PHA. These are necessarily minimum estimates of the fraction of clones capable of lectin-dependent ^{51}Cr-release, since the activity varied with the culture conditions and since rare clones were active with Con A but not with PHA. Table 1 shows examples of the three types of positive clones we have observed. The lectin preference of a clone was a stable character over a couple of months in culture (15).

Table 2 shows that T-depleted spleen cells stimulated with LPS or in an allogeneic MLC caused no ^{51}Cr-release from P815. On the other hand, cells treated with anti-Lyt-2 and complement before stimulation in MLC, which generated no specific CTL, still killed P815 in the presence of PHA. Other experiments have shown that spleen cells fresh from a mouse or cultured for only one day with Con A were inactive in this assay.

Table 2. Only T cells cause lectin-dependent ^{51}Cr-release

Pretreatment of C57BL/6 spleen cells	Stimulus	Net % of total ^{51}Cr released No lectin	PHA
None	Con A	10.0	46.5
C' alone	DBA/2 spleen	51.8	52.1
Anti-Thy-1.2 + C'	DBA/2 spleen	1.9	0.5
Anti-Thy-1.2 + C'	LPS	0.7	2.4
Anti-Lyt-2 + C'	DBA/2 spleen	4.1	35.2

Ten effector cells per P815 target cell. Spontaneous release was 8.5% of total.

Hybridomas. We have tested 30 hybridomas from two fusions between BW5147 and EA-14, a clone which itself scored negative twice for PHA-dependent ^{51}Cr-release. None of the hybridomas were as positive as the best helper clones but about 15 had significant PHA-dependent activity. Fig. 1 shows a test of 21 hybridomas from the second fusion and of BW5147, which was consistently inactive. Con A barely caused detectable ^{51}Cr-release with any of the hybridomas.

Table 3 compares the PHA-dependent ^{51}Cr-release by a selected set of clones with their ability to release TCGF and nonspecific helper factors for an in vitro antibody response of T-depleted spleen cells to SRBC (12). There is no obvious correlation. The most active effector cells produce both activities, but 14/41 makes very little TCGF and 14/48 very little bystander help, and there are clones which produce both, yet are negative for lectin-dependent ^{51}Cr-release. Clones 14/19 and 14/61 are negative for both factor production and PHA-dependent ^{51}Cr-release. We tested whether these clones could be activated to become effector cells by preculturing them for two days in supernatants from the active ones. That was not the case; however, we used only noninduced supernatants. We have also compared the PHA-dependent ^{51}Cr-release of some clones after preculturing them in normal medium and in medium supplemented with partially purified TCGF from a supernatant of Con A-induced rat spleen cells. Again, we found no effect of exogenous factors.

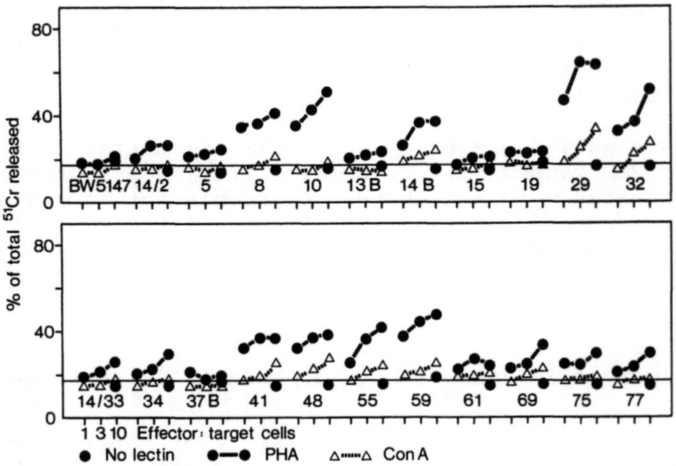

Fig. 1. Lectin-dependent ^{51}Cr release from P815 target cells by helper hybridomas. Spontaneous release is indicated by the horizontal lines.

Table 4 lists the surface markers which we measured. The results with BW5147 and Con A blasts from C57BL/6 are given for comparison with the three most active hybridomas; the same results were obtained with all hybridomas tested. Together with complement, both anti-Thy-1.1 and anti-Thy-1.2 killed more than 90% of 14/10 and 14/29 cells, proving their hybrid nature. As with the cloned helper cells, we found different levels of Lyt-1 on individual hybridomas but no Lyt-2.

The PHA-dependent ^{51}Cr-release requires the presence of the effector cells. Hybridomas 14/10 and 14/29 were cultured for 3.5 hours at $5x10^5$/ml with 10 µg/ml PHA, and the supernatants were then immediately added to labeled P815 cells with or without additional PHA. Whereas the hybridoma cells caused the release of chromium, the supernatants had no effect (Table 5).

The activity of some hybridomas depended on the density at which they were cultured before the assay and each behaved differently. A low density was optimal for 14/32, a high density for 14/10 (Table 6). 14/10 cells precultured at a low density in the supernatant from a culture grown at high density were no more active than if grown in normal medium; thus we have no evidence that the cells secreted factors required for induction of their own effector function.

Table 3. Functional characteristics of selected hybridomas

Hybridoma	^{51}Cr-release[a)] net %	TCGF production[b)] cpm ^{3}H-TdR ($x10^{-3}$)	Bystander help[c)] PFC ($x10^{-2}$)
14/10	35	17	288
14/15	10	31	34
14/19	9	2	7
14/29	43	49	269
14/32	40	108	78
14/37B	11	29	86
14/41	40	3	14
14/48	26	24	5
14/55	45	49	105
14/59	43	40	182
14/61	1	0	0

a) Ten effector cells per P815 target cell in the presence of PHA; spontaneous release (31% of total) was subtracted. Without lectin the ^{51}Cr release did not exceed 3% over background.

b) Two x 10^5 hybridoma cells were cultured for 48 hrs in 1 ml serum-free medium in the presence of 10^6 C57BL/6 *nu/nu* spleen cells and the antigen EA (100 µg/ml). The cell-free supernatant was diluted 1:2 into cultures of a TCGF-dependent killer clone (10^4 cells/0.2 ml). Growth after 48 hrs of culture was measured by ^{3}H-TdR incorporation and expressed as cpm/culture.

c) One hundred µl of the same cell-free 48-hr supernatants were added to 100 µl serum-free medium containing $2x10^5$ C57BL/6J *nu/nu* spleen cells and $5x10^5$ SRBC (11). The number of specific PFC was enumerated after 5 days of culture and is expressed as the mean of four replicate cultures.

Table 4. Surface antigens of C57BL/6 helper hybridomas

Specificity	Antibody (reference)	C57BL/6 Con A blasts	BW5147	14/10	14/29	14/32
$H-2K^k$	H100-5/28 (19)	-	+	+	+	+
$H-2K^b$	B8-24-3 (20)	+	-	+	+	+
Thy-1.1	HO-22-1 (21)	-	+	+	+	+
Thy-1.2	HO-13-4-9 (21)	+	-	+	+	+
Lyt-1	53-7.313 (22)	+	+	+	+	+
Lyt-2.2	HO-2.2 (23)	+	-	-	-	-
Lyt-2	53-6.72 (22)	+	-	-	-	-
Lyt-2	3.168.8 (24)	+	-	-	-	-

Susceptibility and Survival of Target Cells. Many cell lines are susceptible to lectin-dependent killing by CTL, but P815 was the only target which was consistently attacked by the helper cells and the hybridomas. Low levels of ^{51}Cr-release from AKR/A thymoma or EL4 lymphoma targets were occasionally observed, always by effector cells that on the same day were particularly active against P815. Table 6 shows that, like the helper cell lines and clones, hybridomas had no natural killer activity on YAC-1 target cells.

Table 5. Lectin-dependent ^{51}Cr-release cannot be mediated by a hybridoma supernatant

Hybridoma product	Net % of total ^{51}Cr released No lectin	PHA
14/10 PHA-induced supernatant	-1.6	-3.4
14/10 cells	-1.9	24.9
14/29 PHA-induced supernatant	-2.7	-4.8
14/29 cells	-2.7	35.9
BW5147 cells	-1.7	-3.6

Supernatants were prepared from the equivalent of the ten effector cells per target cell used. Spontaneous release from P815 was 11.0 and 12.9% of total.

Table 6. Effect of cell density and target cell susceptibility on lectin-dependent ^{51}Cr-release by hybridoma cells

Effector cells	Density before assay (cells/ml x 10^{-6})	Net % of total ^{51}Cr released P815		EL4		YAC-1	
		-	+PHA	-	+PHA	-	+PHA
14/10	2.1	-5	28	1	1	0	13
	0.20	-1	16	0	-2	-1	0
14/29	1.5	-1	32	-1	7	6	12
	0.15	0	32	2	12	1	8
14/32	3.4	-9	0	0	-2	0	-2
	0.23	-3	23	2	2	-1	7
14/19	0.92	-1	8	2	1	4	2
14/61	0.55	-1	0	-1	-1	1	0
C57BL/6 Con A blasts	-	16	49	5	42	10	39
BALB/c spleen cells	-	-5	-1	0	-2	39	27

The normal spleen cells were assayed at 100, all other cells at 10 effector cells per target cell. The spontaneous release was 36% from P815, 11% from EL4, and 12% from YAC-1.

To test whether the P815 target cells were killed by the helper cells or merely suffered a repairable injury, we assessed their viability after the incubation with effector cells and lectins, both by counting colonies in soft agar and by measuring thymidine uptake. The effector cells were irradiated to prevent them from growing; this did not significantly affect their activity. We confirmed that ^{51}Cr-release with specific CTL signals the death of the target cells (16). Using either PHA or Con A and effector cells from MLC or Con A blasts, the number of surviving P815 cells always dropped as the ^{51}Cr-release increased, consistent with other studies (1,2). The cloned T helper cells had the same effect, provided Con A was used in the assay. But with PHA, the targets remained fully viable, even though they released up to 60% of their chromium (Table 7). We proposed the term *cytolymic* (from Greek *lymē* = maltreatment) to describe the latter interaction (15).

Table 7. Survival of P815 target cells

		Lectin in assay			
Effector cell	Ratio	PHA		Con A	
		Net % ^{51}Cr release	^{3}H-TdR uptake (cpm x 10^{-3})	Net % ^{51}Cr release	^{3}H-TdR uptake (cpm x 10^{-3})
None	-	0	54	0	32
MLC blasts					
DBA/2 anti-C3H	10	22	3	16	12
	3	8	23	5	25
	1	1	51	1	31
Helper clones					
S26-14-26	10	44	43	56	3
	3	34	44	37	10
	1	16	53	20	20
S26-14-38	10	47	44	66	3
	3	26	51	42	8
	1	8	59	26	22

Spontaneous release was 12.6%.

The effect of the hybridomas has been difficult to characterize, as their activity in the ^{51}Cr-release assay was very sensitive to irradiation. A single experiment suggested that they were cytolytic (Fig. 2).

C. Discussion

An Ia-specific killer clone which could also provide help for allogeneic B cells with the appropriate Ia was recently described (17). Our experiments have now shown that lectin-dependent ^{51}Cr-release may be observed with T helper cells (15) and it is thus not a prerogative of specific CTL. The finding of mutants of CTL clones or hybridomas that no longer kill specifically but still give lectin-dependent ^{51}Cr-release (18) must therefore be interpreted with caution; these cells may have lost more than their receptor and the lytic mechanisms may be different. It may be wise to avoid the use of P815 for such studies and choose instead other target cells, insensitive to T helpers. We do not know whether the particular sensitivity of P815 is due to its mastocytoma origin.

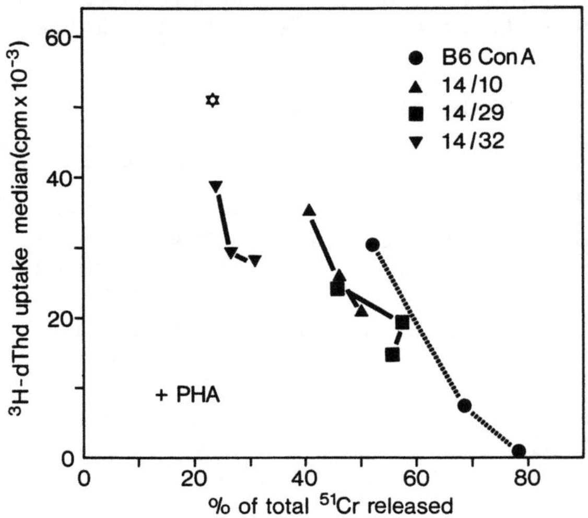

Fig. 2. Survival of P815 target cells after incubation with hybridoma cells or Con A blasts and PHA. The control value with no effector cells is given by the star.

The T helper cells have revealed that ^{51}Cr-release is not synonymous with death, but given Con A they could also kill. Cytolymic interactions of natural killer cells with fibroblast target cells have been observed (M. Cohn, personal communication) as have cytolymic effects of monoclonal antibodies on influenza-infected P815 target cells with up to 80% ^{51}Cr-release but no increased uptake of trypan blue (M. Frankel and W. Gerhard, personal communication). Analysis of cloned helper cells revealed an unexpected heterogeneity; some were effective with Con A, some with PHA and some with both. All the hybridomas, which were derived from a single helper clone, were of the same type, i.e. much more active with PHA than with Con A.

We never observed lectin-dependent killing with BW5147 or with the EA-14 clone used for fusion, and it is possible that the hybridization induced a function displayed by neither parent cell. However, we cannot exclude that EA-14 could become a lectin-dependent killer under the right culture conditions. The lectin-dependent killer activity of the hybridomas is easily recovered after freezing and has been stable over two months in culture; in some cases it even increased with continued culture.

Acknowledgments. We thank B. Hausmann, R. Tees and L. Nordin for their assistance with these experiments and S. Buser-Boyd for preparing the manuscript. The Basel Institute for Immunology was founded and is supported by F. Hoffmann-La Roche and Company, Ltd.

References

1. Zagury, M, Fouchard, M, Morgan, DA, Cerottini, JC (1980) Enumeration of T-effector cells mediating direct and/or lectin-dependent lysis. Immunol Letters 1: 335-339
2. Bradley, TP, Bonavida, B (1981) Mechanism of cell-mediated cytotoxicity at the single cell level. III. Evidence that cytotoxic T lymphocytes lyse both antigen-specific and -nonspecific targets pretreated with lectins or periodate. J Immunol 126: 208-213
3. Bevan MJ, Cohn, M (1975) Cytotoxic effects of antigen- and mitogen-induced T cells on various targets. J Immunol 114: 559-565
4. Rubens, RP, Henney, CS (1977) Studies on the mechanism of lymphocyte-mediated cytolysis. VIII. The use of Con A to delineate a distinctive killer T cell subpopulation. J Immunol 118: 180-186
5. Davignon, DJ, Laux, DC (1978) Lectin-dependent cell-mediated cytotoxicity: Induction of a unique effector cell population. Cell Immunol 41: 294-303
6. Cerottini, JC, MacDonald, HR (1981) Limiting dilution analysis of alloantigen-reactive T lymphocytes. V. Lyt phenotype of cytolytic T lymphocyte precursors reactive against normal and mutant H-2 antigens. J Immunol 126: 490-496
7. Larsson, EL, Fischer Lindahl, K, Langhorne, J, Coutinho, A (1981) Quantitative studies on Concanavalin A-induced, TCGF-reactive T cells. 1. Correlation between proliferation and lectin-dependent cytolytic activity. J Immunol 127: 1081-1085
8. Schreier, MH, Tees, R (1980) Clonal induction of helper T cells: Conversion of specific signals into nonspecific signals. Int Archs Allergy Appl Immun 61: 227-237
9. Schreier, MH, Tees, R, Nordin, AA (1981) Establishment and characterization of helper T cell clones. Their functional heterogeneity and effect on the B cell responses to particulate, soluble and T independent antigens. In: Feldmann, M, Schreier, MH (eds) Lymphokines 5: Monoclonal T Cells and Their Products. Academic Press, New York, in press
10. Schreier, MH, Tees, R, Nordin, AA, Benner, R, Bianchi, ATJ, van Zwieten, MJ (1982) Functional aspects of helper T cell clones. In: Smith, KA (ed) The growth and functional aspects of T cells. Immunobiology Supplement, in press
11. Schreier, MH, Tees, R (1982) Long-term culture and cloning of specific helper T cells. In: Pernis, B, Lefkovits, I (eds) Immunological Methods, Vol. II. Academic Press, New York, pp 263-275
12. Schreier, MH, Tees, R, Nordin, L, Nordin, AA (1982) Functional characteristics of T cell hybridomas obtained by fusion of TCGF-dependent helper T cell clones with BW5147. (This volume.)
13. Fischer Lindahl, K, Hausmann, B (1980) Qed-1 - a target for unrestricted killing by T cells. Eur J Immunol 10: 289-298
14. Fazekas de St.Groth, S, Scheidegger, D (1980) Production of monoclonal antibodies: Strategy and tactics. J Imm Meths 35: 1-21
15. Fischer Lindahl, K, Schreier, MH (1981) Submitted for publication.
16. Lees, RK, MacDonald, HR, Sinclair, NR (1977) Inhibition of clone formation as an assay for T cell-mediated cytotoxicity: Short-term kinetics and comparison with ^{51}Cr release. J Imm Meths 16: 233-244

17. Dennert, G, Weiss, S, Warner, JF (1981) T cells may express multiple activities: specific allohelp, cytolysis, and delayed-type hypersensitivity are expressed by a cloned T-cell line. Proc Natl Acad Sci USA 78: 4540-4543
18. Dialynas, DP, Loken, MR, Glasebrook, AL, Fitch, FW (1981) Lyt-2^-/Lyt-3^- variants of a cloned cytolytic T cell line lack an antigen receptor functional in cytolysis. J Exp Med 153: 595-604
19. Lemke, H, Hämmerling, GJ, Höhmann, C, Rajewsky, K (1978) Hybrid cell lines secreting monoclonal antibody specific for major histocompatibility antigens of the mouse. Nature 271: 249-251
20. Köhler, G, Fischer Lindahl, K, Heusser, C (1981) Characterization of a monoclonal anti-H-$2K^b$ antibody. In: Steinberg, CM, Lefkovits, I (eds) The Immune System, 2. S. Karger, Basel, pp 202-208
21. Marshak-Rothstein, A, Fink, P, Gridley, T, Raulet, DH, Bevan, MJ, Gefter, ML (1979) Properties and applications of monoclonal antibodies directed against determinants of the Thy-1 locus. J Immunol 122: 2491-2497
22. Ledbetter, JA, Herzenberg, LA (1979) Xenogeneic monoclonal antibodies to mouse lymphoid differentiation antigens. Immunol Rev 47: 63-90
23. Gottlieb, PD, Marshak-Rothstein, A, Auditore-Hargreaves, K, Berkoben, DB, August, DA, Rosche, RM, Benedetto, JD (1980) Construction and properties of new Lyt-congenic strains and anti-Lyt-2.2 and anti-Lyt 3.1 monoclonal antibodies. Immunogenetics 10: 545-555
24. Sarmiento, M, Glasebrook, AL, Fitch, FW (1980) IgG or IgM monoclonal antibodies reactive with different determinants on the molecular complex bearing Lyt 2 antigen block T cell-mediated cytolysis in the absence of complement. J Immunol 125: 2665-2672

Cytotoxic T Cell Hybridomas: Generation and Characterization

Z. Eshhar, T. Waks, T. Oren, G. Berke, Y. Kaufmann

A. Introduction

Previous attempts to immortalize cytotoxic T-lymphocytes (CTL) by somatic cell hybridization were unsuccessful (1,2), possibly due to CTL-induced nonspecific lysis of the fusion partner during the fusion process. On the other hand, the functional CTL hybridomas generated by Nabholz and colleagues (3) appeared to be T cell growth factor (TCGF) dependent, because TCGF dependent CTL lines had been used for fusion.

In order to improve the chances of successful fusions and to circumvent nonspecific lysis, we have exposed enriched CTL populations to controlled trypsinization prior to fusion. This procedure reversibly inhibits the lytic capacity of CTL without affecting their fusion capacity. Recently, by using this approach we have established stable, TCGF independent and target cell specific CTL hybridomas (4,5). Here we present further serological and functional characterization of two of the cytolytic hybrid lines that specifically lyse allogeneic leukemia cells.

B. Results and Discussion

I. Generation of CTL Hybridomas and Their Growth Characteristics

The killer cell hybridomas were generated by fusion of the AKR/J thymoma cell line BW5147 with secondary BALB/c anti-EL4 CTL. The protocol used for priming and *in vivo* (PEL) or *in vitro* (MLC) restimulation of the CTL, the trypsinization of the enriched CTL populations prior to hybridization and the fusion conditions have been described in detail previously (4,5). Figure 1 schematically illustrates the protocol we employed for the generation of the CTL hybridomas. The transient inhibition of the lytic capacity of the CTL by trypsin treatment appeared to be effective; thus, in four fusion experiments we have established seven stable CTL hybridomas that specifically lysed EL4 target cells (TC) in the absence of PHA (Table 1). The two hybrid cultures that were obtained from untrypsinized CTL (fusion IV, Table 1) appeared to be unstable: their lectin independent lytic activity was low and declining, disappearing after a few weeks of growth in culture.

The CTL reactive clones from fusions I and II that we studied in more detail are true hybrid cells as verified by their aneuploid number (60-70) of chromosomes, size, pleomorphic appearance displaying uropod structure (6) and coexpression of H-2 and Thy-1 surface antigens of both parental cells. We could not detect any staining of

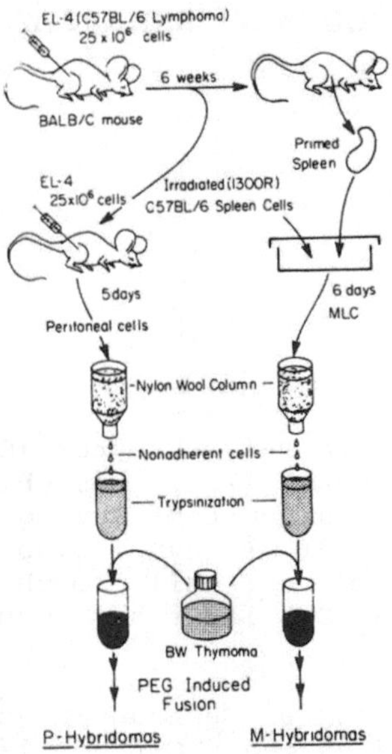

Fig. 1. Schematic experimental protocol used for the generation of functional CTL hybridomas

Table 1. Frequencies of functional hybridomas obtained from fusions of trypsinized and untreated CTL

Fusion number	Source of CTL[1]	Trypsin pre-treatment[2]	Number of CTL fused[3]	Frequency of growing cultures	Number of positive hybrid cultures[4]	
					-PHA	+PHA
I	BALB/c anti-EL4 (MLC)	+	4×10^7	86/96	2	7
II	BALB/c anti-EL4 (PEL)	+	4×10^7	83/96	2	6
III	CBA anti-EL4 (MLC)	+	5×10^7	42/96	3	10
IV	CBA anti-EL4 (MLC)	-	10×10^7	102/144	2[5]	23

[1]Secondary CTL primed in the peritoneum by 25×10^6 EL4 cells and restimulated either in the peritoneum (PEL) by EL4 cells or *in vitro* in mixed lymphocyte culture with C57BL/6 splenocytes (MLC).

[2]Restimulated, nylon wool purified CTL ($5\text{-}10 \times 10^6$/ml) in PBS-10% FCS were treated with trypsin (1.8 mg/ml) for 30 min at 37°. Cells were washed thrice and kept on ice until used for fusion.

[3]Fusion of CTL with BW5147 cells at 2:1 ratio was induced with 41% PEG 1500 (5).

[4]Cytolytic activity was determined against ^{51}Cr-labeled EL4 target cells in the absence or presence of 10 μg/ml phytohemagglutinin M (PHA). Cultures were considered positive when they caused 10% TC lysis or more

[5]Hybrid cells appeared to be less active and unstable and lost their lytic activity after a few weeks in culture.

the CTL hybrid cells with anti-Ig or anti-V_H and anti-V_L antibodies. The most prominent advantages of the CTL hybridoma over continuous CTL lines, which are dependent on antigenic stimulus or TCGF, are the rapid growth capacity (doubling time of about 10 hr in tissue culture conditions), high cell density (up to 4×10^6/ml culture) and ability

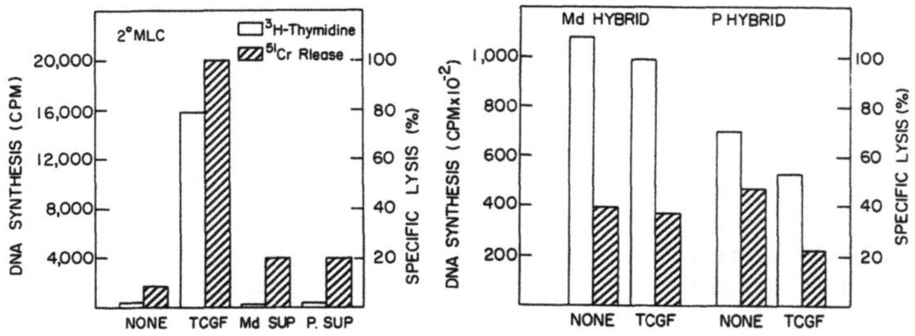

Fig. 2. Analysis for the presence of TCGF-like activity in culture supernatants of the MLC and PEL hybridomas (left panel) and the effect of externally added TCGF (Con A conditioned medium) on the cytolytic activity of the hybridomas (right panel). As indicator cells, secondary MLC memory cells (CBA anti-C57BL/6) have been used 16 days following second stimulation. The TCGF conditioned medium was obtained from spleen cells that were incubated with 2 μg/ml of Con A for 48 hr. After harvesting the supernatant, Con A was saturated with 4 mg/ml methyl-α-D-mannopyranoside. Hybrid culture supernatant was obtained from hybridomas that grew to 10^{6} cells/ml. After incubation of the conditioned medium at 1:2 dilution (5 days with the indicator cells and 3 days with the hybridomas) the ^{3}H-thymidine incorporation (open bars) and the cytolytic activity (^{51}Cr-release from EL4 TC) (striped bars) were determined.

to develop a large mass of tumor in (BALB/c x AKR/J)F_1 mice when inoculated intraperitoneally or subcutaneously.

Another characteristic feature of the CTL hybridomas we have established, which distinguishes them from the hybridomas described by Nabholz *et al.* (3,7), is the independence of the cytotoxicity of the hybrid lines described herein from TCGF. Based on the strict relationship between the cytolytic activity and the TCGF dependence of the continuous CTL line and the hybridomas derived therefrom, it was suggested by Nabholz *et al.* (6,7) that there might be a common linked gene or genes controlling the cytolytic activity and TCGF dependence. Yet the hybridomas derived by us manifest specific cytotoxicity without the addition of any known stimuli. As can be seen in Figure 2, these hybridomas also failed to secrete any detectable growth factor(s). Thus, memory MLC cells that respond to TCGF by increased proliferation and cytotoxicity did not respond to the hybridoma's conditioned medium. In addition, as shown by the second experiment depicted in Figure 2, the addition of potent TCGF preparation to the CTL hybridomas did not enhance the cytolytic activity of the hybridomas and even partially suppressed the lytic activity of the P-hybridoma. Likewise, no effect was observed on the proliferative capacity of the CTL hybridoma.

II. Functional and Serological Characterization

By most criteria tested, the hybrid clones derived from BALB/c anti-EL4 CTL exhibited a pattern of behaviour similar to that of parental CTL, although some differences have been observed (4,5). Thus, for lysis a direct contact between effector cells and TC is required and soluble components were not involved in the lysis. Treatments which have been known to inhibit CTL activity, such as low temperature (21°C), trypsinization, trypsin inhibitor TLCK, cytochalasin-B and EDTA, also inhibited the hybridoma mediated lysis. On the other hand, treatment with neuraminidase or the addition of lectin to the lytic mixture enhanced the TC lysis (4,5). Like the parental CTL population, both PEL and MLC derived hybridomas lysed EL4 cells and several other $H-2^b$ leukemic target cells and did not lyse non $H-2^b$ tumors. However, unlike the parental polyclonal CTL population, the monoclonal hybridomas did not lyse normal and Con A or LPS stimulated lymphocytes (5). To determine whether this pattern does not reflect different levels of sensitivity to lysis of some normal TC and tumor cells, the cytolytic capacity of the hybridomas against the various target cells was determined in the presence of PHA, which promotes nonspecific TC lysis. Figure 3 compares the degree of TC lysis (^{51}Cr release) obtained by CTL clone P-47 as tested on neuraminidase treated TC in the absence and presence of 10 µg/ml PHA in the assay.

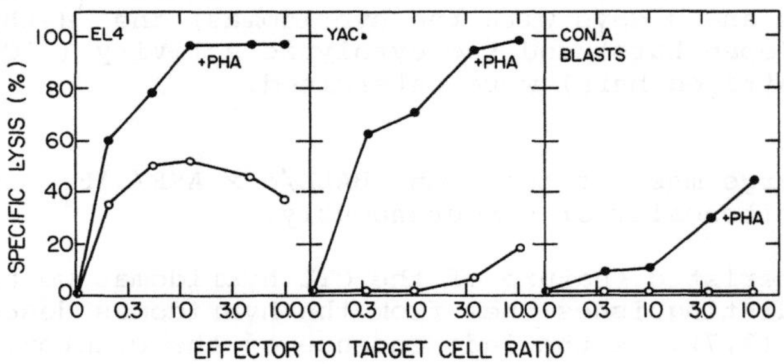

Fig. 3. Specific and PHA promoted nonspecific target cell lysis mediated by CTL hybridomas. Different amounts of hybrid clone P-47 were added to neuraminidase treated (75 units/ml, 15 min 37°C) ^{51}Cr-labeled target cells. Con A lymphoblasts (after 3 days culture with 2 µg/ml Con A) were treated with 20 mg/ml of α-methylmannoside. ^{51}Cr release was determined after 4 hr incubation with EL4 cells and 5 hr with YAC or blasts in the absence (○) or presence of 10 µg/ml PHA (●) in the assay.

The fact that in the presence of PHA the hybrid cells also lysed the $H-2^b$ lymphoblasts or the non $H-2^b$ tumor cells strongly supports our suggestion (5) that the cytotoxicity of the CTL hybridomas is specific for tumor or viral antigens in EL4 TC and is $H-2^b$ restricted.

Monoclonal antibodies against cell surface antigens provide an ideal tool to identify the cellular components of the effector cells that participate (directly or indirectly) in the process of target cell recognition and/or lysis. In addition, antibodies directed towards TC antigens can help us to analyze the fine specificity of the response. Table 2 summarizes the results of experiments where we examined the inhibitory effect of various monoclonal antibodies on the lytic activity of the CTL hybridomas. In the first group, different antibodies against the TC H-2^b region were analysed. As can be seen from the table, antibodies against the H-$2K^b$ region completely inhibited cytotoxicity of both hybrid clones while antibodies against the H-$2K^b$ region did not have any significant effect on the hybridoma mediated lysis, although they have a partial effect on the lysis of EL4 TC by the parental CTL.

The inhibition of alloreactive responses by antibodies against the target cell H-2K or D subregions has been amply reported. Since the CTL hybrid clones do not react against H-2^b determinants *per se* (Figure 3), it is possible that the inhibition observed by the anti-H-$2D^b$ antibodies is due to blockage of allomodified antigen on the target cell in a mechanism similar to the inhibition of self-modified CTL responses by anti-H-2K or H-2D antibodies (15). These results coincide with the allorestriction of the CTL hybridomas as observed by their restricted specificity (Figure 2). Such specificity is minor within the parental CTL population and reflects one of the advantages of the hybridization approach that enables the immortalization and selection of rare specificities.

The second group of monoclonal antibodies contains antibodies against Lyt 2 antigens that have been shown to inhibit T cell mediated cytotoxicity (13,14), implicating a relationship between the T cell receptor and Lyt 2 antigen. As demonstrated in Table 2, anti-Lyt 2 antibodies which inhibited the cytotoxicity of primary MLC did not inhibit the cytolytic activity of the CTL hybridomas. In addition, we could not stain the hybrid cells with the monoclonal antibodies directed against the constant determinants of the murine Lyt 2. The absence of Lyt 2 antigen from the surface of the CTL hybridomas and the insensitivity of the lytic activity to the anti-Lyt 2 treatment imply that the relationship suggested between the killing ability and Lyt 2 is not a necessary requirement. Similar results have been obtained with a long-term I-A specific alloreactive CTL line (16), which in addition to the insensitivity of its reactivity to anti-Lyt 2 is also TCGF independent.

The last group of monoclonal antibodies described in Table 2 includes antibodies that were raised in rats against murine CTL and were selected for their ability to inhibit cell mediated cytolysis (11,12). These antibodies recognize surface structure common to CTL and other cell types which is distinct from Lyt 2. Functionally it was found that H35.27.9 inhibits CTL but not NK mediated lysis. H35.89.9 inhibits both specific T cell proliferation and cytolysis (12,18) and it reacts with polypeptide chains of 180K and 95K similar to the LFA-1 (17). A more detailed report about the effect of these antibodies on the CTL hybridomas described herein has been published elsewhere (18). In summary, it was found that antibody number H35.89.9 inhibited the lytic activity of both PEL and MLR derived hybridomas while antibody number H35.27.9 inhibited only the activity of the PEL derived hybridoma. Interestingly, H35.89.9 antibody did not inhibit the proliferation of the CTL hybridomas. The functional data coincide with the presence on the cell surface of the relevant antigenic structures; thus, antibody H35.89.9 bound to both CTL hybridomas and antibody H35.27.9 reacted only with the PEL derived hybridoma clones. It is still too early to draw conclusions about the association between H35.89.9 and the CTL structure involved in its function. It seems that in different cell types the relationship of a function and a given structure varies considerably, and the mere fact that a certain structure is involved in the inhibition of a function does not necessarily reflect its participation in this function.

Table 2. Effect of various anti-target cells and anti-CTL monoclonal antibodies on the cytolytic activity of the hybridomas

Monoclonal antibody (Ref.)		Serological specificity	Inhibition of anti-EL4 cytolysis			
			Polyclonal CTL		Monoclonal hybridomas	
			MLC	PEL	Md.26	P.47
B22-249.R1	(8)	D^{b};H-2.m2	+	+	+	+
H141-31	(8)	D^{b};H-2.m2	±	+	+	+
B8-24	(9)	K^{b}	±	ND	-	-
H141-11	(8)	$K^{k,b},D^{b}$;H-2.m6	ND	ND	+	+
H142-23	(8)	$K^{k,b,s},D^{k}$;H-2.m9	-	+	-	-
H142-45	(8)	$K^{k,b,q},D^{b,k}$;H-2.m10	+	+	+	+
53-7.313	(10)	Lyt 1	-	-	-	-
53-6.72	(10)	Lyt 2	+	±	-	-
30-H.12	(10)	Thy 1.2	-	-	-	-
H35.17.2	(11,12)	Lyt 2	+	ND	-	-
H35.27.9	(11,12)	Undefined	+	ND	-	+
H35.89.9	(11,12)	LFA-1	+	ND	+	+

The effect of serial dilutions of monoclonal antibodies from culture supernatants or ascitic fluids was determined during all the cytolytic assay (anti-H-2 group) or after preincubation with the effector cells alone (anti-Lyt 2 and LFA-1). The results represent experiments where the degree of inhibition of 50-100% was considered (+), 20-50% inhibition as (±) and 0-20% as (-).

C. Concluding Remarks

Cytolytically active hybridomas are instrumental for our understanding of the following aspects:

1. Constructive hybrid clones provide a homogeneous and stable source for functional analysis of the mechanism of T cell mediated lysis. Minor specificities, such as the allorestricted tumor target cell recognition described in this report, can be revealed and characterized.
2. Large amounts of homogeneous material are available for biochemical studies aimed at the isolation and characterization of the CTL's recognition unit for TC antigen. The generation of additional monoclonal antibodies should help to elucidate the structure-function relationship between a given surface component and the cytolytic activity. So far it appears that Lyt 2 is not required for the lytic activity.
3. The tendency of somatic cell hybrids to lose chromosomes and the

ability of the partner tumor cell to influence the reactivity of the resulting hybrid cell provide a useful model for genetic analysis. The ability to separate between the cytolytic activity and TCGF dependency is one example of such potential studies. The derivation of different nonfunctional variants by subcloning of the CTL hybridomas that we have recently accomplished is a further step toward this end.

Acknowledgements. We are indebted to Drs. P. Golstein, G. Hämmerling and G. Köhler for providing us with the different antibodies. This research was supported by the United States-Israel Binational Science Foundation. Z.E. is an incumbent of the Recanati Career Development Chair in Cancer Research.

References

1. Köhler G, Lefkovits I, Elliott B, Coutinho A (1977) Derivation of hybrids between a thymoma line and spleen cells activated in a mixed leukocyte reaction. Eur J Immunol 7: 758-761
2. Grützmann R, Hämmerling G (1978) Characterization and functional analysis of T cell hybrids. Curr Top Microbiol Immunol 81: 188-191
3. Nabholz M, Cianfriglia M, Acuto O, Conzelmann A, Haas W, von Boehmer H, MacDonald HR, Johnson JP (1980) Cytolytically active murine T cell hybridomas. Nature 287: 437-440
4. Kaufmann Y, Berke G, Eshhar Z (1981) Functional cytotoxic T lymphocyte hybridomas. Transpl Proc 13: 1171-1175
5. Kaufmann Y, Berke G, Eshhar Z (1981) Cytotoxic T lymphocyte hybridomas which mediate specific tumor cell lysis *in vitro*. Proc Natl Acad Sci USA 78: 2502-2506
6. Kaufmann Y, Berke G, Eshhar Z (1982) Cytolytic T cell hybridomas In: Feldmann M and Schreier M (eds) Lymphokines. Vol 5, Academic Press, New York, p 277
7. Nabholz M, Cianfriglia M, Acuto O, Conzelmann A, Weiss A, Haas W, von Boehmer H (1981) The production of murine cytolytic T-cell hybrids. In: Hämmerling GJ, Hämmerling U, Kearney JF (eds) Monoclonal antibodies and T-cell hybridomas. Elsevier/North-Holland, p 539
8. Lemke H, Hämmerling GJ, Hämmerling U (1979) Fine specificity analysis with monoclonal antibodies of antigens controlled by the major histocompatibility complex and by the Qa/TL region in mice. Immunol Rev 47: 176-206
9. A generous present from Dr. G. Köhler, The Basel Institute for Immunology
10. Ledbetter, JA, Herzenberg LA (1979) Xenogeneic monoclonal antibodies to mouse lymphoid differentiation antigens. Immunol Rev 47: 362-378
11. Pierres M, Goridis C, Golstein P (1981) Inhibition of murine T cell mediated cytolysis and T cell proliferation by a rat monoclonal antibody immunoprecipitating two lymphoid cell surface polypeptides of 94000 and 100000 molecular weight. Eur J Immunol in press
12. Golstein P, Pierres M (1981) Monoclonal antibodies as probes to study the mechanism of T cell mediated cytolysis. In: Resch K, Kirchner H (eds) Mechanisms of lymphocyte activation. Elsevier/North-Holland, p 442

13. Sarmiento A, Glasebrook AL, Fitch FW (1980) Monoclonal antibodies block cytolysis by T cells. J Immunol 125: 2665-2672
14. Hollander N, Pillemer E, Weissman I (1980) Blocking effect of Lyt 2 antibodies on T cell functions. J Exp Med 152: 674-687
15. Schmitt-Verhulst A-M, Sachs DH, Shearer GM (1976) Cell mediated lympholysis of TNP-modified autologous lymphocytes: Confirmation of genetic control of response to TNP-modified H-2 antigens by the use of anti-H-2 and anti-Ia antibodies. J Exp Med 143: 211-223
16. Swain SL, Dennert G, Wormsley S, Dutton RW (1981) The Lyt phenotype of a long-term allospecific T cell line. Both helper and killer activities to IA are mediated by Lyt 1 cells. Eur J Immunol 11: 175-180
17. Kürzinger K, Reynolds T, Germain NR, Davignon D, Martz E, Springer TA (1981) A novel lymphocyte function-associated antigen (LFA-1): Cellular distribution, quantitative expression and structure. J Immunol 127: 596-602
18. Golstein P, Pierres M, Schmitt-Verhulst A-M, Luciani M-F, Buferne M, Eshhar Z, Kaufmann Y (1982) Proceedings of the First International Workshop on Mechanism of Cell Mediated Cytotoxicity. Plenum Press, in press

Definition of Function-Related Isotypic Markers on T Cells

E.J. Culbert, S. Kontiainen, L.M. Douglas, M. Feldmann

Introduction

Investigation of the structure and function of antigen-specific molecules synthesized by T cells is a major aspect of current immunological research, since such molecules have an important role in immune regulation. A partial understanding of the structure of T cell factors and receptors has resulted from analysis of antigen-binding materials derived from relatively crude T cell sources, such as activated T cells (e.g. Feldmann and Basten, 1972; Binz and Wigzell, 1975), and this has been confirmed using more homogeneous T cell sources (T cell hybrids, TCGF-dependent lines, virus tranformed lines etc). This understanding has been enhanced by the use of various antisera which recognise T cell factors and receptors.

Early reports demonstrated that some anti-IgM antisera raised in rabbits and chickens also react with T cell molecules (Feldmann and Basten, 1972; Taniguchi and Tada, 1974; Howie and Feldmann, 1977; Marchalonis et al, 1979; Zanders et al, 1980). However, many similar reagents do not react with T cell products, and the precise significance of cross-reactivity between IgM and T cell molecules is unclear at present.

Antisera raised against the variable region of antibody molecules, such as anti-V_H (Puri et al, 1980) and anti-idiotype (reviewed Eichmann, 1978) reagents, are widely accepted as having reactivity with membrane-bound and secreted T cell molecules. However, antisera raised against V_L determinants are unreactive towards T cell factors and receptors, which suggests profound differences between the antigen-binding sites of T cell and B cell antigen recognition structures.

Many T cells and their products react with antisera raised against MHC-coded determinants, principally those against I-region products (Taussig and Munro, 1974; Taniguchi et al, 1976; Howie et al, 1979), and there are reports that some T cell factors have determinants cross-reactive with β_2-microglobulin, a molecule associated with MHC products (Lamb et al, 1981).

Thus there are many antisera available which can be used in the analysis of T cell functions and products. However (with the possible exception of anti-I-J sera), all of the antisera mentioned above have been raised against products of cells other than T cells, and so none are truly specific for T cell molecules. Thus one cannot be certain whether T cell reactivities, especially those present in heteroantisera, are due to shared identities, due to fortuitous cross-reactivity, or perhaps due to an unsuspected minor contaminant in the antiserum. Use of monoclonal antibodies may partially resolve such problems, and there are reports that some monoclonal anti-idiotypes react with both T cells and B cells (Pacifico et al, 1981). However, various monoclonal anti-V_H reagents have been unreactive with T cells which stain with heterologous anti-V_H antisera (Culbert and Kubagawa, unpublished observations), which raises important questions as to the nature of the T cell reactive component in the heteroantisera.

We have attempted to overcome such problems of specificity by raising antisera directly against T cell antigen specific helper and suppressor factors. Various anti-factor antisera raised in rabbits and mice have previously been shown to react with secreted helper factors

(HF) and suppressor factors (SF; Kontiainen and Feldmann, 1979; Feldmann et al, 1980) derived from T cell hybrids and activated T cells (ATC). In this report we show that determinants on T cells recognised by rabbit anti-factor sera define isotypic markers on activated T_H and T_S cells. Similar results have been obtained using monoclonal reagents raised against HF and SF.

Materials and Methods

Animals and Antigens

CBA/Ca, C57BL/10, B10.D_2, AKR, BALB/C, and (B6 x BALB/c) F_1 mice were obtained from the Imperial Cancer Research Fund Breeding Unit. The antigens used were keyhole limpet haemocyanin (KLH), and a copolymer of L-glutamic acid (60)-L-alanine (30)-L-tyrosine (10) (GAT), a gift from Dr. P. Maurer, Philadelphia. Trinitrophenylated KLH had 8 groups of TNP per 100,000 daltons (Rittenberg and Amkraut, 1966). 4-hydroxy-5 nitrophenyl acetic acid (NP) was coupled to GAT by Professor O. Makela (Helsinki) as previously described (Kontiainen and Feldmann, 1980). NP was coupled to coliphage T_4 as previously described (Makela, 1966).

T Cell Tumour Line

The tumour line used for the fusions was BW5147. This was obtained from Dr. Robert Hyman, La Jolla, via Professor L.A. Herzenberg of Stanford in September 1976. It was HGPRT negative, thus allowing for suppression of its growth in HAT medium (see below) following fusion with a source of normal T cells: non-fused T cells die out in a matter of days, and only hybrids between these cells and BW 5147 grow in HAT.

Generation of Helper and Suppressor Cells and Factors

HC and SC induction, production of HF and SF, and co-operative cultures were performed using Marbrook flasks as previously described (Kontiainen and Feldmann, 1976, 1977 and 1979). The optimal doses of KLH or NP required for induction of helper and suppressor cells (HC and SC) and production of helper and suppressor factors (HF and SF) were titrated in preliminary experiments (data not shown). These conditions can be summarised as follows:

HC_{KLH}: 15×10^6 spleen cells + 0.1 μg KLH in 1 ml cultured for 4 days
SC_{KLH}: 15×10^6 spleen cells + 100 μg KLH in 1 ml cultured for 4 days
HF_{KLH}: 5×10^6 spleen cells + 0.1 μg KLH in 1 ml cultured for 24 hours
SF_{KLH}: 5×10^6 spleen cells + 1 μg KLH in 1 ml cultured for 24 hours
HC_{NP}: 15×10^6 spleen cells + 10^5 NP-T_4 in 1 ml cultured for 4 days
SC_{NP}: 15×10^6 spleen cells + 10^7 NP-T_4 in 1 ml cultured for 4 days
HF_{NP}: 5×10^6 HC_{NP} + 10^5 NP-T_4 in 1 ml cultured for 24 hours
SF_{NP}: 5×10^6 SC_{NP} + 10^5 NP-T_4 in 1 ml cultured for 24 hours

HF preparations were tested at a range of dilutions (5%-0.005%) in co-operative cultures containing HF, 15×10^6/ml normal or immunised spleen cells (± anti Thy-1 treatment) plus 0.01 ug/ml TNP-KLH, 1 μg/ml NP-GAT or 0.1 μg/ml DNP-CGG.

SF preparations were tested similarly in co-operative cultures containing SF, 1.5×10^6/ml normal, unimmunised spleen cells and 1 μg/ml TNP-KLH or 1 μg/ml NP-GAT or 1 μg/ml DNP-CGG.

Cultures were performed in a volume of 200 μl in a Mini Marbrook system described in detail elsewhere (Erb et al, 1978). After 4 days of culture the antibody forming cell (AFC) cultures were assayed using the Cunningham modification of the Jerne plaque assay (Cunningham, 1965) utilising DNP (Stausbauch et al, 1970) or GAT (Howie et al, 1979) coupled to sheep red blood cells (SRBC). All the co-operative cultures were performed in triplicate, and results are expressed as AFC/10^6 spleen cells added at the beginning of the co-operative culture.

Cell Hybridization

The cell hybridizations were performed as previously described (Kontiainen et al, 1978) using polyethylene glycol (PEG, BDH, MW 1500). 10^8 in vitro primed cells and 10^7 BW5147 cells were washed twice in serum free BSS and pelleted together at 400 g. 0.5 ml of PEG was added slowly over a period of 1-2 minutes as the cells were gently shaken into suspension, and 0.5 ml of serum-free BSS was then added at the same rate. A further 5 ml were then added dropwise before slowly filling the tube to 20 ml with BSS. The cells were spun at 400g, the supernatant discarded and the cells resuspended in 100ml of MEM with 20% FCS. The cell suspension was dispensed in 2 ml aliquots into 48 wells of two 24 well Linbro trays (Cat. No. Flow FM 1624TC) and incubated at 37°C in a humidified atmosphere of 5% CO_2/95% air. 24 hours after the fusion 1.0 ml of medium was removed from each well and replaced by 1.0 ml HAT (MEM plus 20% FCS, $1x10^{-4}$M hypoxanthine, $1.6x10^{-5}$M thymidine and $4x10^{-7}$M aminopterin) and this procedure was repeated 48 and 72 hours after fusion. On the sixth, eighth and tenth day the medium was changed to HT (MEM plus 20% FCS, 1×10^{-4} M hypoxanthine, $1.6x10^{-5}$M thymidine). Thereafter (day 13 after fusion), the medium was changed to MEM plus 10% FCS, and the contents of each Linbro well which started to grow within the next 1-2 weeks were subcultured in Linbro wells and then transferred to Nunc tissue culture flasks (50 ml, Nunclon-Delta 1461). Aliquots of supernatants were tested for function when the cells were growing in the Linbro plates, and subsequently in flasks.

Functional Assays of Hybridoma Products

Hybridoma supernatants were tested in co-operative cultures exactly as described above for HF and SF preparations (Kontiainen and Feldmann, 1977).

Anti-Factor Antisera

To obtain antibodies against KLH-specific suppressor factor (SF) produced in vitro by KLH-specific suppressor cells of CBA origin, suppressor factor derived from serum-free cultures was first absorbed using KLH immunoadsorbents (Porath et al, 1967) and the neutralized acid eluates (using Sorensen's glycine, pH 2.4) of KLH columns were used to immunize rabbits and CBA mice. Rabbits received 0.5 ml of SF emulsified in 0.5 ml of complete Freund's adjuvant subcutaneously into four sites in the flanks. This is the equivalent of material from $2.5x10^6$ viable cells from SF-producing cultures per injection. The injections were repeated at weekly intervals for about 6 months. CBA mice received 0.2 ml of KLH column eluate emulsified with 0.2 ml of complete Freund's adjuvant intra peritoneally. The injections were repeated weekly for 6-10 weeks. At the end of the immunization period the rabbit serum and serum and ascites from CBA mice were collected (Kontiainen and Feldmann, 1979). The sera were inactivated (56°C for 30 min), Millipore-filtered (0.45μm filter) and stored at -20°C. The mouse ascites was difficult to filter and was either filtered at 1/10 dilution or 'sterilized' by 5000-rad irradiation (using a Co^{60} source) and stored at -20°C.

Maintenance of Stable Function in Hybridoma Cultures

One notorious problem in the study of T cell hybridomas and their products is the phenotype instability of these cells (see Taniguchi and Miller, 1978) due to spontaneous loss of chromosomes. We have found that continual re-selection (every 3-4 weeks) of hybrids by an anti-factor rosetting technique (Culbert et al, 1982) is a quick and easy method of maintaining the functional activity of our T cell hybridomas.

Sheep red cells (SRCs) were washed thrice in saline and 50 µl of appropriately diluted antiserum plus 500 µl $CrCl_3$ (10^{-4}M) was added to 50 µl of packed SRCs for 45 minutes at 37°C. The SRCs were washed three times in saline, resuspended to 10% in medium (+ 10% FCS, $5x10^{-5}$M 2-ME), and added to 1 ml of hybridoma cells (10^5 cells). These were pelleted together by gentle centrifugation (100-200 rpm for 5 minutes), and incubated for at least 6 hours at 4°C.

For separation of RFCs, the mixture was made up to 1 ml, layered on top of 2.5 ml of Ficoll-Hypaque solution (density = $1.075gm.l^{-1}$), and centrifuged at 3000 rpm for 15 minutes. The resulting pellet obtained was treated with NH_4Cl to lyse the SRCs. The hybridoma cells were washed three times, resuspended in complete medium ($2\text{-}3x10^5 ml^{-1}$) and cultured at 37°C in a 5% CO_2/95% air humidified atmosphere.

Cell Surface Antigens

Monoclonal anti-Thyl.1 and anti-Thyl.2 were kindly donated by Dr. P. Lake.

Monoclonal anti-Lyl.1, anti-Ly2.2, anti-D^b, anti-K^k and anti-I-A^k were kindly donated by Drs. G. and U. Hammerling. Monoclonal anti Ly 2.1 and anti -Ia^k and anti-Ia^b antisera were kindly donated by Dr I.F.C. McKenzie.

Indirect Immunofluorescence Technique

Cells (10^5 - 5 x 10^5) were washed thoroughly in serum-free PBS, incubated with 20 µl of appropriately diluted antiserum for 30 mins at 4°C, and then washed twice. 20 µl of appropriately diluted TRITC-conjugated goat anti-mouse, rat, or rabbit Ig (Nordic Diagnostics, Antwerp) were added for 30 mins at 4°C, and the cells washed thrice. Samples were air dried onto a 13 mm dia. coverslip (Chance Propper Ltd, Warley, England), fixed in 5% ethanoic acid/95% ethanol at -20°C for 30 mins, and the coverslips were mounted onto slides in glycerol containing 20% PBS. Fluorescence was evaluated using a x63 objective on a Zeiss Universal fluorescence microscope equipped with phase contrast, fluorescein and rhodamine optics and epi-illumination. At least 200 cells were scored for each sample.

Results

Functional Specificity of Cell Lines

(i) E1.6

This clone was derived from the hybridisation of in vitro activated CBA HC_{KLH} and BW 5147, and has the following phenotype:

Thy 1.1^+; Thy 1.2^+; H-$2K^{k+}$; $Lyl.1^-$; $Ly2.1^-$; Ia^{k-}.

E1.6 secretes an antigen specific HF which augments an in vitro antibody response to TNP-KLH, but not DNP-CGG (Fig. 1).

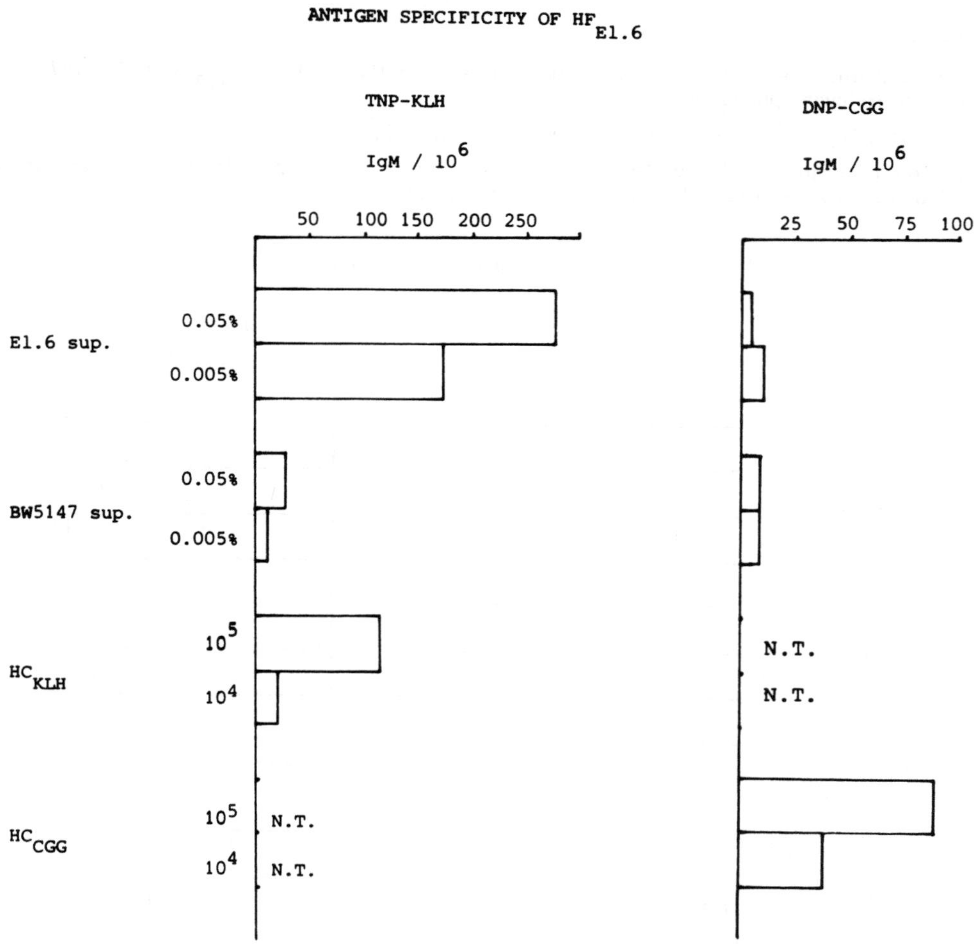

Fig. 1 Functional Specificity of $HF_{E1.6}$

Supernatants or HC were added to 5×10^5 anti-Thy 1.2 + C^1 treated DNP-OA primed spleen cells in the presence of the appropriate antigen (TNP-KLH or DNP-CGG) in microwells (200 µl). Direct PFCs were enumerated on d4 using DNP-coupled SRBCs.

(ii) E2.5

This cell line was derived from a fusion of in vitro activated B10 SC_{NP} with BW5147, and has the following phenotype:

Thy 1.1^+; Thy 1.2^+; $H\text{-}2K^{k+}$; $H\text{-}2D^{b+}$; Ly 1.1^-; Ly 2.2^+; Ia^{b+}.

As shown in Fig 2, E2.5 secretes an antigen specific SF which specifically suppresses an in vitro antibody response to NP-GAT but not TNP-KLH.

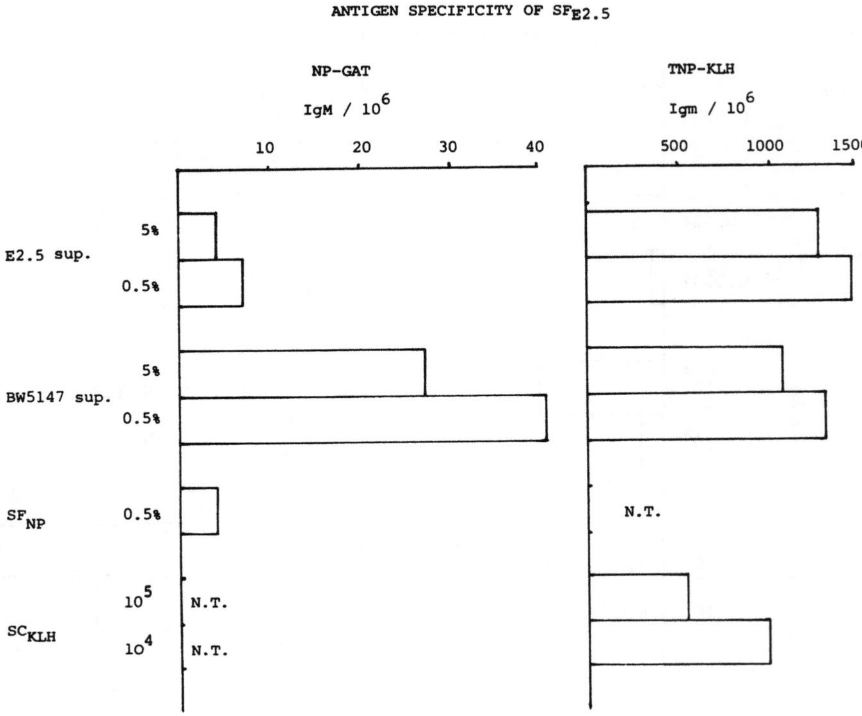

Fig. 2 Functional Specificity of $SF_{E2.5}$

Supernatants or SC were added to 5 x 10^5 B10 spleen cells plus 10^5 HC_{NP} or 10^5 HC_{KLH} in the presence of the appropriate antigen (NP-GAT or TNP-KLH) in microwells (200 µl). Direct PFCs were enumerated on d4 using GAT or DNP-coupled SRBCs.

(iii) K2S

This line was estabished in vitro, in collaboration with Dr P. Ricciardi-Castagnoli, from a thymoma induced by transformation of in vitro activated B10 SC_{KLH} with radiation leukaemia virus (see Ricciardi-Castagnoli et al, this volume). A cloned line, K2S, established from this thymoma suppresses an in vitro antibody response to TNP-KLH, but not DNP-CGG (Feldmann et al, manuscript in preparation).

(iv) $T'_{EA}7$

This TCGF-dependent ovalbumin-specific helper T cell line was kindly provided by Dr Max Schreier. (Schreier et al, 1982)

Rabbit Anti-helper and Anti-suppressor Factor Sera React with T Cell Hybrids

Antisera raised by repeated immunisation of rabbits with antigen-column eluates of in vitro HC or SC supernatants have been previously shown to remove helper or suppressor activities from active supernatants. Rabbit anti-helper factor (R anti-F_H) reacts with HF's, but not SF's, regardless of antigen-specificity, whereas rabbit anti-suppressor factor (R anti-F_S) reacts with SF's, but not HF's, regardless of antigen-specificity. These results have been interpreted by postulating the existence of 'constant' regions on HF's and SF's, such that R anti-F_H recognises a determinant(s) shared by most HF's, but not SF's, whereas R anti-F_S recognises a determinant(s) shared by most SF's, but not HF's (Kontiainen and Feldmann, 1979; Feldmann et al, 1980).

We investigated whether T cells and T cell lines expressed such 'constant' region isotype determinants on their membranes, since it has been suggested that HF and SF are perhaps secreted analogues of the membrane T cell receptor for antigen (Feldmann and Basten, 1972, Culbert et al, 1982). Thus, we tested the reactivity of various R anti-F sera toward various established hybridomas. Data for two sera, one raised against 'HF' (R anti-F_H # 121) and another raised against 'SF' (R anti-F_S # 111), are shown in Table 1.

Antiserum		Binding to: E1.6 (HC_{KLH})	E2.5 (SC_{NP})
R anti-F_H #121	1/5	+*	+
"	1/10	+	+
"	1/25	+	±**
"	1/50	+	-***
"	1/100	±	-
R anti-F_s #111	1/5	+	+
"	1/10	±	+
"	1/25	-	+
"	1/50	-	±
"	1/100	-	±

Table 1 Binding of rabbit anti-factor antisera to helper and suppressor hybridomas

Binding was assayed in a two stage immunofluorescence assay.
+* = > 70% cells stained
±** = 30-70% cells stained
-*** = < 10% cells stained.

As can be seen from Table 1, the R anti-F_H serum stained both helper and suppressor hybrids at higher concentrations (1/10 - 1/25), but stained only the helper hybrid (E1.6) at lower concentrations (1/50-1/100). The R anti-F_S serum stained both hybrids at higher concentrations (1/5-1/10), but stained only the suppressor hybrid (E2.5) at lower concentrations (1/50-1/100). This reciprocal staining pattern suggested the possibility that R anti-F_H #121 contains a species of antibody specific for the helper hybrid, E1.6, and that R anti-F_S #111 contains a species of antibody specific for the suppressor hybrid, E2.5.

Rabbit Anti-factor Sera Define T Cell Isotypes

The reactivity with various helper and suppressor T cell lines of R anti-F sera absorbed with helper or suppressor hybrids was investigated. The absorption protocol is shown in Fig. 3.

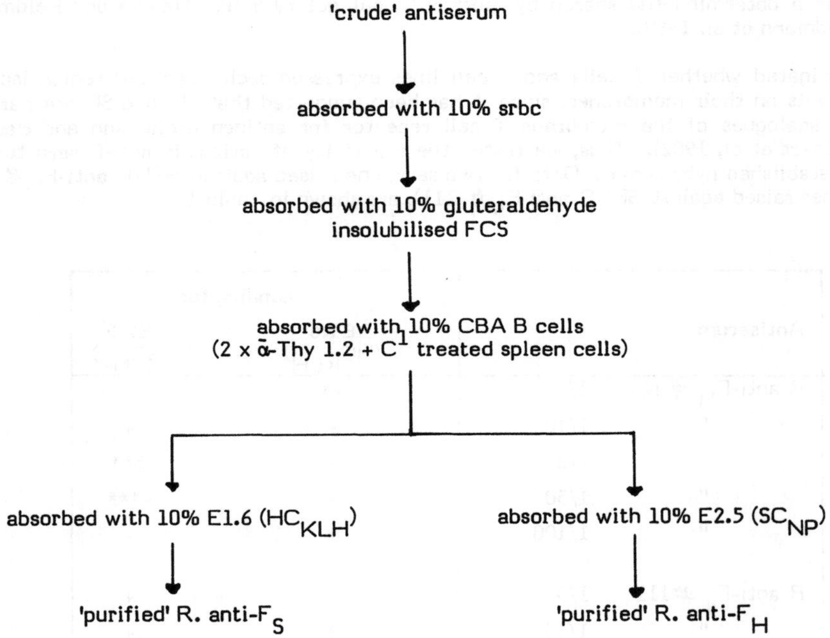

Fig 3 Absorption of R anti-F sera to produce 'purified' R anti-F_H and R anti-F_S

The staining patterns of two antisera before ('crude') and after ('purified') absorption are shown in Table 2.

Antiserum	Binding to:			
	E1.6 (HC_{KLH})	T'_{EA}-7 (HC_{OA})	E2.5 (SC_{NP})	K2S (SC_{KLH})
Rabbit serum	-	-	-	-
R.anti-F_H (# 121) 'crude'	+*	N.T.	+	N.T.
R.anti-F_H(# 121) 'purified'	+	+	-**	-
R.anti-F_S(#111) 'crude'	+	N.T.	+	N.T.
R.anti-F_S(#111) 'purified'	-	-	+	+
Mc.Rat anti-F_H (44.4)	+	+	-	-
Mc.M.anti-F_S (EB8)	-	-	+	+

Table 2 Binding of rabbit anti-factor sera to helper and suppressor T cell lines

Binding was assayed in a two stage immunofluorescence assay.
+* = > 70% stained
-** = < 5% stained
N.T. = Not tested

As shown in Table 2, after absorption with E2.5, R anti-F_H #121 reacted with the helper lines E1.6 and T_{EA}-7, but not the suppressor line, K2S.

After absorption with E1.6, R anti-F_S # 111 stained the suppressor lines E2.5 and K2S, but not the helper line T'_{EA}-7. Thus, R anti-F sera which recognise isotype determinants on helper or suppressor factors also recognise isotype markers on helper or suppressor T cell lines. We find it encouraging that a monoclonal (Mc) anti-F_H (James et al, manuscript in preparation) and monoclonal anti-F_S (Culbert et al, manuscript in preparation), which have opposing effects on in vitro antibody responses, show similar staining patterns to the 'purified' R anti-F_H and R anti-F_S sera, respectively. Binding of Mc anti-F_H to E1.6 is inhibited by R anti-F_H but not R anti-F_S, and binding of Mc anti-F_S is inhibited by R anti-F_S but not R anti-F_H (data not shown), which suggests that the determinants recognised by the monoclonal antisera are closely related, or identical to, determinants recognised by the R anti-factor sera.

Rabbit Anti-factor Sera React with Activated T Cells

'Purified' R anti-F_H and R anti-F_S sera stain less than 10% of virgin spleen cells (data not shown), but were shown to react with a significant number of in vitro activated T cells (ATC), as shown in Tables 3 and 4.

Antiserum	% Fluorescent Cells	
	Expt. 1	Expt. 2
NRS	1	3
R. anti-F_H (#121)	27	30
R. anti-F_S (#111)	12	19
R. anti-F_H + R. anti-F_S	37	48

Table 3 Binding of rabbit anti-factor sera to in vitro CBA HC_{KLH}

CBA spleen cells were cultured with 1 μgml^{-1} KLH for 4d in Marbrook cultures. Cells were eluted from nylon wool columns (Julius et al, 1973) before staining in a two-stage immunofluorescence assay.

As can be seen from Table 3, both R anti-F_H and R anti-F_S stain significant numbers of in vitro activated HC_{KLH}, with consistently more cells stained by R anti-F_H than by R anti-F_S.

The number of cells stained by both antisera is the sum of the totals stained by either antiserum alone, which suggests that the antisera are reacting with distinct subpopulations of cells.

Antiserum	% Fluorescent Cells	
	Expt. 1	Expt. 2
NRS	1	0
R. anti-F_H (#121)	9	5
R. anti-F_S (#111)	58	79
R. anti-F_H + R. anti-F_S	66	87

Table 4 Binding of rabbit anti-factor sera to in vitro CBA SC_{KLH}

CBA spleen cells were cultured with 100 μgml^{-1} KLH for 4d in Marbrook flasks. Cells were eluted from nylon wool columns (Julius et al, 1973) before staining in a two-stage immunofluorescence assay.

Table 4 shows that both R anti-F_H and R anti-F_S stain significant numbers of in vitro activated SC_{KLH}. In contrast to the data for HC^S(Table 3), significantly more cells are stained by R anti-F_S than by R anti-F_H. Again, the number of cells stained in the presence of both antisera is apporoximately the sum of the totals stained by either antiserum alone, suggesting reactivity of each antiserum with distinct subpopulations of cells. Thus, R anti-F_H and R anti-F_S recognise distinct subpopulations of ATC's. R anti-F_H binds to the majority of anti-factor positive in vitro HC, whereas R anti-F_S binds to the majority of anti-factor positive in vitro SC obtained from Marbrook cultures.

Discussion

We have shown that rabbit antisera reactive with T cell derived helper or suppressor factors contain reactivity for cell surface determinants on helper or suppressor cell lines, as well as mouse ATC's. Although these sera initially contain antibodies reactive with both helper and suppressor T cell lines, absorption with the appropriate helper or suppressor cell type results in loss of reactivity for one cell type, while reactivity with the other type remains. Thus R anti-F_H absorbed with E2.5, a SF_{NP}-secreting hybrid, reacts with the helper lines E1.6 and T'_{EA}-7, but not the suppressor line K2S. In contrast, R anti-F_S absorbed with the HF_{KLH}-secreting line, E1.6, reacts with the suppressor lines E2.5 and K2S, but not with the helper line T'_{EA}-7.

The precise nature of the membrane molecule carrying the determinants recognised by the 'purified' R anti-factor sera is unknown. However, by analogy with B cell products, where classes of Ig are termed 'isotypes', we have chosen to refer to these function-related determinants on T cell membrane bound and secreted molecules as 'isotypic determinants'.

Clearly, the determinants are unlikely to be related to the accepted allotypic MHC Class I and Class II determinants, since R anti-F_H reacts with cell lines from both H-2^k (E1.6) and H-2^b (T'_{EA}-7) haplotypes. Similarly, R anti-F_S reacts with both H-2^b (K2S; E2.5) and H-2^k (CBA SC_{KLH}) haplotypes. Additionally, any reactivities against 'conventional' MHC Class I and Class II antigens would have been expected to be removed during the absorbtion process with anti Thy 1 + C^1 treated spleen cells.

Reactivity with conventional Ly1 and Ly2, 3 antigens is excluded for the following reasons:

1. E1.6, which binds R anti-F_H, does not stain with antisera raised in mice and rats against mouse Ly1 and Ly2 antigens (Culbert et al, manuscript in preparation);

2. Two cell lines, WEHI 22 and EL-4, which express Ly1 and Ly2 antigens, respectively, do not bind R anti-factor sera (Culbert et al, manuscript in preparation); and

3. The number of anti-Ly1 and anti-Ly2 reactive cells in populations of in vitro induced HC and SC show significant differences between the numbers of cell stained by R anti-F_H and R anti-F_S (Culbert et al, manuscript in preparation).

The isotypic determinants are unlikely to be related to antigen-specificity, since R anti-F_H reacts with both KLH and OA specific helper lines, and R anti-F_S reacts with both NP and KLH specific suppressor lines. The sera also distinguish between helper and suppressor lines with the same antigen specificity, KLH.

Other reports of rabbit antisera raised against antigen-binding T cell materials (Cone et al, 1981; Binz and Wigzell, 1981) have suggested the presence of a 'constant region' on the T cell antigen receptor. These sera, however, bind to the majority of ATC, which suggests they may be directed against determinants common to T cell molecules, rather than the isotypic determinants recognised by our own antisera (see below).

Preliminary experiments using biosynthetically labelled E1.6 cells show that R anti-F_H precipitates a molecule(s) of MW 60-70K which breaks down on storage (-20°C) to give species of MW 40-50K and 20-30K (W. Cushley, personal communication). These characteristics are similar to those obtained for antigen-binding molecules in other laboratories (Taniguchi et al, this volume; Binz and Wigzell, 1981).

The isotype determinants recognised by R anti-F sera are not restricted to cell lines, but are also present on T cells activated in vitro (Tables 3 and 4) and in vivo (data not shown). Thus, R anti-F_H binds to a majority of HC induced in vitro, and R anti-F_S binds to a majority of SC induced in vitro. The antisera recognise distinct subpopulations in both sets of cultures, since the number of cells stained in the presence of both R anti-F_H and R anti-F_S is approximately the sum of the numbers of cells stained by either antiserum alone. These data are consistent with the notion that in vitro activated 'HC' or 'SC' cultures contain a mixture of both helper and suppressor cells, with the observed function of each culture due to the balance of helper vs suppression. Thus, if helper (= R anti-F_H reactive) cells outnumber suppressor (= R anti-F_S reactive) cells, help is the observed net product, with suppression the net result if suppressor (= R anti-F_S reactive) cells outnumber helper (= R anti-F_H reactive) cells.

Thus, we predict that 'purified' R anti-F_H (or R anti-F_S) reactive ATC's would show enriched helper (or suppressor) function, and experiments are planned to test the validity of this hypothesis.

'Contaminating' anti-SC (in R anti-F_H) and anti-HC (in R anti-F_S)activities may have arisen for two reasons:

1. The antigen eluate used to produce the antisera probably contains both HF's and SF's, since no attempt was made to purify HC or SC before production of factors. Thus, R anti-F_H would be expected to contain antibodies raised against the contaminant SF's, and R anti-F_S would similarly be expected to contain antibodies raised against the contaminant HF's.

2. There may be reactivity to a determinant(s) common to all mouse helper and suppressor factors present in the antisera, distinct from the defined helper or suppressor isotypic determinant(s).

An investigation of the full spectrum of T cell reactive specificities contained in the 'crude' and 'purified' R anti-F sera is being vigorously pursued. There is accumulating evidence for the existence of distinct SC subpopulations which secrete SF's distinguishable on the basis of serology and target cell type (see Germain and Benacerraf, 1981). It will be of great interest to know whether such SC subpopulations can be distinguished by different anti-F_S preparations, and whether similary heterogeneity exists within HC populations as defined by R anti-F_H sera.

We are attempting to resolve these issues by (i) the production, and subsequent characterization of further T cell lines using anti-factor antibodies, and (ii) the production of monoclonal anti-factor antibodies. This latter appears a particularly fruitful approach as various anti-factor antibodies already produced stain helper and suppressor cells in a similar fashion to the R anti-F_H and R anti-F_S sera (Table 2). A monoclonal anti-F_H which enhances an in vitro antibody response by augmenting HC specific for the antigen present in culture (James et al, in preparation) shows a similar staining pattern to R anti-F_H. This binding to HC is blocked by the addition of R anti-F_H, but not R anti-F_S, suggesting that the determinant recognised by the monoclonal anti-F_H is also recognised by R anti-F_H, but not by R anti-F_S. Analogous results have been obtained for a monoclonal anti-F_S which increases suppression in vitro (Culbert et al, in preparation). This antibody binds to SC, and the binding is blocked by R anti-F_S, but not R anti-F_H, suggesting that the determinant recognised by the monoclonal anti-F_S is also recognised by R anti-F_S, but not R anti-F_H.

The existence of antisera against T cell surface determinants directly related to function is an exciting prospect for future study. Such antisera will enable direct manipulations of T cell isotypes, an important advance in the investigation of the complex network of T cell interactions. Identification of analogous determinants in humans would not only aid experimental investigation in man, but might also have important clinical applications.

Acknowledgements

This work was supported by the Imperial Cancer Research Fund, National Institute of Health Grant Nos AI-13145-05, AI-15653-03, A1-15636-03, and the Sigrid Juselius Foundation. EJC is in receipt of a Medical Research Council Training Scholarship. LMD is in receipt of an SERC Training Scholarship.

Professor M.B. Rittenberg is thanked for his gift of KLH, and Professor O. Makela for NP-T_4 and NP-GAT preparations.

We are grateful to Philippa Wells for her help in the preparation of this manuscript.

References

Binz H, Wigzell H (1975) Shared idiotypic determinants on B and T lymphocytes reactive against the same antigenic determinants. I. Demonstration of similar or identical idiotypes on IgG molecules and T cell receptors with specificity for the same alloantigen. J Exp Med **142:** 197-211.

Binz H, Wigzell H (1981) T cell receptors with allo-major histocompatibility complex specificity from rat and mouse: Similarity of size, plasmin susceptibility, and localisation of antigen-binding region. J Exp Med **154:** 1261-1278.

Cone RE, Rosenstein RW, Murry JH, Iverson GM, Ptak W, Gershon RK (1981) Characterisation of T cell surface protein bound by heterologous antisera to antigen-specific T cell products. Proc Natl Acad Sci USA. **78:** 6411-6418.

Culbert EJ, Cecka JM, Erb P, James RFL, Kontiainen S, Ramilla G, Simpson E, Todd I, Feldmann M (1982) Antigen-specific T cell factors derived from monoclonal sources. Lymphokines **5:** (in press).

Cunningham AJ, Szenberg A (1968) Further improvements in plaque techniques for detecting single antibody forming cells. Immunol **14:** 599-580.

Eichmann K (1978) Expression and function of idiotypes on lymphocytes. Adv in Immunol **26:** 195-258.

Feldmann M, Basten A (1972) Cell interactions in the immune response in vitro. III. Specific collaboration across a cell impermeable membrane. J. Exp. Med. **136:** 49-67.

Feldmann M, Cecka JM, James RFL, Todd I, Kontiainen S (1980) Production of antisera to antigen-specific factors. Contemp Topics Immunobiol **11:** 275-288.

Howie S, Feldmann M (1978) Immune response genes expressed at macrophage-B lymphocyte interactions. Nature **273:** 664-666.

Howie S, Parish CR, David CS, McKenzie IFC, Maurer PH, Feldmann M (1979) Serological analysis of antigen-specific helper factors specific for (T,G)-A--L and GAT. Eur. J Immunol **9:** 501-506.

Julius MH, Simpson E, Herzenberg LA (1973) A one-stage method for the purification of T cells. Eur J Immunol **3:** 645-651.

Kontiainen S, Feldmann M (1976) Suppressor cell induction in vitro. I. Kinetics of induction of antigen-specific suppressor cells. Eur J Immunol. **6:** 296-301.

Kontiainen S, Feldmann M (1977) Suppressor cell induction in vitro. III. Antigen-specific suppression by supernatants of suppressor cells. Eur J Immunol **7:** 310-314.

Kontiainen S, Simpson E, Bohrer E, Beverley PCL, Herzenberg LA, Fitzpatrick WC, Vogt P, Torano A, Mckenzie IFC and Feldmann M (1978) T cell lines producing antigen specific suppressor factor. Nature **274:** 477-480.

Kontiainen S, Feldmann M (1979) Structural characteristics of antigen-specific suppressor factors: Definition of constant and variable regions. Thymus **1:** 59-66.
Lamb JR, Zanders ED, Sanderson AR, Ward PJ, Feldmann M, Kontiainen S, Lehner T, Woody JN (1981) Antigen-specific helper factor reacts with antibodies to human β_2-microglobulin. J Immunol **127:** 231-234.
Makela O (1966) Assay of anti-hapten antibody with the aid of hapten-coupled bacteriophage. Immunol **10:** 81-86.
Marchalonis JJ, Warr GW, Bucana C, Hoyer LC (1979) The immunoglobulin-like T cell receptor. I. In situ demonstration of immunoglobulin Fab-region determinants of rhodent T-and B-lymphocytes using chicken antibodies. J. Immunogenet. **6:** 289-310.
Porath J, Axen R, Ernback S (1967) Chemical coupling of proteins to agorose. Nature **215:** 1491-1493.
Puri J, Ben-Neriah Y, Givol D, Lonai P (1980) Antibodies to immunoglobulin heavy chain variable regions protect helper cells from specific suicide with radiolabelled antigen. Eur J. Immunol **10:** 281-284.
Rittenberg MB, Amkraut AA (1966) Immunogenicity of trinitrophenylhaemocyanin: Production of primary and secondary anti-hapten precipitins. J Immunol **97:** 421-430.
Schrier MH, Tees R, Nordin AA (1981) Establishment and characterisation of helper T cell clones: Their functional heterogeneity and effect on the B cell responses to particulate soluble and T-independent antigens. Lymphokines **5:** (in press).
Strausbauch P, Sulica A, Givol D (1970) General method for the detection of cells producing antibodies against haptens and proteins. Nature **227:** 68-69.
Taniguchi M, Tada T (1974) Regulation of homocytotropic antibody formation in the rat. X IgT-like molecule of the induction of homocytotropic antibody response. J Immunol **113:** 1757-1769.
Taniguchi M, Tada T (1976) Properties of antigen-specific suppressive T cell factor in the regulation of antibody responses in the mouse. II. In vitro activity and evidence for the I region gene product. J Immunol **116:** 542-548.
Taniguchi M, Miller JFAP (1978) Specific suppressive factors produced by hybridomas derived from the fusion of enriched suppressor T cells and a T lymphoma line. J Exp Med **148:** 373-382.
Taniguchi M, Saito T, Takei I, Kanno M, Takuhisha T, Tomioka H (1982) Antigen-specific suppressive factors derived from T cell hybridoma: Their structure and interactions. Lymphokines **5:** (in press).
Taussig M, Munro A (1974) Removal of specific co-operative T cell factor by anti-H-2 but not anti-Ig sera. Nature **251:** 63-65.
Zanders ED, Lamb JR, Kontiainen S, Lehner T (1980) Partial characterisation of murine and monkey helper factor to a streptococcal antigen. Immunol. **41:** 587-596.

An Antigen-Specific Suppressor T Cell Factor Controlled by Two Genes in the Immunoglobulin Heavy Chain Linkage Group and in the I-J Subregion of the H-2 Complex

M. Taniguchi, T. Tokuhisa, M. Kanno, T. Honjo

A. Introduction

An antigen-specific suppression of antibody response has been known to be mediated by the factor derived from suppressor T cells (Ts). An antigen-specific suppressor T cell factor (TsF) has been found to possess the antigen-binding moiety and the products of genes in the I-J subregion of the H-2 complex (1). No known constant region determinant of the conventional immunoglobulins (Igs) has been found on suppressor T cells and their factors. It is, however, reported that TsF carries the structure analogous to the variable region of the Ig heavy chain (Igh) (1,2). Therefore, at least two gene products may be involved in the structural entity of the antigen-specific TsF: one encoded by genes in the I-J subregion and the other by genes analogous or identical to the Igh-V genes.

Despite the information mentioned above, the nature of the antigen-binding structure on suppressor T cells or the antigen-specific TsF is largely unknown. To solve these questions, many attempts have been made during the past few years to establish functional T cell lines by somatic cell hybridization techniques. Several reports on T cell hybridomas with antigen-specific suppressor activities have been published (3-6). Such T cell lines are useful in biochemical and molecular analyses of the antigen-recognition units on T cells or TsF. In this report, we will summarize our recent studies on the functional and structural properties of the antigen-specific TsF. Furthermore, the functional activity of the translation products of the fractionated messenger RNA coding for the antigen-specific suppressor T cell factor will also be described.

I. Antigen-Specific Suppressor T Cell Factor Composed of Two Distinct Polypeptide Chains

Suppressor T cell hybridomas specific for keyhole limpet hemocyanin (KLH) were established by the fusion of the AKR-derived thymoma cell line (BW5147) and suppressor T cells from KLH-primed C57BL/6 mice which were enriched by binding to KLH-coated Petri dishes. Several T cell hybrid lines with KLH-specific suppressor function have so far been maintained in our laboratory for about three years (5).

The hybridoma cells express $I\text{-}J^b$ products and KLH-binding activity on their cell surface. Furthermore, the KLH-specific TsF was obtained in the cell-free supernatant after freeze-thaw treatment of the hybridoma cells (extracted TsF) or in the ascitic fluid from hybridoma-bearing mice (secreted TsF) (5). Both the extracted and secreted TsF suppressed the anti-DNP IgG secondary antibody responses in an antigen-specific manner when they were added to the culture of DNP-KLH-primed

C57BL/6 spleen cells. The immunochemical properties of the KLH-TsF derived from the I-J^+ T cell hybridoma were characterized by using immunoadsorbent columns of various antigens and antibodies. In the experiments the extracts or ascitic fluid were applied to and incubated with the column at 0°C for 1 hr. The effluent or the acid eluate with 0.175 M glycine-HCl buffer, pH 3.2, from immunoadsorbent columns was added to the culture and tested for suppressor activity. The results clearly showed that the suppressor activity in the hybridoma extracts or in the ascitic fluid was absorbed with the columns composed of KLH and conventional or monoclonal anti-I-J^b antibodies [B10.A(5R) anti-B10.A(3R)] but not with those of irrelevant antigens (i.e., OVA or Ascaris suum extract), anti-I-J^k or rabbit antibodies against mouse Igs or Fab of mouse Igs. Therefore, it is concluded that KLH-TsF carries two distinct characteristics, i.e., the KLH-binding moiety and the products of genes in the I-J subregion of the H-2 complex.

The next experiments were carried out to determine whether these two determinants are on the same molecule (7,8). If two distinct determinants are on the same molecule, the effluent of the TsF from the KLH or the anti-I-J^b column should not contain any components of the factor. In this case, no suppressor activity would be obtained when the effluents from the KLH and anti-I-J^b are mixed. On the other hand, if the two determinants are present on the two distinct molecules in non-covalent association the mixture of the two effluents from the columns could be reassociated with the KLH-binding and I-J-encoded molecules, which would reconstitute the suppressor activity.

Table 1. Requirement of KLH-binding and I-J-bearing molecules for the expression of TsF activity

Materials added to the culture a)	Suppressor Activity in Anti-DNP IgG PFC Response: Extracted TsF(TsF) b)	Secreted TsF(s-TsF) c)	Reduced s-TsF d)	Alkylated s-TsF d)	Red. + Alkyl. s-TsF d)
Unfractionated TsF	yes	yes	yes	yes	no
TsF absorbed with KLH	no	no	no	-	-
TsF absorbed with anti-I-J^b	no	no	no	-	-
Mixture of 1/2 KLH-absorbed and 1/2 anti-I-J^b absorbed	yes	no	yes	-	-

a) KLH-TsF absorbed with or without immunadsorbent columns was added to the culture of DNP-KLH-primed spleen cells.
b) Extracted TsF: KLH-TsF in freeze-thaw materials from the hybridoma cells.
c) Secreted TsF: KLH-TsF in ascitic fluid from hybridoma-bearing mice.
d) Secreted TsF (s-TsF) purified with the KLH column was treated with DTT (reduction) and/or iodoacetamide (alkylation).

To resolve this question, the hybridoma extract or the ascites was absorbed with the KLH or the anti-I-J^b column; the effluents from the columns were then admixed and tested for suppressor activity. The results are shown in Table 1. The suppressor activity in the extract or in the ascites was completely absorbed out with the KLH or anti-I-J^b column. However, the mixture of the effluents of the extracts from the KLH and anti-I-J^b columns produced a strong suppression, even though twice the amount of either effluent alone had no detectable suppressor activity. Therefore, the KLH-binding and I-J-encoded products in the extracts seem to be present in non-associated form. In this case, the KLH-binding molecules could be recovered from the effluent from the anti-I-J^b column.

Similarly, the effluent from the KLH column contained the nonassociated I-J-encoded products. Hence, the mixture of the two effluents resulted in the reassociation of the KLH-binding and I-J-encoded molecules, which reconstituted the suppressor activity. Moreover, the TsF eluted from either the KLH or the anti-I-J^b column seems to be composed of two chains in noncovalent association since the suppressor activity was elutable from either the KLH or the anti-I-J^b column. It is, therefore, conceivable that a state of equilibrium exists between the associated and dissociated forms of TsF in the extract.

On the other hand, the secreted TsF, unlike the extracted TsF, was found to be composed of KLH-binding and I-J-encoded chains in covalent (disulfide bond) association (8). In this experiment, the secreted TsF purified with the KLH column was reduced with dithiothreitol (DTT) to cleave into two polypeptide chains. The KLH-binding and the I-J-encoded products in the reduced TsF were subsequently separated on the KLH column under reducing conditions. The mixture of the effluent and the acid eluate from the column was tested for suppressor activity. Under these conditions the I-J products are supposed to be present in the effluent. However, the KLH-binding molecules should be recovered in the acid eluate from the column. As expected, suppressor activity was observed in the mixture of the effluent and acid eluate, neither of which showed suppressor effects. It is thus strongly suggested that KLH-binding and I-J-encoded polypeptide chains of the secreted TsF are linked in covalent association with disulfide bonds, and that the two cleaved chains obtained by the reduction of TsF could reconstitute the active suppressor factor.

From the above results we concluded that the KLH-TsF is composed of two distinct polypeptide chains of the KLH-binding and the I-J-encoded products, and that the association of these two chains is essential for the expression of TsF activity. It is also likely that the KLH-binding and I-J-encoded chains are independently synthesized in cytoplasm and secreted as an active suppressor factor after their covalent association with disulfide bonds.

II. Genes Coding for the Constant Region (Ct) of the Antigen-Binding Molecule of KLH-TsF

Recent reports from various laboratories have indicated that the antigen-specific T cell factors or the isolated T cell receptor molecules carry the determinants analogous to the V region of the immunoglobulin heavy chain (Igh). Hence, the antigen-binding structure on T cell antigen-recognition units seems to share the V-region markers with conventional Ig. In this sense, the antigen-recognition unit on T cells appears to possess a constant region, just as do Igs. However, the constant region determinants on T cell antigen receptors or the factors have so far not been defined. Judging by the similarities between the V-region structures on T and B cell antigen receptors, it is possible that T cells utilize similar or the same gene pools as those coding for the V region of Igh, and that the constant region determinants on T cell antigen receptors are encoded by genes on the same chromosome as those coding for the antigen-binding structure.

If this hypothesis is correct, the immunization of mice with T cells from Igh allotype congeneic mice would be able to provide antibodies against the allotypic determinants on the antigen-recognition units on T cells. In this respect, the combination of BALB/c and CB-20 mice is ideal for preparing antibodies against the products of genes linked to the Igh gene cluster. The Igh allotype congeneic mice, CB-20, are derived from a cross between BALB/c ($Igh\text{-}1^a$) and C57BL/6 ($Igh\text{-}1^b$), and carry the $Igh\text{-}1^b$-linked genes on the chromosomal segment from C57BL/6 origin with BALB/c background. In fact, antibodies reactive to the allotypic determinants on the antigen-binding molecules of the TsF from $Igh\text{-}1^b$ mice have been obtained by the repeated immunization of BALB/c with Con A-stimulated CB-20 spleen cells (9,10). Furthermore, monoclonal antibodies have also been established by the hybridization of immunized BALB/c mice with BALB/c-derived P3U1 myeloma (11).

Table 2. Genetic specificity of BALB/c anti-CB-20

Mouse strain	H-2 complex	Igh-1 allotype	Reactivity to thymocytes
C57BL/6	b	b	+
C3H	k	j	-
C3H.SW(CSW)	b	j	-
CWB	b	b	+
BALB/c	d	a	-
BAB-14	d	b	+
SJL	s	b	+
SJA	s	a	-
AKR/J	k	d	-
A/J	a	e	-

The genetic specificty of the conventional and monoclonal BALB/c anti-CB-20 antibodies was investigated by quantitative absorption studies. Some of the results are shown in Table 2. The cytotoxic activity of the antibody on the KLH-Ts hybridoma was absorbed with thymocytes from C57BL/6, CWB and SJL but not with those from BALB/c, C3H, CSW, SJA, AKR, and A/J. Therefore, $Igh\text{-}1^b$ thymocytes, but not those from mice with other Igh allotypes, always removed the antibody activity. Furthermore, the T cell specificity of the antibody was established, since the BALB/c anti-CB-20 could react with B6 spleen cells or thymocytes but not with B6 spleen cells treated twice with anti-Thy 1 and C (B6 B cells), BALB/c thymocytes or various classes of $Igh\text{-}1^b$ immunoglobulins. It is therefore concluded that the antibody recognizes the allotypic determinants on $Igh-1^b$ T cells.

Moreover, the BALB/c anti-CB-20 reacted with BAB-14 thymocytes. The recombinant strain, BAB-14, is derived from a cross between BALB/c and C57BL/6 mice, like CB-20 mice. However, BAB-14 mice carry the Igh-V genes from BALB/c mice and the Igh-C genes from C57BL/6 mice. Therefore, the crossover point of the BALB/c and C57BL/6 chromosomes seems to be in between the Igh-V and the Igh-C genes. As the activity of the monoclonal antibody (7C5) specific for the allotypic determinants on $Igh-1^b$ T cells was absorbed with BAB-14 thymocytes, the gene coding for the T cell allotypic determinants defined by 7C5 is located somewhere to the right side of the Igh-V genes.

Owen et al. have also reported that the BALB/c anti-CAL-20 antiserum could detect the allotypic determinants (Tsu^d) on Ly 2^+ suppressor T cells and $Tind^d$ on Ly 1^+ suppressor-inducer T cells (12). The determinants have been demonstrated to be coded for by genes on the right side of the Igh-C gene cluster. No direct evidence has, however, been demonstrated that the anti-Tsu^d or anti-$Tind^d$ recognizes the constant region determinants on T cell antigen receptors.

Table 3. Adsorption of KLH-specific and OVA-specific TsF with monoclonal BALB/c anti-CB-20 (7C5)

TsF absorbed with	Materials added	Anti-DNP IgG PFC per culture	
		KLH-TsF	OVA-TsF
	None	1430	1010
None	Unabsorbed TsF	160	130
7C5	Effluent	1500	1200
7C5	Eluate	180	60

Table 4. Evidence that the monoclonal BALB/c anti-CB-20 (7C5) recognizes the antigen-binding molecule of KLH-TsF

Materials added to the culture				Anti-DNP IgG PFC per culture
	Effluent of TsF from the column			
Unabsorbed TSF	KLH	7C5	anti-I-J^b	
	None			1750
+	-	-	-	150
-	+	-	-	1640
-	-	+	-	1700
-	-	-	+	1950
-	+*	+*	-	1810
-	+*	-	+*	150
-	-	+*	+*	110

*Half of the effluent of the extracted KLH-TsF from the column was added to the culture.

The results shown in Tables 3 and 4 provide explicit evidence that the 7C5 antibody recognized the determinant on the constant region (Ct) of the antigen-binding molecules of KLH-TsF. In the experiment shown in Table 3 the conventional KLH-TsF or OVA-TsF obtained from primed C57BL/6 thymocytes was absorbed with immunoadsorbent columns of monoclonal 7C5 antibodies. The effluent or the acid eluate from the column was tested for suppressor activity. As shown in Table 3 the same monoclonal 7C5 column was able to absorb the two distinct TsF with different antigen specificity (i.e., KLH and OVA), strongly suggesting that the antibody recognizes the constant region on the TsF. Furthermore, as the monoclonal antibody was prepared by immunization with unprimed Con A-stimulated CB-20 spleen cells, the specificity of the antibody should not be directed against idiotypic determinants on the antigen-binding structure.

Furthermore, the 7C5 was found to recognize the determinants present on the antigen-binding molecule of the KLH-TsF. The KLH-TsF extracted from the T cell hybridoma cells was absorbed with the columns of KLH, anti-I-J^b or 7C5. The effluents from the columns were admixed to test the TsF activity. The suppressor activity of the KLH-specific hybridoma extract was completely absorbed either with the column of KLH, anti-I-J^b or 7C5. However, strong suppressor activity was observed when the effluents of the KLH and anti-I-J^b or the effluents of the 7C5 and anti-I-J^b were mixed. On the other hand, the mixture of the effluents from the KLH and 7C5 could not reconstitute the suppressor activity. Taking into account the results demonstrated in the previous section, showing that the association of the KLH-binding and the I-J products is essential for the expression of the TsF activity, the results in Table 4 are clearly explained by the following: The 7C5 absorbed the KLH-binding molecule so that the I-J-encoded molecule of the extracted TsF could recover in the effluent. Therefore, the mixture of the effluents from the KLH and 7C5 resulted in reconstituting the TsF activity.

The above findings demonstrate that the allotypic marker defined by the monoclonal BALB/c anti-CB-20 (7C5) represents the constant region determinant on the antigen-binding molecule of the suppressor T cell factors. Furthermore, it is also demonstrated that the genes coding for the constant region determinants on the T cell receptor are located on the right side of the Igh-V gene cluster on the twelfth chromosome.

III. Translation of mRNA Coding for the I-J or the KLH-Binding Polypeptide Chain and the Function of Translation Products

Many attempts have been made to characterize biochemical properties of the antigen-specific TsF. However, the data on the chemical structure of TsF are still controversial. This might be due to the lability of TsF, which is very easily degraded, and also to the difficulties of obtaining enough materials to characterize the primary sequence of TsF biochemically. Therefore, we decided to circumvent problems of cleavage and amount by approaching the functional and biochemical determination of the antigen-specific TsF by analysis at the gene level.

The messenger RNA (mRNA) coding for the KLH-binding and I-J-encoded products was isolated. Suppressor Ts hybridoma cells (34S-704) were homogenized in 100 mM Tris-HCl buffer, pH 9.0, containing 250 mM sucrose, 100 mM NaCl, sodium dodecyl sulfate (SDS, 0.3% final concentration), and rat liver extract as RNase inhibitor. The total RNA was extracted from homogenized materials with equal volume of the mixture of phenol : metacresol : 8-hydroxyquinoline (9:1:0.1, w/v). RNA from the aqueous phase was recovered by ethanol precipitation. Poly(A) containing RNA was isolated by chromatography on oligo(dT)-cellulose. Elution was achieved with 10 mM Tris-HCl, pH 7.5, and 0.1 mM EDTA. Poly(A)$^+$-RNA represents 1-2% of the total RNA fraction loaded on the oligo(dT) column and then fractionated on a 5-22% sucrose gradient containing 10 mM Tris-HCl, pH 7.5, and 2 mM EDTA. Poly(A)$^+$-RNA in the fractions was precipitated with ethanol and resuspended in 50 µl of distilled water.

An aliquot (0.05-0.1 µl) of each fraction was injected and translated in *Xenopus laevis* oocytes. Ten injected oocytes were incubated in 100 µl of sterile modified Barth's medium with or without 20 µCi ^{35}methionine at room temperature (19-21°C) for 36 hrs. After translation, a part of the translation products of each fraction was immunoprecipitated with monoclonal antibodies against the I-J products (monoclonal anti-I-J^b: E10, F4 and H6) or against the constant region of the antigen-binding molecule (mixture of monoclonal BALB/c anti-CB-20: 7C5 and 7D1). Furthermore, for the detection of the I-J or the antigen-binding molecule in the translation products, the inhibition of the cytotoxicity of monoclonal antibodies on the T cell hybridoma was carried out. The cytotoxicity of the monoclonal anti-I-J^b on the KLH-specific suppressor T cell hybridoma (KLH-8C28) was completely blocked by addition of the translation products of Fr 13 but not by those of any other fractions. Similarly, the cytotoxic activity of the 7C5 antibody was blocked by the Fr 15 product and also by the Fr 20 product, indicating that the antigen-binding polypeptide chains would have two different molecular sizes. The next experiments were carried out to determine whether the translation products would have suppressor activity when the two distinct translation products were added together in the in vitro culture system.

Table 5. Reconstitution of TsF activity with two distinct translation products of mRNA from KLH-Ts hybridoma (34S-704)

Translation products of fractionated mRNA	Products defined by	Anti-DNP IgG PFC per culture
None		1470
Fr 13	Anti-I-J^{b*}	1800
Fr 15	Anti-Ct^{b**}	1450
Fr 20	Anti-Ct^b	1620
1/2 Fr 13 and 1/2 Fr 15		210
1/2 Fr 13 and 1/2 Fr 20		630

*Monoclonal B10.A(5R) anti-B10.A(3R) antibodies (E10, F4 and H6).
**Monoclonal BALB/c anti-CB-20 antibody (7C5).

Combinations of Fr 13 and Fr 15 products or of Fr 13 and Fr 20 products were added to the culture and tested for their activity. The results are shown in Table 5. The translation products of Fr 13, Fr 15 and Fr 20 mRNA did not exert any suppressor activity by themselves. However, the mixture of the half volume of the materials from Fr 13 and Fr 15 produced strong suppression. Significant but not strong suppression of antibody response was also obtained by mixing Fr 13 and Fr 20. The facts that the Fr 13 mRNA encodes I-J products and the Fr 15 and Fr 20 mRNA encode the antigen-binding polypeptide chains suggest that the mixture of the I-J and the antigen-binding polypeptide chains obtained by the translation of the fractionated mRNA successfully reconstitutes the active form of the antigen-specific suppressor T cell factor. The suppression was found not to be nonspecific, because the mRNA products suppressed the response of C57BL/6 but not that of BALB/c mice to DNP-KLH. The results are in agreement with our previous data showing that the hybridoma-derived KLH-TsF (34S-704) suppresses the responses of syngeneic or semisyngeneic mice [C57BL/6 or (C57BL/6 x C3H)F_1] but not those of allogeneic mice (C3H or BALB/c). Furthermore, the suppressor activity mediated by the mixture of the translation products of the Fr 13 and Fr 15 mRNA was found to be KLH-specific, since the materials suppressed anti-DNP IgG PFC response against DNP-KLH but not against DNP-OVA.

B. Concluding Remarks

Our results presented here clearly demonstrate that an antigen-specific TsF is composed of two gene products, i.e., the I-J and the antigen-binding polypeptide chains, one of which is controlled by a gene in the I-J subregion of the H-2 complex on the seventeenth chromosome while the other is controlled by a gene on the right side of the Igh-V gene cluster on the twelfth chromosome. The two gene products were successfully translated in *Xenopus laevis* oocytes by the fractionated messenger RNA having sizes of about 19S (Fr 20), 14S (Fr 15) and 11.5S (Fr 13), obtained from the KLH-specific suppressor T cell hybridoma (34S-704).

The translation product of the Fr 13 mRNA was determined to be the I-J molecule in the inhibition of the cytotoxic assay or in the biochemical precipitation analysis with monoclonal anti-I-J^b antibodies. Similarly, the products of the Fr 15 and Fr 20 mRNA were demonstrated to be the antigen-binding molecule, since they could inhibit the cytotoxic activity of the anti-Ct (7C5). Therefore, it is conceivable that these two products of the Fr 15 and the Fr 20 mRNA are the antigen-binding molecule in different forms, i.e., membrane and secretory types.

Most intriguing is the finding that the combination of the translation products of mRNA coding for the I-J and the antigen-binding polypeptide chains, neither of which had detectable functional activities by itself, could reconstitute the strong suppressor function. The results are quite in agreement with our previous data (8,9) demonstrating that the mixture of the KLH-binding and the I-J-encoded molecules obtained from the hybridoma extract or from the reduced secreted TsF was able to reconstitute the antigen-specific suppressor activity.

In order to achieve further biochemical characterization and gene organization of the antigen-specific TsF, cloning and sequencing of the corresponding DNA has been undertaken.

Acknowledgments. We thank Miss Hisano Nakajima and Miss Chie Ono for their excellent secretarial assistance. The work was partly supported by grants from The Ministry of Education, Culture and Science, Japan.

References

1. Taniguchi M, Saito T, Takei I, Kanno M, Tokuhisa T, Tomioka H (1981) Suppressor T cell hybridomas and their soluble products. In: Pick E (ed) Lymphokines, Vol. 15. Academic Press, New York, p 77
2. Tada T, Hayakawa K, Okumura K, Taniguchi M (1980) Coexistence of the variable region of immunoglobulin heavy-chain and I-region gene products on antigen-specific suppressor T cells and suppressor T cell factors - a minimal model of a functional antigen receptor of T cells. Mol Immunol 17: 867
3. Kontiainen S, Simpson E, Bohrer E, Beverley PCL, Herzenberg LA, Fitzpatrick WC, Vogt P, Torano A, McKenzie IFC, Feldmann M (1978) Hybrid cell lines with T-cell characteristics. Nature 274: 477
4. Taussig M, Covalan JRF, Binns RM, Holliman A (1979) Production of an H-2-related suppressor factor by a hybrid T cell line. Nature 277: 305
5. Taniguchi M, Saito T, Tada T (1979) Antigen-specific suppressive factor produced by a transplantable I-J-bearing T cell hybridoma. Nature 278: 555
6. Okuda K, Minami M, Ju ST, Dorf ME (1981) Functional association of idiotypic and I-J determinants on the antigen receptor of suppressor T cells. Proc Natl Acad Sci USA 78: 4557
7. Taniguchi M, Takei I, Tada T (1980) Functional and molecular organization of an antigen-specific suppresor factor derived from a T cell hybridoma. Nature 283: 227
8. Taniguchi M, Saito T, Takei I, Tokuhisa T (1981) Presence of interchain disulfide bonds between two gene products that compose the secreted form of an antigen-specific suppressor factor. J Exp Med 153: 1672
9. Tokuhisa T, Taniguchi M (1982) Two distinct allotypic determinants on the antigen-specific suppressor and enhancing T cell factors that are encoded by genes linked to the immunoglobulin heavy-chain locus. J Exp Med, in press
10. Tokuhisa T, Taniguchi M (1982) Constant region determinants on the antigen-binding chain of the antigen-specific suppressor T cell factor. Submitted for publication.
11. Tokuhisa T, Komatsu S, Uchida Y, Taniguchi M (1982) Monclonal antibodies specific for the constant region of the T cell antigen receptors. Submitted for publication.
12. Owen FL, Riblet R, Taylor BA (1981) The suppressor cell alloantigen Tsu^d maps near immunoglobulin allotype genes and may be a heavy-chain constant-region marker on a T cell receptor. J Exp Med 153: 801

Structural and Functional Studies on Antigen-Specific Suppressor Factors from T Cells and T Cell Hybrids

M.J. Taussig, G.L. Asherson, A. Holliman, N. Richardson, G. Skibinski

Introduction

T cell hybrids are a potentially valuable source of antigen-specific helper and suppressor factors, molecules characterised by their ability to bind specifically to antigen, their lack of immunoglobulin (Ig) determinants, the presence of Ia specificities and the ability to mediate the biological effects of the T cells from which they originate (1-3, and this volume). In the first part of this paper we describe an approach to the structure of these molecules which we are applying to factors in supernatants both of T cells primed to antigen *in vivo* and of T cell hybrids. The method consists of binding the factors to antigen adsorbents and radioiodination *in situ*, followed by elution, characterisation by SDS-polyacrylamide gel electrophoresis (PAGE) and 2-dimensional thin-layer peptide mapping. The second part of this paper describes some of the biological effects of a specific suppressor factor produced by a T-hybrid line.

Hapten-specific non-Ig (factor) molecules in supernatants of suppressor T-cell cultures.

CBA mice were treated with the oxazolone (Ox) or picryl (Pic) haptens using a protocol which generates suppressor T cells capable of inhibiting the passive transfer of contact sensitivity (4,5). Culture supernatants of these cells contain hapten-specific suppressor factors which similarly inhibit contact sensitivity (4,5). These molecules were isolated and labelled as follows.

(a) Oxazolone-specific suppressor factor

CBA mice received oxazolone thioglycolic acid (10mg) intravenously, and a second dose 2 days later; on day 6 they were skin painted with ethanolic oxazolone, a step needed for the production of biologically active suppressor factor (6). Lymphocytes enriched for T cells by passage over nylon wool were prepared from spleens and lymph nodes and the cells cultured for 48 hours. Their supernatants were harvested and Ig removed by adsorption with Sepharose-coupled sheep anti-mouse IgM(λ) and anti-IgG(κ). Oxazolone-specific molecules in the supernatants were then adsorbed onto oxazolonated microcrystalline cellulose beads (Ox beads), using 25μl packed beads to 50ml supernatant; proteins adherent to the beads after washing were radioiodinated *in situ* by the chloramine T method using 100-200μCi ^{125}I for every 10μ. aliquot of packed beads. After washing the beads with a total of 500ml PBS, specific elution was carried out with oxazolone-ε-aminocaproic acid (Ox-EACA) at 10mg/ml for a total of 45 minutes, followed by nonspecific elution with 3.5M potassium thiocyanate (KSCN). The dialysed eluates were then tested for their ability to bind back to Ox beads or Pic (picrylated) beads (control). Table 1 shows that the radioiodinated product had considerable specificity for Ox beads, and the ratio of binding of Ox vs. Pic beads was 4-6 to 1. Specific material was not all eluted by hapten, since hapten-binding activity was also present in the KSCN eluate.

Table 1. Specific binding activities of radioiodinated anti-oxazolone and anti-picryl factors.

Factor	Elution conditions	%cpm binding back to Ox beads	Pic beads	Difference	Ratio
Anti-Ox	Ox-EACA	34.5(±0.9)	6.0(±1.5)	28.5%	5.7
	KSCN	46.6(±0.1)	11.7(±0.1)	35.0%	4.0
Anti-Pic	Pic-EACA	11.3(±1.1)	49.6(±0.1)	38.3%	4.4
	KSCN	20.2(±0.3)	26.3(±0.6)	6.1%	1.3

(b) Picryl-specific suppressor factor

Non-Ig factor molecules of a second specificity were prepared from nylon-wool separated T cells of mice injected with picrylsulphonic acid and skin painted with picryl chloride (5,7). The protocol described above was used, except that Pic beads and picryl-ε-aminocaproic acid (Pic-EACA) were used instead of their oxazolone counterparts. Radioiodinated material eluted with 1mg/ml Pic-EACA rebound specifically to Pic beads (Table 1); in this case sequential elution with KSCN failed to elute further significant amounts of specific material.

The binding of radioiodinated anti-oxazolone and anti-picryl factors could be specifically blocked by preincubation with the corresponding hapten-EACA at 1-20 mg/ml. About 80% specific inhibition was seen at the highest antigen concentration.

SDS-PAGE and peptide mapping

(a) Oxazolone-specific factor

The radiolabelled material rebinding to Ox beads was studied by SDS-PAGE. Ox beads carrying labelled material were heated in SDS sample buffer at 100°C for 2 minutes and electrophoresed according to the discontinuous system of Laemmli (8). Fig. 1a-c shows the 10% gel profiles of the material assayed in Table 1. Under reducing conditions (Fig. 1a) there were peaks at apparent molecular weights of 75K, 48K and 27K; under nonreducing conditions (Fig. 1b,c) the 27K peak was the major one, with some material too large to enter the separating gel. The specific nature of the 27K protein is seen in Fig. 1c, which compares the nonreduced gels of 125I-anti-Ox factor bound to Ox beads or Pic (control) beads; there was a 10 to 1 ratio of preferential binding based on the areas under the peak. In other experiments, SDS-gels showed two main peaks, both under reducing and nonreducing conditions, with molecular weights of 48K and 27K (Fig. 1d,e); the gels again indicated a high degree of antigen specificity.

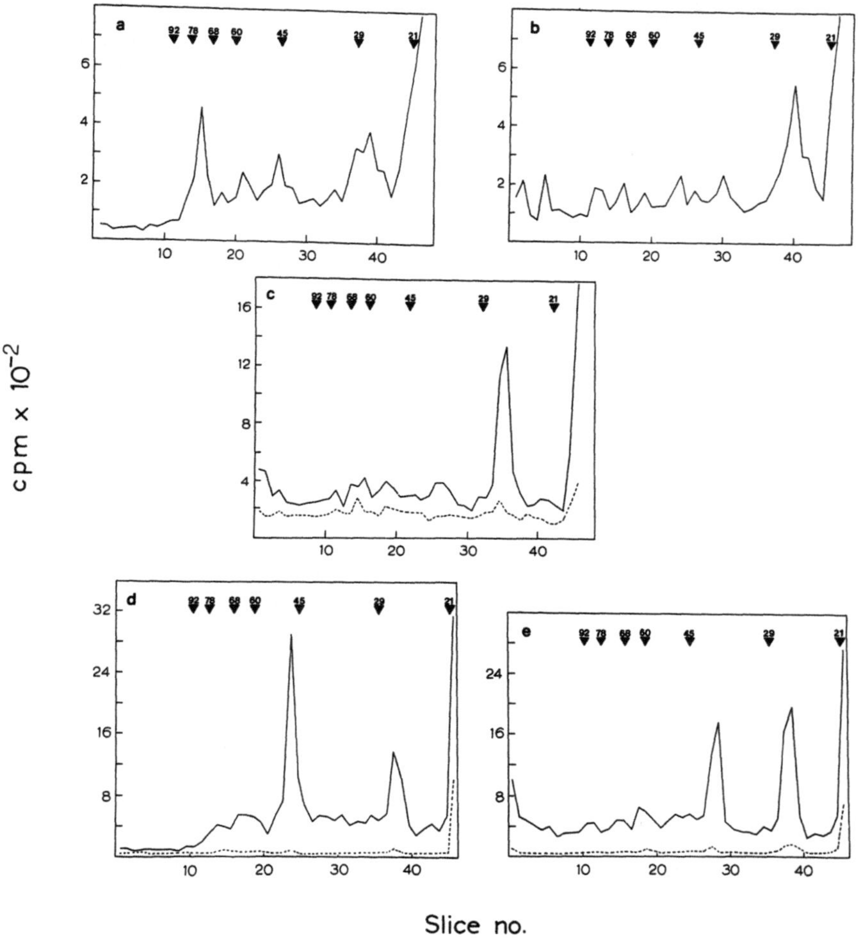

Fig. 1. SDS-polyacrylamide gel electrophoresis of specific anti-oxazolone T cell factor. (a) Reduced and (b) nonreduced 10% gels of oxazolone-specific factor bound to Ox beads; (c) nonreduced 10% gel of oxazolone-specific factor bound to Ox beads (———) or Pic beads (-----); (d) reduced and (e) nonreduced gels of oxazolone-specific factor bound to Ox beads (———) or Pic beads (-----).

Peptide maps were prepared from peaks separated on SDS-gels under nonreducing conditions, the latter being a further precaution against the possibility of Ig contamination. Two-dimensional thin-layer peptide mapping was performed after Feinstein et al. (9), soluble peptides being separated in the first dimension by electrophoresis at pH 3.5, followed by chromatography in the second dimension and autoradiography (10). For comparison with the peptide maps of the specific factors, maps were prepared in parallel of heavy and light chains of radioiodinated myeloma proteins, namely μ and λ chains of MOPC 104E and γ_{2a} and κ chains of Adj PC5; BSA was also used as a further comparison. Figs. 2a and d show peptide maps of the 27K and 48K polypeptides of the oxazolone-specific factor; κ and γ chains of Ig are shown in Figs. 2b and e. Detailed comparisons showed no more than a chance similarity (25-30%) between maps of Ig chains (μ,γ,κ,λ) and those of polypeptides from the factor (i.e. no greater than the similarity between Ig chains with each other or with BSA). Nor was there significant similarity between the maps of 27K and 48K factor polypeptides themselves.

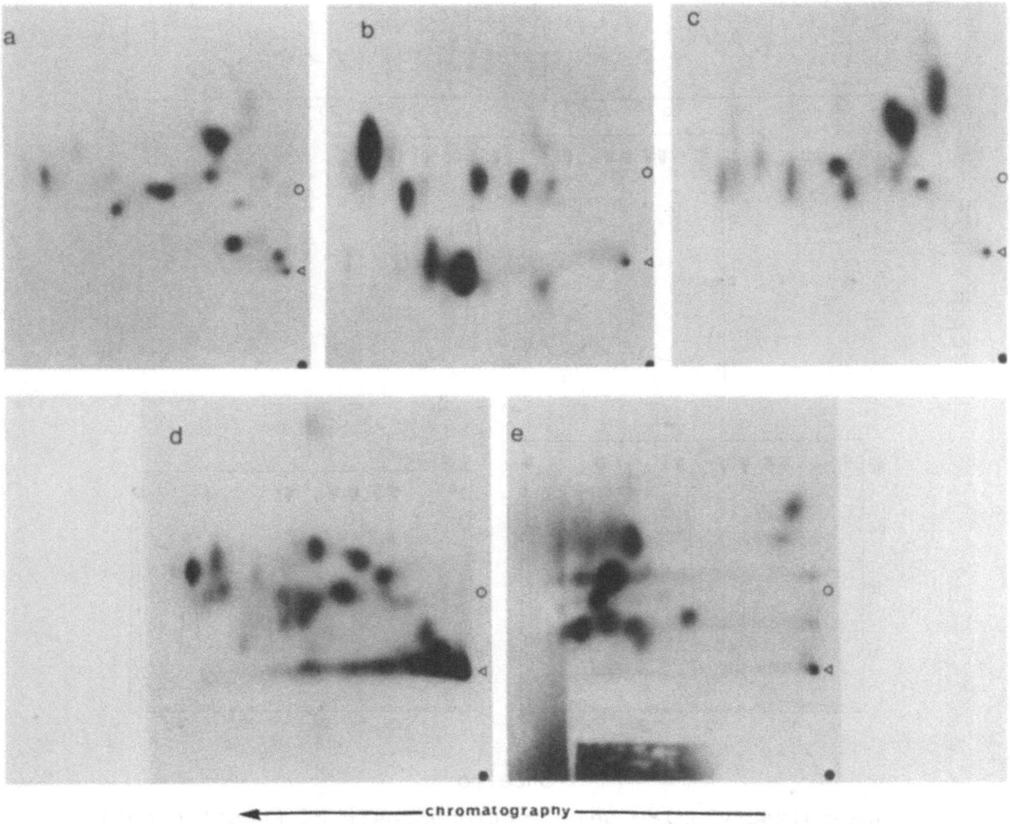

Fig. 2. Two-dimensional peptide maps of specific anti-oxazolone and anti-picryl T cell factors and Ig heavy and light chains. (a) 27K polypeptide of oxazolone-specific factor from nonreduced SDS-PAGE; (b) κ-light chain of mouse IgG_{2a} myeloma Adj PC5 mapped concurrently with (a); (c) 27K polypeptide of picryl-specific factor from nonreduced SDS-PAGE; (d) 48K polypeptide of oxazolone-specific factor from nonreduced SDS-PAGE; (e) γ-chain of mouse IgG_{2a} myeloma Adj PC5 mapped concurrently with (d). ○ ● indicate the electrophoretic movement of markers ε-DNP-lysine and thymol FF which were run concurrently. ◁ indicates the origin.

(b) Picryl-specific factor

^{125}I-labelled picryl-specific material rebinding to Pic beads was also analysed by SDS-PAGE; Fig. 3 shows that the profile of the material was similar to that of the anti-oxazolone product, with a major peak again at 27K and a 48K peak in minor amount. A high degree of specificity for picryl vs. oxazolone (10 to 1) is again indicated from the areas under the 27K peak. The peptide map prepared from this polypeptide (Fig. 2c) showed 55-60% similarity with its 27K anti-oxazolone counterpart (Fig. 2a), but no significant similarity with the peptide maps of any of the Ig chains examined.

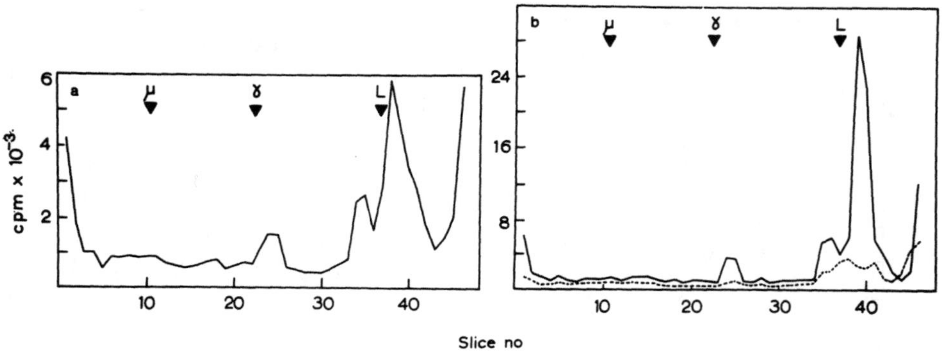

Fig. 3. SDS-PAGE of specific anti-picryl T cell factor.
(a) Reduced and (b) nonreduced 10% gels of picryl-specific factor bound to Pic beads (———) or Ox beads (-----). For these gels the markers were the heavy (μ and γ) and light (L) chains of mouse IgM (MOPC 104E) and IgG (Adj PC5) myeloma proteins.

These results indicate that the antigen-specific T cell factors in supernatants of suppressor T cells are structurally distinct from the 'classical' immunoglobulin chains used as comparisons and also suggest that factors of different specificities may share a common constant region (see also Discussion).

SRBC-specific factor from a T-hybrid line.

The origin of the T-hybrid line designated Al was by fusion between SRBC-primed mouse (C57Bl) spleen cells and the thymoma BW5147 (11,12).The ability of the supernatant of this line to suppress specifically the anti-SRBC antibody response has been described and reviewed in detail (13,14).

(a) Structural studies

We have previously reported the results of internal labelling of the Al suppressor factor, which indicated that it was composed of polypeptide chains of about 80K and 25K, and that the SRBC-binding site was present on the larger chain while the smaller carried MHC (Ia) determinants (15). Recently we have attempted radioiodination of the factor using a similar protocol to that described above, with a view to peptide mapping and comparison with the anti-hapten factors. Al supernatant of known biological activity was adsorbed onto glutaraldehyde-fixed SRBC and, after washing, was iodinated on the red cells. After washing the red cells, radiolabelled material was eluted with 3.5M KSCN, and following dialysis this material was able to bind back specifically to SRBC. Figure 4a shows a 10% SDS-PAGE profile of the labelled factor bound to SRBC or pig (control) RBC; three peaks are evident at apparent molecular weights of 75K, 50K and 30K, a picture quite similar to that of anti-oxazolone factor (Fig. 1a). Peptide mapping of these polypeptides is in progress.

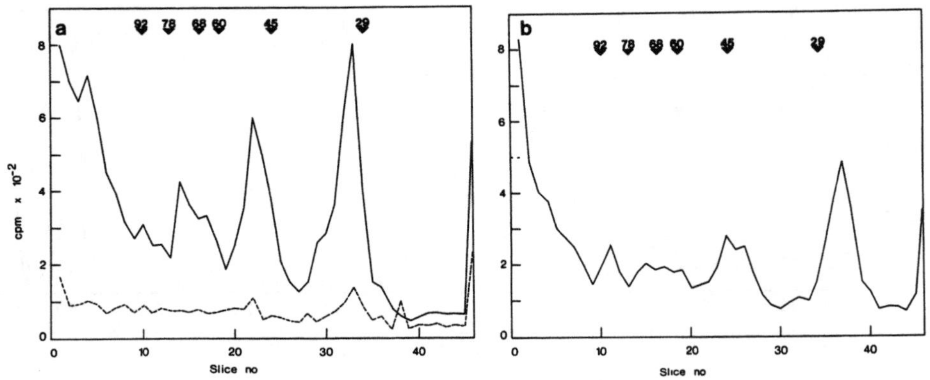

Fig.4. (a) Reduced and (b) nonreduced 10% SDS-gel profiles of ^{125}I labelled T-hybrid Al factor bound to SRBC (———) or pig RBC (----).

(b) Functional studies: an interaction with monoclonal anti-SRBC antibodies

As described elsewhere, the Al factor appears to suppress the anti-SRBC response by binding to an acceptor site at the surface of B cells (13); this reaction is under the control of two I-region genes (13). Recently we have found that the action of the factor is modified by the presence of monoclonal IgG and IgM anti-SRBC antibodies. Figure 5a shows that in the presence of an IgG antibody (D3) suppression produced by the factor was markedly enhanced, and that this was a synergistic effect which occurred at levels of antibody which alone were too low to significantly suppress the response. In contrast, an IgM antibody (Spl) tended to abolish, in a concentration dependent manner, the suppressive activity of Al factor (Fig. 5b). For both monoclonal antibodies, absorption with SRBC negated the effect. It is easy to envisage that this type of interaction between factor and antibody may have physiological importance, since it would be advantageous for the activity of T cell factors to be responsive to antibody levels and class. Our observation also suggests that assays for suppressor factors, such as when screening T-hybrid lines, might with advantage be carried out in the presence of a small amount of added IgG antibody.

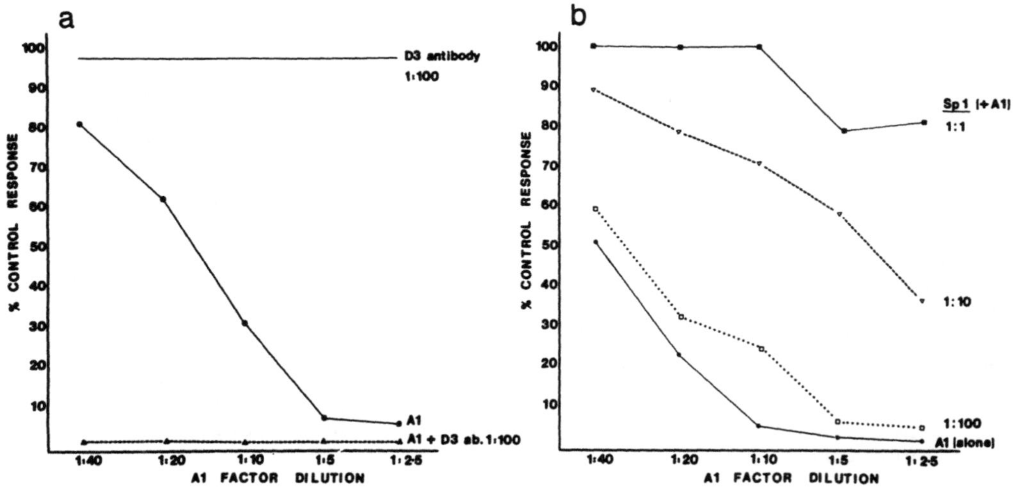

Fig. 5. Effect on suppression by T-hybrid Al supernatant of monoclonal anti-SRBC antibodies. (a) Al supernatant was titrated for suppressor activity in the presence of 'D3' IgG_2 monoclonal anti-SRBC antibody, by addition to microcultures of mouse (C57Bl) spleen cells responding to SRBC; the topmost line shows the effect of D3 alone. (b) Al activity was titrated in the presence of 'Spl' IgM monoclonal anti-SRBC antibody at dilutions of Spl from 1:1 to 1:100. Spl alone produced about 65% inhibition of response at 1:1, but at 1:10 and 1:100 was without effect. Positive control response (factor and antibodies omitted) was 342 pfc/culture ($5x10^5$ spleen cells).

Discussion

(a) Factor structure

The basic approach described here was to absorb factors onto their corresponding immobilised antigen (hapten or SRBC), label in situ with ^{125}I and elute with specific hapten or KSCN. The method enables small amounts of material to be retrieved from supernatants and labelled, and the combining site is protected during iodination. Several precautions were taken to exclude antibody from the supernatants of cultures containing suppressor T cells, including: nylon-wool enrichment of T cells; absorption of supernatants before iodination with anti-mIg adsorbents; and the use of SDS-gels run under nonreducing conditions in obtaining material for peptide mapping, which should exclude Ig chains from the bulk of the gel. The peptide maps confirmed the absence of Ig chains ($\mu,\gamma,\kappa,\lambda$) from the final preparatons. Thus the results support the existence of antigen-recognition molecules which are distinct from Ig in the structure of their polypeptide chains.

The most prominent feature of the labelled hapten-specific molecules on reduced and nonreduced gels was a 27K peak, while a second peak at 48K appeared in most gels. The 48K and 27K components were not linked by disulphide bonds, nor did the 27K chain appear (from peptide maps) to be a breakdown product of the 48K moiety. The presence of a 75K peak in some reduced gels raised the possibility that the 27K and 48K polypeptides were degradation fragments of this larger chain. The variability in yield of the three peaks may reflect variable proteolytic degradation during the 48 hour culture; subsequent degradation may have been limited by the presence of PMSF.

Our preliminary result with ^{125}I-labelled A1 factor showing 3 peaks on SDS-gel at molecular weights of 75K, 50K and 30K is similar to the above. However, previous studies with internally labelled factor showed only 2 peaks in material binding specifically to SRBC, at about 80K and 25K(15). Proteolytic breakdown of the largest chain may partly account for the discrepancy, particularly the appearance of the 50K polypeptide in iodinated factor. The fact that the supernatant used for radioiodination had been stored at -20°C and freeze-thawed a few times before labelling (in the course of assaying biological activity) might also have encouraged fragmentation.

The results described here are broadly in agreement with a number of other studies on antigen-specific products and receptors of suppressor T cells, T cell clones and T-hybrid lines (16-19). The general indication is that these molecules have a large chain in the 70-90K range, with proteolytic degradation products of 40-50K and 20-30K.

(b) <u>Biological properties of T-hybrid (A1) suppressor factor</u>

It is to be expected that the active products of cloned T-hybrid lines, while recognisably factor-like in their major properties, should have idiosyncrasies which may not have been noticed in the products of antigen-primed T cells; in other words, they may represent subclasses with hitherto unrecognised characteristics. This is illustrated by the suppressor factor produced by the A1 line. The combination of SRBC-specificity in both binding and activity, its non-Ig nature and MHC derivation all suggest that its parent cell was an SRBC-specific suppressor T cell. On the other hand, it shows several differences from many of the factors obtained from suppressor T cells and other clones or hybrid lines. These include non-I-J coding, lack of strict self-H-2 restriction in activity, ability to interact with B cells rather than T cells, and a relatively high molecular weight of about 200K (13). Hence there is some reason to suggest that the A1 molecule is a member of a subclass of suppressor factors, the properties of which have not been defined using more conventional material.

An important question, probably as yet unanswerable, is to what extent are the products of T-hybrid lines reported in this volume and elsewhere representative of the molecules produced by normal primed T cells? For example, which are major and which minor T cell products? and are the products of T hybrids influenced in an unknown way by gene products of the parent tumour? The most secure approach must be to study and compare in parallel the products of purified T cell populations primed <u>in vivo</u>, and the products of lines, both hybrid and nonhybrid T cell clones.

Acknowledgements

G. Skibinski was supported by the Royal Society under the exchange agreement with the Polish Academy of Sciences. We thank Dr. A. Feinstein for helpful discussions, Dr. D.B.A. Symons for immunofluorescence, Dr. M. Glennie and Miss A. Froggatt for some peptide mapping, and Dr. J. Vidal-Gomez for help in some of the preparations.

References

1. Taussig MJ (1980) Antigen-specific T-cell factors. Immunology 41: 759-787.
2. Tada T, Okumura K (1979) The role of antigen-specific T-cell factors in the immune response. Adv. Immunol. 28: 1-87.
3. Germain R, Benacerraf B (1980) Helper and suppressor T-cell factors. Springer Seminars in Immunopathology 3: 93-127.
4. Asherson GL, Zembala M, Thomas WR, Perera MACC (1980) Suppressor cells and the handling of antigen. Immunol. Revs. 50: 3-45.
5. Asherson GL, Zembala M (1980) T suppressor cells and suppressor factor which act at the efferent stage of the contact sensitivity skin reaction: their production by mice injected with water-soluble, chemically reactive derivatives of oxazolone and picryl chloride. Immunology 41: 1005-1013.
6. Zembala M, Asherson GL (1974) T-cell suppression of contact sensitivity in the mouse. III. The role of macrophages and the specific triggering of nonspecific suppression. Europ. J. Immunol. 4:799-804.
7. Asherson GL, Zembala M, Noworolski J (1978) The purification of specific anti-picryl T suppressor factor which depresses the passive transfer of contact sensitivity: affinity chromatography on antigen and concanavalin-A sepharose and specific elution with hapten and α methyl-mannoside. Immunology 35: 1051-1056.
8. Laemmli UK (1970) Cleavage of structural proteins during the assembly of the head of bacteriophage T4. Nature 227: 680-681.
9. Feinstein A, Richardson NE, McIlhinney RAJ (1979) Structure of lymphocyte membrane immunoglobulin. In Pernis B, Vogel HJ (eds) Cells of Immunoglobulin Synthesis, Academic Press, New York, p 165-188.
10. Laskey RA, Mills AD (1977) Enhanced autoradiographic detection of 32P and 125I using intensifying screens and hypersensitised film.FEBS Letters 82: 314-316.
11. Taussig MJ, Corvalan JRF, Binns RM, Roser B, Holliman A (1979) Production of an H-2-related suppressor factor by a hybrid T cell line. Nature 277: 305-307.
12. Taussig MJ, Corvalan JRF, Binns RM, Roser B, Holliman A (1979) Immunological activity of a T-hybrid line. I. Production of an H-2 related suppressor factor with specificity for sheep red blood cells. Europ. J. Immunol. 9: 768-776.
13. Taussig MJ (1982) Studies on an antigen-specific suppressor factor produced by a T-hybrid line. Lymphokines 5: 161-185.
14. Taussig MJ, Corvalan JRF, Holliman A (1979) Characterisation of an antigen-specific factor from a hybrid T cell line. Ann N.Y. Acad Sci. 332: 316-335.
15. Taussig MJ, Holliman A (1979) Structure of an antigen-specific suppressor factor produced by a hybrid T-cell line. Nature 277: 308-310.
16. Goodman JW, Lewis GK, Primi D, Hornbeck P, Ruddle NH (1980)

Antigen-specific molecules from murine T lymphocytes and T cell hybridomas. Molecular Immunol. 17: 933-946 .
17. O'Connor S, Eardley D, Shen FW, Gershon RK, Cone RE (1980) Isolation and partial characterisation of antigen-binding molecules produced by in vitro educated T cells. Molecular Immunol. 17: 913-924
18. Spurll GM, Owen FL (1981) A family of T-cell alloantigens linked to Igh-1. Nature 293: 742-744.
19. Fresno M, Boudreau LM, Nabel G, Cantor H (1981) Antigen-specific T lymphocyte clones. II. Purification and biological characterisation of an antigen-specific suppressive protein synthesized by cloned T cells. J. Exp. Med. 153: 1260-1274

Purification and Biochemical Analysis of Antigen-Specific Suppressor Factors Isolated from T-Cell Hybridomas

D.R. Webb, B.A. Araneo, C. Healy, J.A. Kapp, K. Krupen, I. Nowowiejski, C.W. Pierce, C.M. Sorensen, S. Stein, K.J. Wieder

A. Introduction

Recent advances in technology have made the understanding of the molecular and biochemical aspects of immune regulation an attainable goal. We have used some of these advances in somatic cell hybridization, microanalytical peptide chemistry and recombinant DNA technology, to study the molecular aspects of antigen-specific suppression of immune responses. These studies have used a group of related polypeptide antigens; L-glutamic acid^{60}-L-alanine^{30}-L-tyrosine^{10} (GAT), L-glutamic acid^{50}-L-tyrosine^{50} (GT), and L-glutamic acid^{50}-L-alanine^{50} (GA). The immune responses in mice to these polymers are controlled by immune response (Ir) genes which map to discrete loci within the I region of the major histocompatability complex (H-2). In the case of GAT for example, immunization of mice bearing H-$2^{a,b,d,f,k}$ haplotypes results in antibody production to GAT and primes lymph node T cells for subsequent in vitro proliferation in response to GAT. In mice bearing the H-$2^{p,q,s}$ haplotypes immunization with GAT neither stimulates antibody-forming cells nor does it prime lymph node T cells for proliferation unless the GAT is complexed with an immunogenic carrier such as MBSA. Earlier studies suggested that the reason for this lack of responsiveness in mice bearing H-$2^{p,q,s}$ haplotypes was the development of GAT-specific suppressor T cells (1-5). More recently it has been shown that extracts from GAT-specific suppressor T cells from non-responder mice contain soluble proteins which can specifically suppress the response to GAT. These soluble proteins are called GAT-specific suppressor T-cell factors (GAT-TsF) and may also be found in the supernatant of cultures or cell extracts from responder mouse strains exposed to GAT under the appropriate conditions (6,7). Like GAT, GT stimulates suppressor T-cells that produce suppressor factors in certain strains of mice. GAT-TsF and GT-TsF are antigen binding proteins that bear determinants coded by the I-J region of the H-2 complex and idiotypic determinants that cross-react with murine antibodies of the same specificity (8,9).

To investigate the molecular properties of these suppressor factors we decided to take advantage of somatic hybridization methodology to make T-cell hybrids between antigen-specific suppressor T-cells and the AKR thymoma BW5147. In the following sections we summarize our studies to date on the purification and characterization of the suppressor polypeptides which we have obtained from a collection of T-cell hybrids.

I. Construction of T Cell Hybrid Cell Lines

Initially mice of the appropriate non-responder haplotype were injected i.p. with 10 μg GAT (Vega Biochemical Co., Tucson, AZ) or 100 μg GT (Miles Laboratories,

Elkhart, IN) mixed with Maalox (Wm. H. Rorer, Inc., Fort Washington, PA). Four days later the mice were given an i.p. injection of 2.5 mg cortisone acetate (Merck, Sharp and Dohme, West Point, PA). It has been found that cortisone-treated mice have 3-4 times the number of GAT or GT specific suppressor cells in the spleen compared to untreated mice (unpublished observation). Three days after cortisone treatment the mice were sacrificed and nylon wool non-adherent T cells prepared from spleen cell suspensions. These were fused to the $HGPRT^{(-)}$ AKR thymoma, BW5147 (Ouabain resistant), using polyethylene glycol (Carbowax 1500, Fisher Scientific Co., St. Louis, MO) according to previously published techniques (10,11). Hybrid cells were selected using HAT medium containing 10% gammaglobulin-free horse serum and 5% calf serum. Once detectable growth occurred and the cultures became confluent (3-6 weeks) the supernatants were assayed for suppressor activity as previously described (11). Cells from primary cultures causing $\geq$50% suppression of the GAT-MBSA response and no suppression of sheep erythrocyte responses at a dilution of 1/1000 were cloned by limiting dilution in soft agar over feeder layers of 3T3 fibroblasts. Isolated colonies were picked, grown in liquid cultures and the supernatant reassessed for suppressive activity. Selected clones have been maintained in culture and aliquots of each clone have been frozen in liquid nitrogen.

To obtain suppressor T-cells for somatic fusion from responder strains, a different induction protocol was used. Syngeneic macrophages (1×10^7) bearing GAT (0.3-0.5 μg) were injected into neonatal mice (8-10 hr old). After 12 weeks the mice were sacrificed and the spleen cells cultured for 24 hr with 2μg GAT. The spleen cells were then fused with BW5147 as outlined above. The selection procedure was identical to that described previously.

Table I lists the hybridomas which have been characterized to date. The list does not include all hybridomas obtained from the various fusions, rather we have listed those where the hybridoma and its product have been partially characterized. Several points are worth noting. Hybridomas which represent fusions between BW5147 and non-responders to GAT or GT seem to be exclusively of the single chain type with a molecular weight below 30,000 daltons. We have termed these GAT-TsF as TsF_1 since recent experiments have shown that such factors are capable of inducing a second set of suppressor cells, Ts_2 which produce a TsF with different molecular properties (Kapp and Araneo, manuscript submitted). These factors termed TsF_2 are composed of two chains, have a higher molecular weight and appear to be similar to suppressor factors described by Taniguchi et al. (12). We have recently produced hybrid lines from non-responders which secrete TsF_2 specific for GT (469B5.A5) as well as GAT (not listed). In responder strains, hybridomas of both TsF_1 (372B3.B5) and TsF_2 class (372D6.D5) have been obtained. Future studies will include attempts to produce hybridomas representing other possible Ts subclasses.

II. Purification to Chemical Homogeneity of Antigen-Specific TsF Obtained From T-Cell Hybrids

The purpose of making the T-cell hybrids was to obtain TsF in sufficient quantity to study their biology and biochemistry. On a comparison basis at least some of the hybridomas make as much as 40 times the amount of TsF as can be extracted from a comparable number of lymphoid cells from GAT-treated animals.

The first hybridoma product to be purified and characterized was 258C4.4 (C4.4); this was the result of the fusion of GAT-primed non-responder T cells from DBA/1 ($H-2^q$) mice. A complete description of the purification and biochemical analysis of this GAT-TsF has recently been presented (13). Briefly, supernatants from C4.4 were passed over BSA-Sepharose and GAT-Sepharose; the TsF was eluted from the GAT-Sepharose with KCl. This eluate purified from antigen-affinity columns was then subjected to a series of high-pressure, reverse phase liquid chromatographic steps.

Table I. Bank T cell hybridomas fused with BW5147

				Serological Characterization of TsF				
	Spleen Cell					Id[2]		
Clone No.	Donor	1° Ag	S_{50} U/ml[1] TsF	Ia	Ag Binding	CGAT	GA-1	-C4.4
258C4.4	DBA/1	GAT	20,000	I^q	GAT+GA	+	+	+
258C4.6	DBA/1	GAT	20,000	I^q	GAT+GA	NT	NT	NT
342B1.11	B10.S	GAT	40,000	$I-J^s$	GAT+GT	+	-	+
366D3.2	B10	GT	30,000	$I-J^b$	GAT+GT	+	-	-
367A5.4	CBA	GT	30,000	$I-J^k$	GAT+GT	NT	NT	NT
368B1.5	BALB/c	GT	27,000	I^d	GAT+GT	±	-	-
395A4.4	B10.S	GT	30,000		GAT+GT	±	-	-
469B5.A5	B10.A	GT+TsF	80,000	$I-J^k$	GAT+GT	+	-	NT
372D6.D5	B10	GAT-MØ	50,000	$I-J^b$	GAT+GT	+	NT	-
372B3.B5	B10	GAT-MØ	100,000	$I-J^b$	GAT+GT	+	-	+

[1] S_{50} units represent the reciprocal of the dilution which gives 50% inhibition of the response to antigen.

[2] Anti-idiotype antisera from several sources were used to study the TsF protein.

CGAT = guinea pig anti-mouse GAT antibody; GA-1 = guinea pig anti-mouse GAT antibody; anti C4.4 = rabbit anti C4.4 TsF (affinity purified); NT = not tested.

Purification to apparent chemical homogeneity was accomplished with a final chromatography step employing a high-pressure DEAE-nugel column. This chromatograph showed a single protein peak which contained all of the biological activity. Analysis of this protein on 15% polyacrylamide gel (in SDS) by electrophoresis showed a single band at 24,000 daltons. The material eluted from the polyacrylamide contained all of the biological activity of the crude supernatant. Various serological parameters were also assessed using the purified TsF and showed that the 24,000 dalton protein could bind GAT, GA but not GT; and it reacted with anti-I^q alloantisera as well as anti-idiotypic antisera. The observation that GAT-TsF_1 can be bound to and eluted from GAT-Sepharose and anti-I-J-Sepharose sequentially, suggests that it is a single molecule. Purification of TsF_1 in the presence of a reducing agent such as 5 mM dithiothreitol (14) did not alter the sequential absorption by GAT- and anti-I-J-Sepharose which provides evidence that TsF_1 cannot be broken down into smaller constituents by reduction.

Since our initial purification of C4.4, we have begun to study several other T-cell hybridomas. The molecular characteristics of some of these are listed in Table II. Evidence obtained from studies using GAT-specific hybridoma supernatant suggests that C4.4 and B1.11 represent factors produced by Ts_1 type cells (Kapp and Araneo, submitted for publication). In the responder strains this also is true for B3B5 (see Table II). The TsF_1 type factors not only have similar molecular weights but also similar properties on high pressure columns. Factors from both responder and non-responder hybridomas behave as highly hydrophobic proteins eluting from Lichrosorb RP-8 resins in 40% propanol. In addition, these proteins appear to have carbohydrate based on their binding to lentil lectin and peanut agglutinin (unpublished observations). Future studies will focus on the structural relationship between the various factors from different haplotypes (e.g. primary amino acid sequence); the relationship between TsF directed towards the related polymers GAT and GT; the biochemical relationship between TsF derived from responder versus non-responder strains; and the structural analysis of TsF_2.

Yet another aspect of this problem has to do with the relationship between these antigen-specific factors and the T-cell surface antigen receptor. The GAT-TsF molecules from C4.4 and B1.11 exist on the membrane, as shown by membrane extraction and fractionation experiments where as much TsF activity can be extracted from the membrane as exists in the supernatant. Currently we are determining whether the membrane associated TsF has the same molecular properties as TsF isolated from the supernatant. Future studies will be directed at the question of whether TsF can serve as the T-cell antigen receptor.

III. Characterization of the Genes Which Code for GAT-TsF

In addition to studying the TsF protein structure and function we are interested in understanding the structure of gene or genes which code for this family of proteins. For this reason studies were begun to isolate and characterize the GAT-TsF (H-2^q) mRNA in preparation for making a cDNA clone bank in _E. coli_ which would contain the appropriate gene sequences coding for the GAT-TsF.

To isolate the mRNA, cell pellets obtained from 258C4.4 were extracted with phenol according to established procedures. The RNA was chromatographed on oligo(dT) cellulose and the polyA$^+$ RNA isolated. The polyA$^+$ RNA was separated according to size using sucrose density gradient centrifugation. The various size fractions of polyA$^+$ RNA were recovered from the sucrose and translated in a cell-free, rabbit reticulocyte lysate system (Wieder et al., submitted for publication). The translated products were assayed for GAT-TsF activity. PolyA$^+$ RNA with a sedimentation coefficient of 16s contained a mRNA which coded for GAT-TsF. The 16s RNA fraction from the parent BW5147 contained no such message. The cell-free translated GAT-TsF was bound by anti-I^q antisera but not by anti-I-J^s antisera. The cell-free translated GAT-TsF was affinity purified using anti-I^q-Sepharose and then chromatographed on a Lichrosorb RP-8 reverse phase, high pressure liquid

Table II. Molecular characteristics of suppressor factors isolated from T cell hybrids

Clone No.	Ag Specificity	Ia	M.W.	No. of chains	Specific activity (S_{50} units/ g)	HPLC Charac.	CHO
283C4.4	GAT/GA	I^q	24,000	1	7.8×10^7	hydrophobic	+
342B1.11	GAT/GA	$I\text{-}J^s$	24,000	1	2.6×10^7	hydrophobic	NT
372B3.B5	GAT	$I\text{-}J^b$	24,000	1	NT	hydrophobic	+
469B5.A5	GT	$I\text{-}J^k$	50,000	2	NT	NT	NT
372D6.D5	GAT	$I\text{-}J^b$	60,000	2	NT	NT	+

chromatographic column. The biologically active material was recovered in a single fraction in 40% propanol. This fraction was then subjected to polyacrylamide gel electrophoresis using an 8-18% gradient gel (15). All of the biological activity was recovered in a gel slice corresponding to a molecular weight of 19,000 daltons. These data illustrate several points. First, the serological determinant for I^q is not carbohydrate since no glycosylation occurs in cell-free translation (16,17); second, carbohydrate is not necessary for biological activity; and third, the bioactive protein need be no larger than 19,000 daltons and probably does not require more than one chain for activity.

A cDNA clone bank was constructed using total $polyA^+$ RNA from C4.4 and the bacterial plasmid PBR-322 (unpublished data). Using hybrid selection techniques we have established that the clone bank contains cDNA sequences which will bind to GAT-TsF mRNA (unpublished observations). Experiments are in progress to isolate the relevant clones which will allow us to directly examine the genes which code for C4.4 GAT-TsF. This methodology will also allow us to sequence the mRNA and thereby derive the sequence of the GAT-TsF protein. The possibility also exists that we can construct plasmids which will produce GAT-TsF in large quantities facilitating biological and biochemical studies.

B. Concluding Remarks

These studies suggest that by integrating biological and biochemical methodologies real progress can be made in the study of immune regulation. The work summarized in this report suggests that the suppressor T cell pathway can be dissected using somatic cell fusion methodology coupled with biochemical analysis of somatic cell hybrid products. The stability of the T-cell hybridomas and their capacity to continuously produce the TsF as well as other MHC gene products suggests they will continue to be very useful tools in our analysis of T-cell function.

Acknowledgments. The authors wish to thank Drs. Sidney Udenfriend and John Monahan for their encouragement and support, and Drs. Martin Dorf and Shyr-Te Ju for providing the anti-idiotypic reagents. This research was supported in part by USPHS grants AI-15353 and AI-13915. J.A. Kapp is the recipient of a USPHS Research Career Development award AI-00361. K. Krupen is the recipient of a pre-doctoral fellowship from the Roche Institute of Molecular Biology.

References

1. Dunham EF, Dorf ME, Shreffler DC, Benacerraf B (1973) Mapping the H-2 linked genes governing respectively the immune responses to a glutamic acid, alanine, tyrosine copolymer and to limiting doses of ovalbumin. J Immunol 111: 1621-1625
2. Gershon RK, Maurer PH, Merryman CF (1973) A cellular basis for genetically controlled immunologic unresponsiveness in mice: Tolerance induction in T-cells. Proc Nat Acad Sci USA 70: 250-254
3. Kapp JA, Pierce CW, Benacerraf B (1973) Genetic control of immune responses in vitro. I. Development of primary and secondary plaque-forming cell responses to the random terpolymer L-glutamic $acid^{60}$-L-$alanine^{30}$-L-$tyrosine^{10}$. J Exp Med 138: 1107-1120
4. Araneo BA, Kapp JA (1980) H-2 linked Ir gene control of T-cell proliferative responses to the synthetic terpolymer L-glutamic $acid^{60}$-L-$alanine^{30}$-L-$tyrosine^{10}$ (GAT). Requirements for the T-cell activation in responder and non-responder mice. J Immunol 124: 1492-1498
5. Kapp JA, Pierce CW, Schlossman S, Benacerraf B (1974) Genetic control of immune responses in vitro. V. Stimulation of suppressor T cells in non-responder mice by terpolymer L-glutamic $acid^{60}$-L-$alanine^{30}$-L-$tyrosine^{10}$. J Exp Med 140: 648-659

6. Kapp JA, Pierce CW, DeLaCroix F, Benacerraf B (1976) Immunosuppressive factor(s) extracted from lymphoid cells of nonresponder mice primed with L-glutamic acid^{60}-L-alanine^{30}-L-tyrosine^{10}. I. Activity and antigenic specificity. J Immunol 116: 305-309
7. Kapp JA, Pierce CW, Benacerraf B (1977) Immunosuppressive factor(s) extracted from lymphoid cells of non-responder mice primed with L-glutamic acid^{60}-L-alanine^{30}-L-tyrosine^{10}. II. Cellular source and effect on responder and non-responder mice. J Exp Med 145: 828-838
8. Kapp JA (1978) Immunosuppressive factor(s) from lymphoid cells of nonresponder mice primed with L-glutamic acid^{60}-L-alanine^{30}-L-tyrosine^{10}. IV. Lack of strain restrictions among allogeneic nonresponder donors and recipients. J Exp Med 147: 997-1006
9. Germain RN, Ju S-T, Kipps TJ, Benacerraf B, Dorf ME (1979) Shared idiotypic determinants on antibodies and T cell derived suppressor factor specific for the random terpolymer L-glutamic acid^{60}-L-alanine^{30}-L-tyrosine^{10}. J Exp Med 149: 613-622
10. Galfre G, Howe SC, Milstein C, Butcher GW, Howard JC (1977) Antibodies to major histocompatibility antigens produced by hybrid cell lines. Nature 266: 550-552
11. Kapp JA, Araneo BA, Clevinger BL (1980) Suppression of antibody and T cell proliferative responses to L-glutamic acid^{60}-L-alanine^{30}-L-tyrosine^{10} by a specific monoclonal T cell factor. J Exp Med 152: 235-240
12. Taniguchi M, Takei I, Tada T (1980) Functional and molecular organization of an antigen-specific suppressor factor from a T cell hybridoma. Nature 283: 227-228
13. Krupen K, Araneo BA, Brink L, Kapp JA, Stein S, Wieder KJ, Webb DR (1982) Purification and characterization of a monoclonal T cell suppressor factor specific for L-glutamic acid^{60}-L-alanine^{30}-L-tyrosine^{10}. Proc Nat Acad Sci USA (in press)
14. Taniguchi M, Saito T, Takei I, Takuhisa T (1982) Presence of interchain disulfide bonds between two gene products that compose the secreted form of an antigen-specific suppressor factor. J Exp Med 153: 1672-1677
15. Laemmli UK (1970) Cleavage of structural proteins during the assembly of the head of bacteriophage T4. Nature 227: 680-685
16. Lingappa VR, Shields D, Woo SLC, Blobel G (1978) Nascent chicken ovalbumin contains the functional equivalent of a signal sequence. J Cell Biol 79: 567-572
17. Erickson AH, Blobel H (1979) Early events in the biosynthesis of the lysosomal enzyme cathepsin D. J Biol Chem 254: 11771-11774

Dissection of a Suppressor Cell Cascade

M.E. Dorf, K. Okuda, M. Minami

The preparation of monoclonal T cell hybridomas and isolation of biologically active factors derived from these cells, provides a means for isolating the individual T cell subsets that exist within heterogeneous T cell populations. This approach is especially useful when dissecting immune phenomena which require the interactions of several T cell subsets. Immune suppression is one such T cell mediated activity which involves multiple T cell subpopulations and several distinct soluble factors. Although initial characterization of the various suppressor cells (Ts) has been possible with heterogeneous Ts populations, it remains difficult to analyze the contributions of individual Ts subsets and to ensure that a single subset instead of a mixture of subsets is being studied. Furthermore, the quantities of cells or factors obtained from conventional cell preparations are often insufficient for detailed analysis. Thus the establishment of monoclonal Ts hybridomas provides an ideal tool for characterizing the individual Ts subsets, their products, and their interactions.

The regulation of cellular and humoral immune responses to the hapten 4-hydroxy-3-nitrophenyl acetyl (NP) has been well characterized (1,2). Previous studies using conventional analyses have revealed that in the NP suppressor cell pathway, at least three distinct subsets of T cells are required for suppression (Table 1). The first set of Ts, termed Ts_1, are induced by antigen, bear idiotypic receptors, and function early, *i.e.*, during the induction phase of the immune response (3). These Ts_1 cells produce a soluble suppressor factor, termed TsF_1, which can induce a second set of Ts, termed Ts_2 (4). The Ts_2 cells bear anti-idiotypic receptors and function late, *i.e.*, during the effector phase of the immune response (2,5). The Ts_2 cells or factors derived from them, activate an antigen primed Ts_3 population which bears idiotypic determinants (1,6). The Ts_3 population provides still another suppressor signal; however, the target of this signal has not been identified.

To obtain a better understanding of the precise nature and function of Ts, we prepared a series of T cell hybridomas that correspond to the Ts_1, Ts_2, and Ts_3 populations. This goal was only achieved after the Ts subsets could be specifically enriched and distinguished by their surface phenotype. Thus, prior to fusion, the Ts_1 and Ts_3 subsets were enriched by adherence to antigen (NP-BSA) coated plates, while the Ts_2 were enriched by adherence to NP^b idiotype coated plates. This procedure results in a 20 to 50-fold enrichment of Ts activity on a per cell basis (3,5). Following enrichment, the Ts cells were hybridized with the drug marked AKR thymoma BW5147 at a 1:4 ratio using a batch of polyethylene glycol which had been preselected for its ability to fuse cells (7-9). The cells were seeded in 0.2 ml at a concentration of 10^5 tumor cells per ml in 96-well Linbro culture plates. The hybridomas were selected with HAT media. After 14 days, colonies were apparent in 30 to 60% of the wells. Since cells of the Ts_1, Ts_2 and Ts_3 subsets carry the I-J marker on their surface, the hybridomas were screened for the presence of I-J determinants using a dye-exclusion microcytotoxicity test (7-9). Furthermore, since the Ts_1 and Ts_3 cells also carry NP^b-related idiotypic receptors, the $I-J^+$ hybridoma cells were screened with guinea pig anti-idiotypic antisera (7,9). For the first several weeks after hybridization, the levels of I-J and NP^b-related idiotypic determinants were relatively high, 30 to 70% lysis of the hybridoma cells was frequently observed. However, with maintained *in vitro* passage, the percentage of hybridoma cells lysed with these antisera decreased. After the hybridoma cells were cloned, we noted that the individual clones also demonstrated low levels of lysis (0 to 35%) with anti-I-J or anti-idiotypic reagents. This may be due to the differential expression of these markers during various phases of the cell cycle or to other possibilities.

After selecting the hybridomas on the basis of surface phenotype, their ability to constitutively secrete biologically active suppressor factors was assessed. Approximately one-third of the hybridomas prepared from enriched Ts populations and selected by phenotype secreted antigen specific suppressor factors (TsF) into the culture media. The properties of these TsF have been described in detail elsewhere (4,8,9) and are summarized in Table 1.

The genetic restrictions of the hybridoma derived TsF have been well characterized. For instance, B6 or CKB derived hybridoma TsF_1 will only suppress Igh^b bearing mice, but the TsF_1 can induce anti-idiotypic Ts_2 cells in any strain regardless of Igh allotype (4). However, since these Ts_2 cells must interact with NP^b-related idiotypic determinants, the Ts_2 will only cause suppression when adoptively transferred to strains carrying the Igh^b allotype (4). This type of TsF restriction was termed a pseudogenetic restriction (4). These pseudogenetic restrictions are a reflection of the series of idiotypic-anti-idiotypic interactions which occur in the suppressor cell cascade. It should be noted that Ts_1 derived factors do not demonstrate any H-2 genetic restrictions (4). In contrast, the Ts_2 and Ts_3 derived factors are restricted by genes in both the H-2 (I-J) and Igh gene complexes (8,9). Thus, these factors will only suppress recipients which are I-J and Igh compatible with the TsF donor (8,9). This dual genetic restriction may reflect a two signal activation process required for the triggering of Ts_3 or other members of the suppressor cell cascade.

Most of the hybridomas appear to be relatively stable in long-term culture and many have been maintained for over one year in continuous *in vitro* culture. However, to ensure against the loss of these hybridomas, we periodically passage the Ts_1 and Ts_3 hybridomas over NP-BSA coated plates and then expand the adherent cell population. In spite of these efforts, we have sometimes noticed a decrease in the titer of TsF activity over prolonged culture periods from some of the hybridoma lines. Maintenance of the hybridoma cells also appears to be highly dependent on fetal calf serum. Thus, some Ts hybridomas would not grow in certain lots of serum while others grew but lost biological activity. Therefore, we recommend routine monitoring of TsF activity and periodic recloning of Ts hybridomas.

The creation of Ts_1, Ts_2 and Ts_3 hybridomas permitted a thorough comparison of these distinct Ts subsets and allowed analysis of the three types of TsF. It appears that TsF is a soluble form of the T cell receptor. The evidence supporting this hypothesis are: 1) the binding specificity of each subset of Ts cells corresponds with that of the respective TsF (Table 1); 2) the I-J determinants which are

Table 1. Properties of Suppressor T Cells and Factors

Property	Cells			Factors		
	Ts_1	Ts_2	Ts_3	TsF_1	TsF_2	TsF_3
H-2I determinants	I-J	I-J	I-J	I-J	I-J	I-J
Lyt determinants	$1^+,2^-$	2^+	2^+	-	-	-
Igh determinants	Igh-V	?	Igh-V	Igh-V	?	Igh-V
binding specificity	antigen	idiotype	antigen	antigen	idiotype	antigen
phase of action	induction	effector	effector	induction	effector	effector
genetic restriction	-	H-2 + Igh	-	none (Igh pseudo-restriction)	I-J + Igh	I-J + Igh
cellular target	-	-	-	pre-Ts_2	Ts_3	?
activity in cyclophosphamide treated recipients	?	no	yes	?	no	yes

Note: Ts_1 and Ts_3 are induced by antigen presumably in the context of MHC products, whereas Ts_2 are induced by TsF_1. The determinants on the TsF_1 impart genetic restrictions.

also present on TsF appear to be functionally involved with the antigen binding of Ts cells (7); and 3) cell free membrane preparations from Ts_1 or Ts_3 cells have suppressive activity (Table 2).

However, since the Ts hybridomas constitutively secreted TsF, they did not permit analysis of the triggering process required for release of these biologically active mediators. The understanding of how mature Ts cells are triggered is particularly important for the Ts_3 population. Ts_3 cells are generated by conventional antigen immunization and consequently are present concomitant with helper or DTH effector T cells. The Ts_3 population apparently expands and matures following immunization, but remains functionally inactive unless triggered by Ts_2 cells or TsF_2 (1,6,9). This situation permits the primary immune response to proceed normally in the absence of suppression. However, once Ts_2 cells or factors are generated, they will induce the terminal maturation of the Ts_3 population to secrete TsF_3, which in turn rapidly modulates the immune response (2,5,8). Thus, both TsF_2 and TsF_3 function in previously primed animals (*i.e.*, during the effector phase of the immune response) and can mediate suppression of either T or B cell responses within 24 hrs (2,8).

To study the process of Ts_3 activation, we generated a series of inducible TsF_3 secreting cells, termed pTs_3. Such hybridoma lines were prepared by the fusion of NP-O-Su primed lymph node Ts_3 cells enriched from an antigen plate with the BW5147 thymoma. We have prepared pTs_3 hybridomas from both C57BL/6 (B6) and CKB mice. The pTs_3 hybridoma lines were selected from the colonies of $I\text{-}J^+$ cells which also reacted with anti-NP^b idiotype antiserum but failed to constitutively secrete TsF. These cell lines were cultured in the presence of TsF_2 for 1 to 2 days; the cells were then washed extensively and cultured in fresh media for an additional 18-24 hrs. The supernatants were then tested for NP specific suppressive activity in a contact sensitivity system. To ensure that we were not assaying contaminating TsF_2 which may have been carried over into the second culture, suppressive activity was assessed in cyclophosphamide treated mice which are refractory to TsF_2. The data in Table 3 demonstrate the activity and specificity of these pTs_3 derived factors. Furthermore, additional preliminary data indicate that these pTs_3 derived factors have all the properties of TsF_3, including Igh and H-2 genetic restrictions, presence of I-J and NP^b related idiotypic determinants, and activity during the effector phase of the immune response. These inducible hybridomas should permit direct analysis of: 1) non-activated Ts_3 cells, 2) the ability of anti-idiotype or other reagents to activate Ts_3, 3) the analysis of Ts_2 acceptor sites, and 4) characterization of the intracellular activation signals leading to TsF_3 secretion.

Summary

T cell hybridomas can be used to dissect the black box of immune suppression. The use of hybridoma technology permits the series of T-T interactions to be viewed individually. Furthermore, hybridomas can be selected which represent different differentiation or maturation states of a single cell type, *e.g.*, inducible or constitutive TsF_3 secreting cells.

Table 2. Suppressive Activity Obtained from Ts Cell Membranes[a]

TsF source	Cellular fraction	NP-O-Su priming	Specific Footpad swelling ± S.E.
BW5147	-	+	27.0 ± 1.7
CKB-Ts_1-17	-	+	16.6 ± 1.9*
CKB-Ts_1-17	membranes	+	13.8 ± 1.5*
CKB-Ts_1-17	cell sap	+	29.3 ± 2.2
BW5147	-	-	9.5 ± 1.3
BW5147	-	-	21.8 ± 1.5
B6-Ts_3-8	-	+	11.2 ± 3.1*
B6-Ts_3-8	membranes	+	12.4 ± 0.8*
B6-Ts_3-8	cell sap	+	20.7 ± 3.3
BW5147	-	-	4.6 ± 2.2

[a] Groups of 4-5 C57BL/6 mice were primed with 7mg NP-O-Su in the flanks. Animals receiving Ts_1 derived factor were given 0.5 ml i.v. containing 1-5 x 10^6 cell equivalents of culture supernatant, cell membrane or cell sap on day 0. The membranes were prepared by nitrogen bomb cavitation followed by a series of differential centrifugations (20 min at 365g, 15 min at 3,600g, and 30 min at 22,000g). The membranes were washed and the pellet was suspended in DMEM. The Ts_3 derived fractions were administered on day 6. All mice were challenged with NP-O-Su in the hind footpad on day 6. Footpad swelling responses were measured in units of 10^{-3}cm on day 7. An asterisk indicates a significant level of suppression, $p < 0.01$.

Table 3. Antigen Specificity of Activated pTs_3 Hybridoma Derived Factors[a]

Recipient strain	TsF source: pTs Cells	TsF source: TsF_2 used for activation	CY treatment	Antigen priming	Contact Sensitivity Response[b]: NP-O-Su	Contact Sensitivity Response[b]: DNFB
C57BL/6	BW5147	B6-Ts_2-28	+	+	25.8 ± 2.3	9.5 ± 1.6
C57BL/6	B6-Ts_2-28	-	+	+	27.0 ± 3.4	N.D.
C57BL/6	B6-Ts_2-28	-	-	+	16.3 ± 2.8*	N.D.
C57BL/6	B6-pTs_3-9	-	+	+	23.6 ± 1.2	N.D.
C57BL/6	B6-pTs_3-9	B6-Ts_2-28	+	+	9.8 ± 1.0*	9.5 ± 1.2
C57BL/6	-	-	+	-	4.8 ± 1.6	1.3 ± 0.6
B10.A	BW5147	CKB-Ts_2-59	+	+	42.6 ± 2.1	9.3 ± 0.8
B10.A	CKB-Ts_2-59	-	+	+	41.8 ± 4.2	N.D.
B10.A	CKB-Ts_2-59	-	-	+	17.6 ± 1.8*	N.D.
B10.A	CKB-pTs_3-80	-	+	+	39.5 ± 3.1	N.D.
B10.A	CKB-pTs_3-80	CKB-Ts_2-59	+	+	16.5 ± 2.4*	11.2 ± 0.6
B10.A	-	-	+	-	11.8 ± 0.8	0.5 ± 0.3

a Groups of four or five mice were immunized with either NP-O-Su or DNFB. 24 hrs later some groups were given 20 mg/kg cyclophosphamide i.p. On the day before and the day of antigen challenge, mice were given an injection of 0.4 ml i.v. of either control BW5147 or suppressor factors. Activated pTs_3 suppressor factors were obtained by culturing suppressor hybridomas with TsF_2 for 2 days and then in fresh media for 1 day.

b The data are expressed as the increment of footpad or ear swelling ± S.E. in units of 10^{-3} cm. An asterisk (*) indicates significant suppression, $p < 0.01$.

Acknowledgments. We wish to thank Mrs. T. Dinse and N. Axelrod for expert secretarial assistance and Drs. M. Mescher and S-T. Ju for help in preparing the membrane preparations. This work was supported by NIH grants CA 14723 and AI 16677 and a grant from the Cancer Research Institute.

References

1. Sunday ME, Benacerraf B, Dorf ME (1981) Hapten specific T cell responses to 4-hydroxy-3-nitrophenyl acetyl. VIII. Suppressor cell pathways in cutaneous sensitivity responses. J Exp Med 153:811-822
2. Sherr DH, Dorf ME (1981) Hapten specific T cell responses to 4-hydroxy-3-nitrophenyl acetyl. IX. Characterization of idiotype specific effector phase suppressor cells on plaque forming cell responses *in vitro*. J Exp Med 153:1445-1456
3. Weinberger JZ, Germain RN, Ju S-T, Greene MI, Benacerraf B, Dorf ME (1979) Hapten-specific T cell responses to 4-hydroxy-3-nitrophenyl acetyl. II. Demonstration of idiotypic determinants on suppressor T cells. J Exp Med 150:761-776
4. Okuda K, Minami M, Sherr DH, Dorf ME (1981) Hapten-specific T cell responses to 4-hydroxy-3-nitrophenyl acetyl. XI. Pseudogenetic restrictions of hybridoma suppressor factors. J Exp Med 154:468-479
5. Weinberger JZ, Germain RN, Benacerraf B, Dorf ME (1980) Hapten-specific T-cell responses to 4-hydroxy-3-nitrophenyl acetyl. V. Role of idiotypes in the suppressor pathway. J Exp Med 152:161-169
6. Sherr DH, Dorf ME (1982) Hapten specific T cell responses to 4-hydroxy-3-nitrophenyl acetyl. XIII. Characterization of a third T cell population involved in suppression of *in vitro* PFC responses. J Immunol 128:in press
7. Okuda K, Minami M, Ju S-T, Dorf ME (1981) Functional association of idiotypic and I-J determinants on the antigen receptor of suppressor T cell. Proc Natl Acad Sci 78:4557-4561
8. Minami M, Okuda K, Furusawa S, Benacerraf B, Dorf ME (1981) Analysis of T cell hybridomas. I. Characterization of H-2 and Igh restricted monoclonal suppressor factors. J Exp Med 154:1390-1402
9. Okuda K, Minami M, Furusawa S, Dorf ME (1981) Analysis of T cell hybridomas. II. Comparisons among three distinct types of monoclonal suppressor factors. J Exp Med 154:1838-1851

4-Hydroxy-3-nitro-phenylacetyl (NP)-Specific T Cell Hybridomas

M. Cramer, R. Mierau, T. Takemori, G. Suzuki, T. Tada

A. Introduction

T cell hybridomas are a potential source of large amounts of monoclonal, antigen-specific T cell receptor or T cell factor material. However, despite numerous efforts, this goal has not yet been achieved. There may be several reasons for this and we would like to stress three of them.

Firstly, available tests for antigen-binding properties of T hybridoma cells are laborious and time-consuming (1-3) and do not lend themselves as routine screening assays of high numbers of hybridomas. Therefore, screening was done not for antigen binding but for surface components potentially associated with the antigen-specific T cell receptor: I-J (4, 5), idiotype (6), Ig-V_H (7) or a combination of two such markers (8-10). Interesting and partially successful as these indirect approaches are, they nevertheless - at least conceptually - restrict the analyses of T cell hybridomas to certain functional subsets of T cells, to the availability of major idiotypes and in many other ways as well.

Secondly, T cell hybridomas have been screened for anti-idiotype-binding (6) or functionally active (4, 5, 11-14) T cell factors released or secreted into the supernatants of T cell hybridoma cultures. Such a strategy hinges on the assumption that the fused antigen-specific T cell is of the type of a T cell factor producer and that on top of this the ability to produce a factor is also maintained in the hybridoma. These may be very stringent requirements indeed.

Finally, the most crucial drawback of T cell hybridomas seems to be their notorious instability. Despite the fact that some antigen-specific T cell hybridoma lines stabilized to a certain extent (1-3, 5-14) the rule seems to be a gradual loss of the ability to produce antigen-specific molecules. It is unknown whether this loss is due to "non-producers" overgrowing the "producers" or whether every T hybridoma cell is intrinsically unstable, in which case stabilization would need further as yet undefined molecular events in the genome of such cells.

Table 1: Origin, functional activity and surface phenotype of the two T cell hybridomas used in our study

	6NA/58	7C3-13
T lymphoma used for fusion	BW5147	BW5147
NP-sensitized splenic T cells used as fusion partner	C57BL/6	B10.BR
Functional activity	n.t.[a)]	suppressive
Surface phenotype[b)]		
Immunoglobulin	-	-
Thy-1.1	n.t.	+
Thy-1.2	+	-
$I\text{-}J^k$	n.t.	+
Qa-1	n.t.	+
NP^b-idiotype	(+)[c)]	+

a) n.t..: not tested, b) The surface phenotype of 7C3-13 cells was tested in some detail (9, 10). In addition to the data given in this table, 7C3-13 cells typed as follows: $I\text{-}J^{b-}$, $I\text{-}A,B,E,C^{k-}$, $Lyt\text{-}1,2,3^-$, $Ig\text{-}V_H^+$, $Ig\text{-}V_L^-$. c) 6NA cells were stained with rabbit-anti-NP^b antibodies (gift of Dr. T. Imanishi-Kari) and fluorescein-conjugated goat anti-rabbit-Ig and positive cells were selected in the fluorescence activated cell sorter (FACS). Cells selected this way showed a significant titer in the HPI-test on detergent lysates while control cells did not.

B. Our Approach

The aim of our studies was to grow 4-hydroxy-3-nitro-phenylacetyl (NP)-specific T cell hybridomas to numbers of 10^8-10^9 truely NP-positive cells in order to analyse their NP-specific molecules and the genes coding for them. For this reason we developed two methods of screening T cell hybridomas for NP-specific molecules: a) Detergent lysates of cells were assayed for inactivation of hapten-coupled bacteriophages. No special requirements for factor production or surface expression of receptors have to be met in this test system. b) A fast and efficient method is the specific binding of hapten-coupled bacteriophages to T hybridoma cells. This latter technique was used to screen for NP-specific subclones of two independent T cell hybridoma lines. Positive subclones were selected and recloned again in order to get a stabilized line that can be grown to high cell numbers and still expresses the desired properties.

C. The T Cell Hybridomas

Some characteristics of the T cell hybridomas used in this study are summarized in Table 1.

6NA: C57BL/6 mice were sensitized twice intraperitoneally with NP-conjugated C57BL/6-Ig in complete Freund's adjuvant. After two weeks spleens were taken and Ig^+ cells were removed on rabbit-anti-mouse Ig-coated Petri dishes. Ig^- cells were then panned on NP_{17}-bovine serum albumin (BSA). These panned T cells thus enriched for NP-specificity were fused with BW5147 lymphoma cells according to the protocol of Taniguchi et al. (5). The 14 resulting hybrids were screened for NP-specific molecules in detergent lysates by the method described in section D. One of the 3 positive hybrids, clone 6NA/58, was recloned to yield the cell line 6NA/58-44-8-341 (6NA, Table 1), which showed a positive reaction in the test on detergent lysates (compare section D.).

7C3-13: The production and characterization of this hybridoma is described in detail elsewhere (9, 10). Briefly, B10.Br mice were sensitized with NP-coupled gelatine and boosted 2 weeks later with NP-coupled mouse-Ig. Splenic T cells enriched for NP specificity were prepared and fused to BW5147 as described above and in references 9, 10. The resulting hybridomas were screened with anti-I-J^k and anti-NP^b idiotypic antisera in the FACS (9) and in a microcytotoxicity test (10). After two clonings 7C3-13 cells were isolated. The 7C3-13 hybridoma is Ig^- but does not express the Thy-1.2 allele of B10.Br any more. Nevertheless, we take the expression of Qa-1 and I-J molecules on 7C3-13 (Table 1) as indicators that we are dealing with a T cell hybridoma in this case as well (9, 10). 7C3-13 was shown to produce an NP-specific suppressor factor (9).

D. Haptenated Bacteriophage Inactivation (HPI) Tests on Detergent Lysates

Because of its exquisite specificity and sensitivity we successfully used the haptenated bacteriophage inactivation (HPI) assay (15, 16) to analyse isolated hapten-specific T cell receptor material (17-19). In control experiments employing N-(4'-hydroxy-3'-iodo-5'-nitrophenyl-acetyl)-6-amino-caproic acid-coupled T4 bacteriophages (NIP-cap T4) (20) and anti-NP antibodies we now demonstrated that this assay can also be performed in the presence of moderate concentrations of mild detergents (Fig. 1). The sensitivity of the HPI test does, however, decrease in the presence of deoxycholate (Fig. 1) and even traces of sodiumdodecyl-sulfate (SDS) are deleterious for this assay system (data not shown).

Detergent lysates of T hybridoma cells were prepared as follows: Cells are washed 4x in serum free minimal essential medium buffered with 0.01 M HEPES and containing 10 μg DNAse/ml. Up to 10^7 cells are spun down and the pellet is resuspended in 0.5 ml 0.9% NP-40 in 0.01 M phosphate buffered saline, pH 7.2 (PBS) containing 10^{-3} M phenyl-methyl-sulfonyl-fluoride as a protease inhibitor. Lysis is performed for 30 min at 37°C and nuclei are removed by centrifugation at 2000 g for 10 min. For practical reasons 5-15 clones or subclones had to be pooled for the preparation of lysates. Positive pools were then dissected to identify the positive members of the pool.

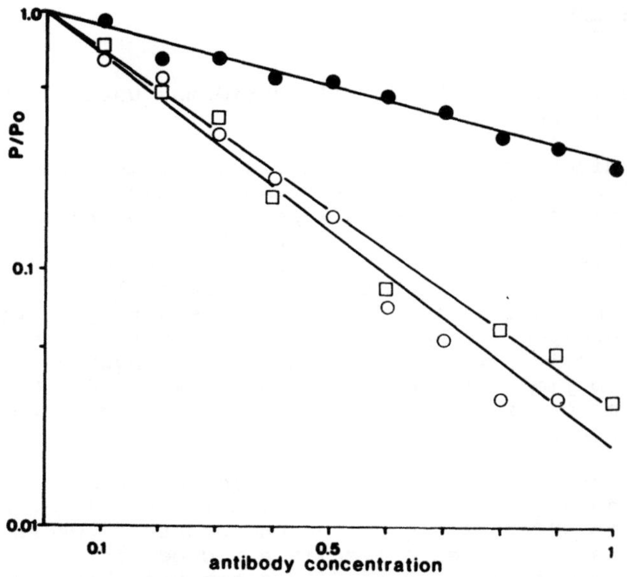

Figure 1. Haptenated bacteriophage inactivation (HPI) assay in the presence of detergent. Primary C57BL/6 anti-NP antibodies are titrated with NIP-cap T4 in the absence of detergent (○ ○ ○) and in the presence of 0.5% NP-40 (□ □ □) and 0.05% NP-40, 0.1% deoxycholate (● ● ●). Plotted is the ratio of surviving plaques P to input plaque number P_o versus a relative concentration of antibodies.

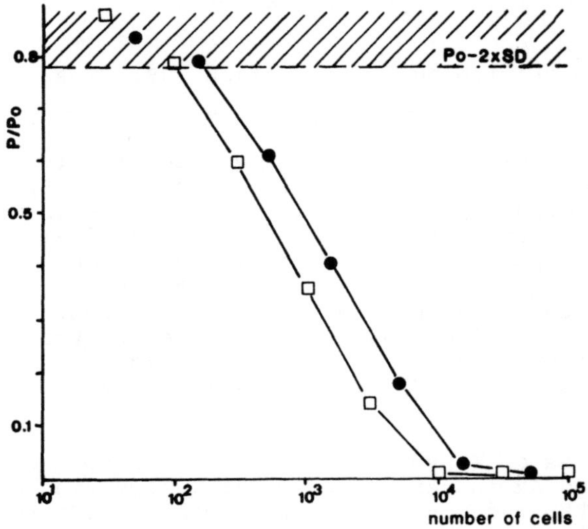

Figure 2. Cellular haptenated bacteriophage (CHP)-binding test with NP-specific B cell hybridoma B1-8δ. B1-8δ cells (□ □ □) were tested in CHP-binding with NIP-cap T4. Every well contained 2 x 10^5 BW5147 cells as fillers. Plotted is the fraction of unbound or surviving phages P/P_o versus the cell number per well. Values above P/P_o = 0.78 are not considered significant as marked by the shaded area. Filled circles (● ● ●) give the HPI results of antibodies produced under CHP-binding conditions by the respective cell numbers per well.

HPI tests lasting 4 hrs were performed at 37°C by serial dilutions of detergent lysate in 10 mg BSA per ml PBS (BSA/PBS). NIP-cap T4 was also diluted in BSA/PBS. Cell free detergent solution served as a medium for determining the input number of plaque forming units (PFU) for every dilution. As an obligatory control, free hapten (NIP-cap, 5×10^{-5} M) was added to an aliquot of the most concentrated lysate dilution in order to verify the specificity of the HPI by inhibition with the hapten. Detergent lysates of various preparations of BW5147 cells did not show any specific HPI titer. Duplicates or triplicates were used throughout.

As mentioned in section C, the cell line 6NA was selected using this method. The original well 6NA/58 showed hapten-specific inactivation values P/P_o of 0.430, 0.662, 0.864, 0.733 and 0.651 at lysate dilutions of 1:3 in five individual lysates. The values for the clone 6NA/58-44-8-341 were 0.372, 0.635, 0.550 and 0.835 at a dilution of 1:4. It has been shown earlier that HPI assays with NIP-cap T4 do not work with monovalent hapten-specific molecules but at least bivalency (or aggregation) is strictly required (19). This may in part explain the rather low but significant (i.e. hapten-specific) P/P_o values obtained with 6NA lysates, since the possibility exists that the hapten-specific molecules we deal with are indeed monovalent and their state of aggregation may be variable.

E. Cellular Haptenated Phage (CHP)-Binding Assay

We therefore developed another assay that uses intact cells to bind/inactivate the hapten-coupled bacteriophage: $1.5\text{-}3 \times 10^5$ cells to be tested are washed two times with 200 µl BSA/PBS/0.1% azide in a well of a 96-well microtiter plate (No. 1-220-25, Cooke, Alexandria, VA/USA). The cell pellet is resuspended in 25 µl bacteriophage suspension containing about 1000 PFU diluted in BSA/PBS/0.1% azide. The plate is kept on ice for 2 hrs and the cells are carefully resuspended on a Vortex every 10 min. Then 100 µl BSA/PBS/0.1% azide are added to every well and the cell/phage suspension is mixed. The microtiter plate is spun at 180 g to pellet the lymphocytes (and to leave the unbound phages in suspension). 50 µl of the supernatants are plated on bacterial plates. Controls with BW5147 lymphoma cells and cell culture medium are performed in 6 individual wells each. Under our experimental conditions BW5147 cells do not bind/inactivate T4 bacteriophage nor a whole battery of hapten-coupled bacteriophages including NIP-cap T4, if BW5147 is grown in fetal calf serum-supplemented culture media (data not shown).

To get an idea of the sensitivity of this CHP-binding assay, control experiments were carried out on surface Ig^+, anti-NP antibody secreting B hybridoma cells (gift of Ms. S. Klein and Dr. A. Radbruch, Cologne). The results appear in Fig. 2. It turns out in repeated experiments on several clones of the B cell hybridoma that 100-300 cells per well can be detected by our assay in the presence of 2×10^5 BW5147 cells (Fig. 2.). In this case a considerable part of the observed effect is, however, due to inactivation of NIP-cap T4 by released antibodies and not to binding of NIP-cap T4 to the cells. This becomes obvious when the B cell hybridoma is cultured under CHP-binding conditions in the absence of NIP-cap T4 and the supernatants are then tested for HPI activity (Fig. 2). In various control experiments 45-80% of the observed effects were due to CHP-binding and not

Table 2: Cellular haptenated bacteriophage (CHP)-binding test with NIP-cap T4 bacteriophages of some subclones of 6NA and 7C3-13 T cell hybridomas

	NIP-cap T4 plaques			Classification c)			
Subclone No. a)	Test A b)	Test B	Test C	Test A	Test B	Test C	Final
6NA-1 - 2	159	n.t. d)	303	+		+	+ e)
4	316	293	n.t.	−	−		−
6	n.t.	313	725		−	−	−
39	165	n.t.	654	+		−	?
43	248	n.t.	832	−		−	−
46	319	n.t.	746	−		−	−
49	129	153	n.t.	+	+		+
68	161	n.t.	486	+		+	+
BW5147 f)	279±40	286±47	752±111				
7C3-13-5-A6	247	410	161	+	−	+	?
B2	n.t.	366	313		−	−	−
D8	330	223	174	+	+	+	+ e)
F7	480	507	358	−	−	−	−
F12	n.t.	119	139		+	+	+
G10	314	334	427	+	−	−	−
H3	394	434	316	−	−	−	−
H10	n.t.	283	309		−	−	−
BW5147	468±39	394±67	422±33				

a) Representative examples of the 1st cloning of 6NA and the 5th cloning of 7C3-13. b) Test A was performed on single wells as soon as the available cell number reached 5-6x10^5 cells/well; Tests B and C followed 3-4 and 6-9 days later. c) Classification: Values lower than the arithmetric means of BW5147 controls minus 2-3×standard deviation of this mean were classified +, higher values as −; discordant results in the tests were finally classified with ?. d) n.t.: not tested. e) Chosen for further cloning. f) BW5147: Arithmetric mean of 6 individual wells ± standard deviation. BW5147 controls did not differ from medium controls, i.e. BW5147 did not bind NIP-cap T4.

Table 3: Increase in the frequency of positive subclones upon successive cloning of positive wells.

	T cell hybridomas	
No. of cloning	6NA	7C3-13
1	9/62 (15%)[a)]	2/77 (3%)
2	12/27 (44%)	4/19 (21%)
3	30/68 (44%)	5/39 (13%)
4		5/25 (20%)
5		5/33 (15%)
6		9/37 (24%)
7		19/49 (39%)
8		17/37 (46%)

a) Given are numbers of subclones reacting positively in several consecutive CHP-binding tests (compare Table 2) versus the total number of subclones tested. In brackets, this frequency of positive subclones is expressed in percent.

to HPI. The question remains open to what extent these B cell hybridoma controls actually reflect the situation of T cell hybridomas (compare section F) as far as receptor densities and affinity/avidity of hapten-binding are concerned.

F. CHP-Binding on T Cell Hybridomas

We now started a program of rapid recloning of 6NA and 7C3-13 cells in order to obtain stable subclones by repeated selections of positive wells for the next subcloning.

As documented in Table 2, individual wells were tested in CHP-binding at least 2-3 times within a period of 7-9 days. The specificity of the binding by positive wells was verified using uncoupled T4 bacteriophages in the CHP-binding test. Even in the relatively short time, some clones turned negative in the CHP-binding assay, e.g. 6NA-1-39 or 7C3-13-5-G10 (Table 2).

The frequency of positive clones in the original cell populations described in section C was surprisingly low, namely 15% and 3%, respectively (Table 3). It took 7 subclonings to increase this frequency to about 40% in the case of the fast growing 7C3-13 line, while 6NA stabilized after the 2nd cloning already (Table 3). Most other series of clonings started from other subclones along this protocol did not show this increase and were thus abandoned.

Parallel to the increase in frequency of CHP-binding subclones we observed an increase in their lifetime as well: While we were unable to keep the earlier clones positive for more than 21 days, we found some wells of the 7th cloning of 7C3-13 to be positive in CHP-binding for longer than 66 days (data not shown).

It was, therefore, feasible to expand 9 clones of the 8th cloning of 7C3-13 to cell numbers of 3-11x10^8 per clone. Three of the clones were negative in CHP-binding after such a massive expansion, 3 clones were just at the limit of significance of the test (mean of control minus 2x standard deviation) and three clones were clearly CHP-binding positive when aliquotes of the 3-11x10^8 cells were tested.

G. Concluding Remarks

With the help of two hapten-specific assay systems we were able to analyse NP-specific T cell hybridomas independently of their functional properties. The CHP-binding test permitted the selection of subclones with a considerably higher frequency of NP-specific cells as compared with original cell populations. In addition these clones express the hapten-specific molecules for a significantly longer time. We thus hope to be able to start a detailed analysis of the gene(s) coding for the NP-specific T cell receptors/factors expressed/produced by T hybridoma cells. We shall try to make use of an NP^b-V_H-DNA probe established some time ago by Bothwell et al. (21). We still feel that there is only one way to get to this stage of experimentation: T cell hybridomas.

Acknowledgements. We thank Ms. S. Klein and Dr. A. Radbruch for the gift of B1-8δ cells, Dr. T. Imanishi-Kari for the rabbit-anti-NP^b antibodies, Ms. C. Schenkel for expert technical help, Drs. U. Krawinkel and A. Radbruch for critical discussions of this manuscript and Ms. A. Böhm for putting all this into the proper frame. This work was supported by the Deutsche Forschungsgemeinschaft through the SFB 74.

References

1. Lonai P, Puri J, Hämmerling G (1981) H-2 restricted antigen binding by a hybridoma clone that produces antigen-specific helper factor. Proc Natl Acad Sci USA 78: 549-553
2. Ruddle NH, Beezley B, Lewis GK, Goodman JW (1980) Antigen specific T cell hybrids - I. T cell hybrids which bind azobenzenearsonate. Mol Immunol 17: 925-931
3. Kaufmann Y, Berke G, Eshhar Z (1981) Cytotoxic T lymphocyte hybridomas that mediate specific tumor cell lysis in vitro. Proc Natl Acad Sci USA 78: 2502-2506
4. Taniguchi M, Miller JFAP (1978) Specific suppressive factors produced by hybridomas derived from the fusion of enriched suppressor T cells and a T lymphoma cell line. J Exp Med 148: 373-382
5. Taniguchi M, Saito T, Tada T (1979) Antigen-specific suppressive factor produced by a transplantable I-J bearing T-cell hybridoma. Nature 278: 555-558
6. Pacificio A, Capra JD (1980) T cell hybrids with arsonate specificity. I. Initial characterization of antigen-specific T cell products that bear a cross-reactive idiotype and determinants encoded by the murine major histocompatibility complex. J Exp Med 152: 1283-1301
7. Eshhar Z, Aple RN, Löwy H, Ben-Neriah Y, Givol D, Mozes E (1980) T cell hybridoma bearing heavy chain variable region determinants producing (T,G)-A--L-specific helper factor. Nature 286: 270-272

8. Okuda K, Minami M, Ju S-T, Dorf ME (1981) Functional association of idiotypic and I-J determinants on the antigen receptor of suppressor T cells. Proc Nat Acad Sci USA 78: 4557-4561
9. Tada T, Okumura K, Hayakawa K, Suzuki G, Abe R, Kumagai Y (1981) Immunological circuitry governed by MHC and V_H gene products. In: Janeway C, Sercarz EE, Wigzell H (eds) Immunoglobulin Idiotypes. ICN-UCLA Symposia on Molecular and Cellular Biology Vol 20. Academic Press, New York, pp 563-572
10. Suzuki G, Hayakawa K, Okumura K, Cramer M, Rajewsky K, Tada T (1982) Heteroclitic fine specificity of monoclonal T cell receptor. Manuscript in preparation.
11. Kontiainen S, Simpson E, Bohrer E, Beverly PCL, Herzenberg LA, Fitzpatrick WC, Vogt P, Torano A, McKenzie IFC, Feldmann M (1978) T cell lines producing antigen-specific suppressor factor. Nature 274: 477-480
12. Taussig MJ, Corvalan JRF, Binns RM, Holliman A (1979). Production of an H-2-related suppressor factor by a hybrid T-cell line. Nature 277: 305-308
13. Hewitt HJ, Liew FY (1979) Antigen-specific suppressor factors produced by T cell hybridomas for delayed type sensitivity. Eur J Immunol 9: 572-575
14. Kapp JA, Aranco BA, Clevinger BL (1980) Suppression of antibody and T cell proliferative responses to L-glutamic acid^{60}-L-alanine^{30}-L-tyrosine^{10} by a specific monoclonal T cell factor. J Exp Med 152: 235-240
15. Haimovich J, Sela M (1966) Inactivation of poly-DL-alanyl bacteriophage T4 with antisera specific towards poly-DL-alanine. J Immunol 97: 338-343
16. Mäkelä O (1966) Assay of anti-hapten antibody with the aid of hapten-coupled bacteriophage. Immunol 10: 81-86
17. Krawinkel U, Rajewsky K (1976) Specific enrichment of antigen-binding T and B lymphocyte surface receptors. Eur J Immunol 6: 529-536
18. Krawinkel U, Cramer M, Kindred B, Rajewsky K (1979) Isolated hapten-binding receptors of sensitized lymphocytes. V. Cellular origin of receptor molecules. Eur J Immunol 9: 815-820
19. Cramer M, Krawinkel U (1980) Immunochemical properties of isolated hapten-specific T cell receptor molecules. In Pernis B, Vogel H-J (eds) Regulatory T Lymphocytes, Academic Press, New York, pp 39-55
20. Becker M, Mäkelä O (1975) Modification of bacteriophage with hapten-ε-aminocaproyl-N-hydroxysuccinimide esters; increased sensitivity for immunoassay. Immunochem/Mol Immunol 12: 329-331
21. Bothwell ALM, Paskind M, Reth M, Imanishi-Kari T, Rajewsky K, Baltimore D (1981) Heavy chain variable region contribution to the NP^b family of antibodies: Somatic mutation evident in a $\gamma 2a$ variable region. Cell 24: 625-637

Regulation of the IgE Response by IgE Class-Specific Suppressor T Hybridomas

M. Suemura, K. Sugimura, A. Ishizaka, S. Kobatake, O. Shiho, H. Deguchi, S. Kishimoto, Y. Yamamura, T. Kishimoto

A. Introduction

Previous studies have demonstrated the presence of IgE class-specific helper and suppressor T cells which show a regulatory effect only on the IgE antibody response (1,2,3). Thus, T cells from mice primed with hapten (DNP or PC)[2]-coupled mycobacterium showed the selective suppressive effect on the in vivo as well as in vitro IgE responses against DNP-antigen or PC-antigen, respectively (2,4,5). The suppressor function was shown to be mediated by antigen-nonspecific, IgE class-specific suppressor factor(s) (IgE-TsF) released from DNP- or PC- mycobacterium (Myc)-primed T cells by antigen-specific stimulation. IgE-TsF had neither Ig determinants nor antigen binding sites (4), but bore the I region gene products (6). Furthermore, IgE-TsF has been proved to have the binding site(s) for IgE molecules as well as I region gene products on the same molecules (7). In the present experiment, we have established IgE class-specific suppressor T hybridomas and studied the function and properties of IgE specific suppressor factors obtained from them.

B. Methods and Results

1. Establishment of IgE class-specific and PC-specific T hybridomas.

In order to characterize and isolate IgE-TsF, we attempted to establish T hybridomas and to obtain monoclonal IgE-TsF from them. In our previous experiments, it was shown that PC-specific suppressor T cells were able to be stained with fluorescent-labeled anti-T15 antibody and enriched by anti-T15-coated Petri dish (5). Thus, in this experiment we employed PC-Myc for the induction of IgE class-specific suppressor T cells, because PC-specific and IgE class-specific suppressor T cells could be compared by employing anti-T15 antibody. Thirty-two clones were established by hybridization of PC-Myc-primed T cells and BW5147 and all culture supernatants derived from them were screened for the activity to suppress the in vitro IgE or IgG response of DNP-KLH-primed cells stimulated with either DNP-KLH or DNP-KLH-PC in Marbrook's culture system. The results of

---

Footnote. 1) This work was supported by a grant from the Ministry of Education, Science and Culture, Japan. 2) Abbreviations used in this paper: DNP, 2.4-dinitrophenyl; PC, phosphorylcholine; Myc, Mycobacterium tuberculosis.

the initial functional screening of the 32 clones were as follows, i.e., 3 clones of a total of 32 hybrid clones showed IgE class-specific but antigen non-specific suppression, and 13 clones showed PC-specific suppression both on the IgE and IgG responses. Four clones showed antigen-non-specific suppressor effect both on the IgE and IgG responses, and the remaining 12 clones did not show any suppression. Among 32 hybrid clones established, clones with IgE

Table 1. Functional analysis of culture supernatants derived from T hybrid cell lines.[a)]

Response	Ag in vitro	Medium	Reverse PFC/culture	% Sup.[b)]
IgE	KLH-DNP	Fresh medium	252 ± 19	-
		BW5147	347 ± 24	< 0
		F18-3-4	225 ± 30	13
		F18-3M16	63 ± 7	87
	PC-KLH-DNP	Fresh medium	151 ± 12	-
		BW5147	221 ± 18	< 0
		F18-3-4	45 ± 9	91
		F18-3M16	51 ± 4	86
	None	Fresh Medium	35 ± 5	-
IgG_1	KLH-DNP	Fresh medium	6,240 ± 384	-
		BW5147	6,720 ± 192	< 0
		F18-3-4	6,676 ± 97	< 0
		F18-3M16	6,630 ± 253	< 0
	PC-KLH-DNP	Fresh medium	9,640 ± 601	-
		BW5147	10,218 ± 432	< 0
		F18-3-4	1,840 ± 72	90
		F18-3M16	9,200 ± 400	5
	None	Fresh medium	1,020 ± 340	-

a) Twenty five x 10^6 DNP-KLH-primed cells were suspended in 1 ml of the supernatants and stimulated with 1 μg of PC-KLH-DNP or KLH-DNP. Four days later, IgE or IgG_1 producing cells were enumerated by the reverse plaque assay.

b) $$\% \text{ suppression} = \left(1 - \frac{\text{experimental group - background}}{\text{positive control - background}} \right) \times 100.$$

specific suppressor function and antigen (PC) specific suppressor function were recloned by the limiting dilution method. Representative results of the function of these hybrid clones are shown in Table 1. The supernatant from the clone with IgE specific suppressor function (F18-3M16) exerted the selective suppression of the IgE antibody response in vitro against either DNP-KLH-PC or DNP-KLH. In contrast, the supernatant from the clone with antigen-specific suppressor activity (F18-3-4) showed almost complete suppression of both IgE and IgG antibody responses, when DNP-KLH primed cells were stimulated in vitro with DNP-KLH-PC but did not show any suppressive effect when stimulated with DNP-KLH.

Surface phenotypes of antigen-specific (F18-3-4) and IgE class-specific (F18-3M16) hybridomas were analysed (Table 2). The results

Table 2. Characteristics of T hybrid clones.

Properties	F18-3-4	F18-3M16	Methods
Suppressor function	PC-specific	IgE-specific	
Karyotype	61 ± 5	58 ± 2	
$H-2^k$	+	+	A[a), B
$H-2^d$	+	+	A , B
Thy 1.2	+	+	A , B
Lyt 1.2	-	-	A , B
Lyt 2.2	+	+	A , B
Ig	-	-	B
Antigen-binding	+ (64K)	-	C
T15 idiotype	+	-	B , C
Fcε receptor	-	+	C

a) A: Cytotoxicity, B: Fluorescence, C: Rosetting.

are summarized as follows, i) Both clones expressed $H-2^k$ and $H-2^d$ products. ii) They also expressed Thy 1.2 and Lyt 2.2, but did not bear any Lyt 1.2 antigen and Ig-determinants. iii) Clone F18-3-4 cells were stained by anti-T15 idiotypic antibody and constantly rosetted with PC-conjugated SRBC at the range of 20-66%, however F18-3M16 cells did not express any PC-binding or T15 idiotype bearing molecules on their surface. iv) PC-binding molecules with the molecular weight of 64K were isolated from PC-specific hybrid clone, F18-3-4, while any PC-binding molecules were detected on IgE-specific suppressor hybridomas, F18-3M16. v) The presence of Fcε receptors was demonstrated in 3-20% of clone F18-3M16 cells, but not in clone F18-3-4 cells.

These results clearly showed that two distinct kinds of hybridomas i.e., antigen-specific suppressor hybridomas, and IgE class-specific, antigen-non-specific suppressor hybridomas, have been established. Although the results did not completely exclude the presence of IgE class-specific suppressor T cells with antigen (PC) specificity, such a possibility seemed to be less likely, since none out of 32 clones tested showed dual specificity. Therefore, it is conceivable to assume that antigen-specific suppressor T cells may be involved in the induction or activation of IgE class-specific suppressor T cells, since the induction of IgE-TsF from PC-Myc-primed T cells required the PC-specific stimulation but the effect of IgE-TsF was antigen-nonspecific but IgE class-specific (see discussion).

2. Properties and functions of IgE specific suppressor factor(s).

Properties and functions of IgE specific suppressor factors were studied by employing the culture supernatant from IgE class-specific suppressor T hybridomas. In order to study whether the factor(s) had the binding sites for IgE, attempts were made to absorb the suppressor activity with IgE-conjugated Sepharose column. Culture supernatant of IgE specific suppressor hybridomas, clone F1819-d, was applied to IgE-, IgM- or IgG-column and the suppressor activity of the column effluent was tested in the in vitro experimental system, in which DNP-KHL-primed cells were cultured with DNP-KLH and IgE and IgG_1 producing cells were enumerated by reverse plaque assay with anti-ε and anti-γ_1 antibody.

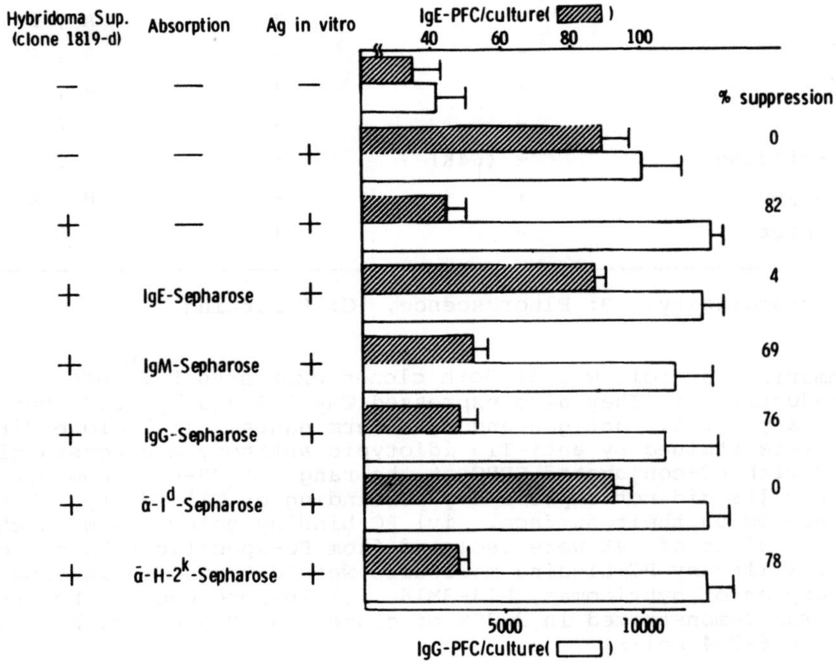

Fig. 1. Absorption of IgE-TsF with IgE-column and anti-I^d-column. The supernatant of IgE class-specific suppressor T hybridoma was applied to IgE-, IgM-, IgG-, anti-I^d- or anti-H-2^k-coupled Sepharose column, and the column effluents were examined for their suppressive activity.

As shown in Fig. 1, the suppressor activity was specifically absorbed with IgE-column, whereas IgM- or IgG-column did not remove any suppressor activity, indicating that the IgE specific suppressor factor derived from T hybridomas has the binding site(s) for IgE molecules. The suppressor activity was also absorbed with anti-I^d-column, but was not absorbed with anti-H-2^k-column although the hybridomas expressed H-2^k products on their surface. This result shows the presence of I region gene products derived from BALB/c mice on the suppressor factors. In the next experiment, functional organization of the IgE binding site(s) and I region gene products on the suppressor factor(s) was studied. The suppressor activity absorbed with IgE-column was eluted with acetate buffer (0.05 M, pH 4.0), and reabsorbed by anti-I^d-column, indicating that the suppressor factor derived from IgE specific suppressor T hybridomas, like the conventional IgE-TsF obtained from PC-Myc-primed T cells, was composed of the binding site(s) for IgE and the I region gene products.

3. Mechanisms of action of IgE class-specific suppressor factor(s).

These results shown above and the previous results (6,7) suggested that IgE-TsF act directly on IgE B cells which bear IgE molecules. In the following experiments, the mechanisms of the action of IgE-TsF derived from T hybridoma were studied by using anti-DNP IgE-producing B hybridoma cells as target cells. IgE-producing B hybridoma cells were incubated for 1 hr at 37°C or 4°C in the presence or absence of the supernatant from IgE specific suppressor hybridomas, and the number of IgE-producing cells were enumerated by reverse PFC assay (Table 3). The results showed that the reduction of IgE-PFC

Table 3. Suppressive effect of the supernatant from IgE-specific suppressor T hybridoma on the development of IgE-PFC from IgE B hybridoma cells.[a)]

Temp.	Incubation (1 hr)	sIgE(+)cells	IgE-PFC $x10^{-2}$/culture	Recovered cells
37°C	Medium	13%	173 ± 29	9.5×10^4
	IgE-Ts hybridoma sup.	4%	56 ± 7	9.2×10^4
4°C	Medium	11%	120 ± 16	9.0×10^4
	IgE-Ts hybridoma sup.	6%	118 ± 20	9.6×10^4

a) 1×10^5 IgE-producing B hybridomas were incubated with or without the supernatant of IgE specific suppressor hybridomas for 1 hr at 37°C or 4°C and IgE-producing cells were counted by reverse plaque assay.

was observed when cells were incubated with the suppressor factor at 37°C, as compared with the control experiment. Although the inhibition was not complete and the reduction of PFC was 40-60% through experiments, inhibition of PFC was consistently observed in repeated experiments. At the same time, surface IgE(+) cells were examined by immunofluorescence. The number of surface IgE(+) cells was significantly decreased in the experimental groups incubated with the suppressor factor at both 37°C and 4°C, however neither capping nor patching of IgE receptors was observed. These results suggest that IgE-TsF suppress the secretion and/or production of IgE after binding with IgE B hybridoma cells via IgE receptors, and that its binding per se with IgE receptors is not enough to reduce IgE-PFC, since the incubation at 4°C did not inhibit the secretion of IgE. The possibility that the suppression might be due to some cytotoxic effect of the culture supernatant was excluded, because the number of the recovered viable cells did not differ with or without the suppressor factor. The results were confirmed by using IgE-TsF purified with IgE-column from the culture supernatant of IgE specific suppressor hybridomas. The numbers of surface IgE(+) cells and the IgE-PFC were dose-dependently suppressed with purified IgE-TsF.

In order to examine the possibility that IgE-TsF suppressed the biosynthesis of IgE, IgE B hybridoma cells were cultured for 24 hrs with the culture supernatant from IgE specific suppressor T hybridomas and the number of cytoplasmic IgE(+) cells was enumerated by immunofluorescence (Table 4).

Table 4. Specific suppression of IgE biosynthesis with the supernatant from IgE-specific suppressor T hybridoma.[a)]

Culture	cIgE (+) cells	Recovered cells
Medium	134/1030 (13.0%)	13.6×10^4
Ag-Ts hybridoma sup.	138/ 927 (14.9%)	11.4×10^4
IgE-Ts hybridoma sup.	44/ 963 (4.6%)	12.2×10^4

a) IgE B hybridomas were cultured with fresh medium or antigen specific suppressor factor or IgE specific suppressor factor for 24 hr and IgE-producing cells were counted by immunofluorescence.

Significant reduction of the number of cytoplasmic IgE(+) cells was observed only in the experimental group cultured with IgE specific suppressor factor(s). The supernatant of PC-specific suppressor T hybridomas did not suppress the development of cytoplasmic IgE(+) cells. As it is also clear from the results, the cell proliferation was not affected, and it was the case even after 48 hr cultivation. This result shows that the suppression of the development of IgE-PFC is attributable mainly to the inhibition of biosynthesis of IgE with IgE-TsF, and IgE-TsF does not affect the proliferation of IgE B cells.

In the next experiment, IgE B hybridomas were cultured for 24 hr with serially diluted IgE-TsF purified from IgE specific hybridoma supernatant and IgE-producing cells were counted by reverse plaque assay or cytoplasmic IgE staining. The results showed that purified

IgE-TsF suppressed dose-dependently the development of IgE-PFC as cytoplasmic IgE(+) cells (see Table 5).

Table 5. Parallel suppression of the induction of IgE-PFC and cytoplasmic IgE with IgE-TsF.[a)]

Culture	IgE-PFC $\times 10^{-2}$/culture	cIgE (+) cells	Recovered cells
Medium	153 ± 7	69/519 (13.3%)	11.1×10^4
IgE-TsF (x 2 dil)	81 ± 4	20/396 (5.1%)	10.9×10^4
(x 4 dil)	117 ± 6	51/790 (6.5%)	9.7×10^4
(x 8 dil)	122 ± 21	62/659 (9.4%)	10.5×10^4

a) 1×10^5 IgE B hybridomas were cultured with or without IgE-TsF for 24 hr and IgE-producing cells were counted by reverse plaque assay or cytoplasmic IgE staining.

The specificity of the suppressive effect of IgE-TsF was demonstrated by employing anti-DNP IgM-producing B hybridoma cells. IgE-TsF did not suppress the biosynthesis of IgM in IgM-producing B hybridomas, indicating the specificity of IgE-TsF in its effector phase.

If IgE-TsF suppress the biosynthesis of IgE by binding with surface IgE receptors, removal of IgE receptors from IgE B hybridomas will make them insensitive to the effect of IgE-TsF. IgE B hybridomas were preincubated for 1 hr with anti-ε antibody (20 μg/ml) to cap surface IgE receptors, washed extensively and cultured with IgE-TsF. The results showed that the capping of IgE receptors abrogated the suppressive effect of IgE-TsF, proving that IgE-TsF exerts its effect by binding with IgE receptors (see Table 6).

Table 6. Failure of IgE-TsF to suppress IgE induction of B hybridoma cells after stripping IgE receptors.[a)]

Pretreatment	Culture	IgE-PFC $\times 10^{-2}$/culture
—	Medium	152 ± 26
—	IgE-Ts hybridoma sup.	84 ± 23
a-ε Ab	Medium	168 ± 24
a-ε Ab	IgE-Ts hybridoma sup.	151 ± 27

a) IgE B hybridomas were treated with 20 μg/ml anti-ε antibody for 1 hr and cultured with or without IgE-TsF for 24 hr.

Finally, we examined the effect of the simultaneous addition of IgE with IgE-TsF on its suppressive effect. It was found that the suppressive effect of IgE-TsF was neutralized with 0.1 μg/ml IgE. Taken together, these results confirm that the target of IgE-TsF is IgE molecules present on the surface of IgE B cells and suggest that IgE-TsF binds to the Fc portion of IgE which is common to the membrane and secretory type of IgE.

C. Discussion

The present study demonstrated the establishment of two distinct types of T hybridomas, i.e., PC-specific suppressor T hybridomas and IgE class-specific, antigen non-specific suppressor T hybridomas, from the PC-Myc-primed T cell populations, which showed PC-specific suppressive function selectively on the IgE antibody response. Phenotypically, IgE class-specific as well as PC-specific suppressor hybridomas had common characters as suppressor T cells. However, there were some clear differences between them as follows. i.e., i) The presence of PC-binding molecules which bore T15 idiotype was demonstrated on the surface of PC-specific hybridomas by the rosette formation and immunofluorescence, whereas the hybrid clone with IgE class-specific suppressive function did not express any PC-binding or T15 idiotype bearing molecules on the surface. ii) The Fc receptors for IgE were found on the surface of IgE class-specific suppressor hybridomas, but not on the surface of PC-specific suppressor hybridomas suggesting that Fcε receptors on IgE specific hybridomas may be intimately related with their IgE class-specific suppressor function. Thus, the nature of the surface phenotypes of the T hybridomas seems to coincide with their functions. iii) The nature of the suppressor factors released from PC-specific and IgE class-specific suppressor T hybridomas was also different, i.e., PC-specific suppressor hybridomas released the suppressor factor(s) which could bind with PC-antigens and suppressed both IgE and IgG responses only when PC-antigens were used as stimulating antigens, while IgE class-specific suppressor hybridoma released the suppressor factor(s) which did not bind with PC-antigens but could bind with IgE molecules and showed antigen non-specific and IgE class-specific suppressor function.

These results strongly suggested that two distinct subsets of T cells, i.e., PC-specific T cells and IgE class-specific but antigen-nonspecific T cells, were present in the PC-Myc-primed T cell populations. In the preliminary experiment, it has been demonstrated that the depletion of T15 positive T cells from PC-Myc-primed T cell population abrogates their IgE class-specific suppressor function. However, their function could be restored by the addition of PC-antigen and PC-specific suppressor factor(s) from PC-specific hybridomas, clone F18-3-4. The result strongly suggests that antigen specific suppressor factor(s) released from antigen-specific T cells may activate IgE class-specific suppressor T cells in collaboration with the antigen.

The previous results with the conventional IgE-TsF suggested that the factor(s) exerted the suppressive function by acting on B cells (6) and showed that the suppressor activity could be absorbed with IgE-producing B hybridomas (7). The present result with the factor(s) from T hybridomas demonstrated the presence of the binding site(s) for IgE in IgE-TsF. Taken together, these results suggest that IgE-TsF act on Bε cells through IgE receptors and inactivate the differentiation or proliferation of Bε cells. This was further confirmed by the experiments with IgE-producing B hybridomas, in which incubation of IgE-producing B hybridomas with the supernatant from IgE specific suppressor T hybridomas or purified IgE-TsF resulted in the reduction of IgE-PFC. The reduction of IgE-PFC was demonstrated to be due to the suppression of IgE biosynthesis in hybridomas. In these experiments, the specificity of the effect of IgE-TsF was proved. Thus, the supernatant from PC-specific suppressor hybridomas did not show any effect on the development of IgE-PFC, and IgE-TsF did not exert any suppressive effect on IgM-producing B hybridomas. Binding of IgE-TsF with IgE B hybridomas through IgE receptors was demonstrated to be essential for the suppression, since IgE-TsF was ineffective after stripping IgE receptors off from the surface of IgE B hybridomas. However, some active metabolism after binding of IgE-TsF with IgE receptors was also required for the suppression, because incubation of IgE B hybridoma cells with IgE-TsF at 4°C did not inhibit the development of IgE-PFC.

The reason why the suppression of the development of IgE-PFC with IgE-TsF was incomplete is not clear, but some explanations may be tenable, i.e., i) IgE-TsF may be neutralized with IgE which is released from IgE B hybridomas during the incubation, ii) the proportion of surface IgE(+) cells in IgE B hybridomas was usually 10-30%, and IgE-TsF may be inaccessible to surface IgE(-) cells in some phase of cell cycle, iii) tumor cell nature of IgE B hybridomas makes them relatively resistant to the effect of IgE-TsF.

Although the suppressive effect of IgE-TsF on IgE-producing B hybridomas was incomplete, the result was reproducible and the suppression was dose-dependent of IgE-TsF. Thus, this system will be applied for the assessment of the activity of IgE-TsF.

D. Summary

In order to characterize and isolate IgE class-specific suppressor T cell factor(s), attempts were made to establish IgE class-specific suppressor T hybridomas. PC-Myc-primed T cells were hybridized with BW5147 and two distinct types of T hybridomas, i.e. antigen (PC)-specific and antigen-nonspecific and IgE class-specific, were obtained. They were clearly different from each other with respect to their surface phenotypes and functions. IgE class-specific suppressor factor(s) (IgE-TsF) derived from T hybridomas was demonstrated to be composed of the binding site(s) for IgE and the I region gene products. In the analysis of the mechanism of the action of IgE-TsF by using IgE producing B hybridomas, IgE-TsF was shown to bind to IgE B cells via IgE receptors on the surface and to suppress the biosynthesis of IgE.

Acknowledgments. We thank Miss K. Kubota for her secretarial assistance.

References

1. Kishimoto T, Ishizaka K (1973) Regulation of antibody response in vitro. VI. Carrier-specific helper cells for IgG and IgE antibody response. J Immunol 111:720-726.
2. Kishimoto T, Hirai Y, Suemura M and Yamamura Y (1976) Regula-

tion of antibody response in different immunoglobulin classes. I. Selective suppression of anti-DNP IgE antibody response by preadministration DNP-coupled Mycobacterium. J Immunol 117:396-404.
3. Kimoto M, Kishimoto T, Noguchi S, Watanabe T and Yamamura Y (1977) Regulation of antibody response in different immunoglobulin classes. II. Induction of in vitro IgE antibody response in murine spleen cells and demonstration of a possible involvement of distinct T helper cells. J Immunol 118:840-845.
4. Suemura M, Kishimoto T, Hirai Y and Yamamura Y (1977) Regulation of antibody response in different immunoglobulin classes. III. In vitro demonstration of "IgE class-specific" suppressor functions of DNP-mycobacterium-primed T cells and the soluble factor released from these cells. J Immunol 119:149-155.
5. Sugimura K, Kishimoto T, Maeda K and Yamamura Y (1981) Demonstration of T15 idiotype-positive effector and suppressor T cells for phosphorylcholine-specific delayed-type hypersensitivity response in CBA/N or (CBA/N x BALB/c) F_1 male mice. Eur J Immunol 11:455-461.
6. Kishimoto T, Hirai Y, Suemura M, Nakanishi K and Yamamura Y (1978) Regulation of antibody response in different immunoglobulin classes. IV. Properties and functions of "IgE class-specific" suppressor factor(s) released from DNP-mycobacterium-primed T cells. J Immunol 121:2106-2112.
7. Suemura M, Shiho O, Deguchi H, Yamamura Y, Böttcher I and Kishimoto T (1981) Characterization and isolation of IgE class-specific suppressor factor (IgE-TsF). I. The presence of the binding site(s) for IgE and of the H-2 gene products in IgE-TsF. J Immunol 127:465-471.

Establishment of Functional, Antigen-Specific T Cell Lines by RadLV-Induced Transformation of Murine T Lymphocytes

P. Ricciardi-Castagnoli, F. Robbiati, E. Barbanti, G. Doria, L. Adorini

A. Introduction

Oncogenic viruses, ionizing radiations or carcinogens have been extensively used to induce experimental T cell lymphomas in mice. These T cell tumors have unknown antigenic specificities and are therefore not very useful for functional and structural analyses of specific T cell products. The availability of a transforming virus able to infect *in vitro* a pre-selected, antigen-specific T cell population should overcome this problem. A retrovirus having such peculiar characteristics is the Radiation Leukemia Virus (RadLV) originally isolated by Lieberman and Kaplan from radiation-induced thymic lymphomas in C57BL mice (1). These retroviruses have a T cell-restricted tropism and T lymphocytes can be easily infected *in vitro* by a short exposure to the virus. Some of the *in vitro*-infected cells become transformed after injection into syngeneic hosts and give rise, a few months later, to donor-type T cell lymphomas, as judged by genetic markers (2).

Using this approach we have been able to establish T cell lines which secrete antigen-specific products and maintain functional activity in spite of their neoplastic transformation. Both helper and suppressor T cells can be transformed and antigen-specific products can be obtained in unlimited amounts for structural, genetic and functional studies.

B. Biological Properties of Radiation Leukemia Virus (RadLV)

The radiation leukemia viruses are endogenous leukemogenic retroviruses of the C57BL/Ka mouse strain. Expression of these endogenous viral genes was originally obtained by whole body X-irradiation of C57BL/Ka mice (1).

From the radiation-induced thymic lymphomas, Lieberman and Kaplan extracted a viral agent, designated RadLV, which has been successfully passed *in vivo* for more than 20 years (3). Recently, several cell lines from RadLV-induced lymphomas have been established *in vitro* (4, 5). These permanent T cell lines are chronically infected with highly oncogenic RadLV and produce large amounts of infectious particles which can be recovered from the culture supernatant. The most used cell lines for the *in vitro* production of RadLV are VL_3, a T cell line derived from RadLV-induced thymic lymphomas of C57BL/Ka mice (6) and Nu_1, a lymphoblastoid cell line derived from RadLV-induced non-thymic lymphomas of nude (nu/nu) mice (5). The biological properties of RadLV are well characterized (6,7). These viruses induce neoplastic transformation in susceptible target cells such as bone marrow, spleen thymus and fetal liver cells (8, 9). The tropism of RadLV for cells of the T lymphocyte lineage is remarkable although not yet fully under stood. Until now, RadLV has never been shown to transform lymphocytes of the B cell lineage, unlike Abelson murine leukemia virus (10). RadLV, continuously produced in vitro by VL_3 or Nu_1 cell lines, are typical C-type particles with biological characteristics similar to the original RadLV passed in vivo. When injected into susceptible hosts they induced 100% T-cell lymphomas within three months. These lymphomas are generally localized in the thymus, but lymph nodes and spleen can also be involved (4, 5).

Antigen-specific T lymphocytes obtained from different lymphoid tissues of antigen-primed mice can also be infected *in vitro*. However, neo plastic transformation occurs only if the infected lymphomas are main tained under conditions which support their proliferation. Until now the easiest procedure to achieve transformation consists in the inocu lation of the infected cells into irradiated syngeneic hosts. Suscepti bility to RadLV-induced transformation is age-dependent. Viral infec tion of lymphocytes from newborn or young mice results in a much higher (40-50%) percentage of donor-type lymphomas as compared to that ob tained from infected lymphocytes of adult mice (5%) (11). However, irradiation can transiently restore the susceptibility of adult mice almost to neonatal levels.

Two different restrictions confer resistance to RadLV. The first is operating at the H-2 complex level. Mice of $H\text{-}2^b$ haplotype are the most susceptible to RadLV-induced leukemogenesis. Mice with different haplo types are generally less susceptible or resistant (11). According to Meruelo (12) the H-2 restriction of leukemogenesis may be assigned to the H-2 D region of the H-2 complex and H-2 D^q and $H\text{-}2^s$ haplotypes would confer susceptibility to RadLV. In our own experience, using RadLV/Nu_1 to infect T cells *in vitro*, we have successfully transformed lymphocytes from $H\text{-}2^b$ (C57BL/6) and from $H\text{-}2^a$ (A/J) mice or from their F_1 hybrids, but we have not been able to transform lymphocytes from BALB/c mice ($H\text{-}2^d$).

A second genetic restriction is determined by the Fv-1 gene (13). Rad LV is a B-tropic virus (14) and replicates up to titer 10^3 fold higher on cells bearing $Fv\text{-}1^{bb}$ genotype than on cells bearing $Fv\text{-}1^{nn}$ genotype. This classification is unrelated to the H-2 type since mice with $H\text{-}2^b$ and $H\text{-}2^k$ haplotypes are present in both groups.

C. RadLV Infection of Enriched T Lymphocytes

The procedure to establish functional, antigen-specific T cell lines by RadLV-induced transformation of T lymphocytes is outlined in fig. 1. To achieve best results, specific T lymphocytes should be highly en riched before viral infection since RadLV may infect any T cell, al though probably mainly activated T cells will be transformed. Several different methods have been described to select antigen-specific sub populations of T lymphocytes. In our hands, a satisfactory enrichment of splenic suppressor T cells has been obtained by a two-step panning procedure. First, erythrocytes are lysed by treatment with Tris-buffered ammonium chloride, then nucleated cells are incubated for one hour at 4°C on 100 mm plastic Petri dishes, previously coated overnight with 5 ml of purified anti-mouse immunoglobulins (100 μg/ml), to eliminate B lymphocytes and adherent cells. Non-adherent cells are then recovered by gentle swirling and incubated for one hour at 4°C on antigen-coated Petri dishes. Unbound cells are discarded by washing the plate several times and the bound cells removed by pipetting. For hen egg-white lysozyme (HEL)-specific suppressor T cells the final recovery was about 5% of the total spleen cells and 80% of the recovered cells were $I-J^+$ (15).

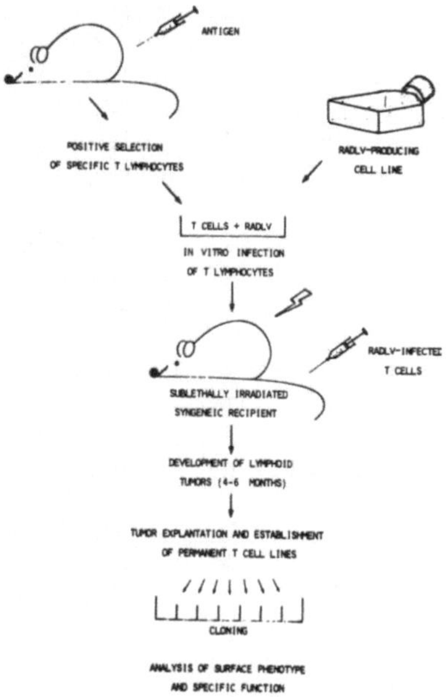

Fig. 1. Procedure for the establishment of functional, antigen-specific T cell lines by RadLV-induced transformation of murine T lymphocytes.

Besides using other cell separation procedures, such as cell sorting, antigen-specific T cell clones maintained *in vitro* by IL-2 and antigen may also represent suitable cell populations for RadLV-induced transformation.

To obtain viral infection of enriched T lymphocytes the following procedure has routinely been used. Enriched T cells ($20x10^6$) in 1 ml of RPMI containing 10% FCS and 4 μg/ml polybrene (Sigma) are incubated in 60 mm Petri dishes with 1 ml of freshly collected cell-free supernatant, harvested from Nu_1 or VL_3 cell lines, for one hour at 37°C in a humidified CO_2 incubator and every 15 min the plates are gently swirled. Then the cells are washed three times in 50 ml of medium to remove free viral particles and immediately injected i.v. ($1-5x10^6$ cells/mouse) into syngeneic or congeneic irradiated (450 R) recipients. Within 6 months a high percentage (about 50%) of lymphomas is usually obtained. Depending on the selected T cells used the tumors are localized either in the thymus or in the lymph nodes and spleen (15).

D. Establishment of Permanent T Cell Lines

Tumors cells obtained from the explanted lymphomas can be immediately injected i.p. into pristane-primed (0.5 ml one week before) syngeneic mice or washed and plated *in vitro* ($4\text{-}5\times10^6$ cells/ml) in RPMI medium supplemented with 20% FCS, 2×10^{-5} 2-mercaptoethanol and 1mM L-glutamin. When the tumor is localized in the thymus or in the lymph nodes, a feeder layer of peritoneal macrophages should be added to cultures. Such feeder layers improve the establishment of a permanent T cell line. Therefore, we routinely wash the mouse peritoneum before explanting the tumor and plate peritoneal cells together with the lymphatic tissue.

Cultures are incubated at 37°C in 5% CO_2-humid air and once a week half medium is removed by very careful suction and replaced by fresh medium. After 1-2 weeks of culture massive cell death may be observed. Nevertheless, in close association with adherent cells, small clusters of large, round cells are frequently observed. Later on, these clusters increase in size as cells start to proliferate. Generally, this occurs 30-40 days after the initial plating and before this time cultures should not be splitted off. During this period syngeneic macrophages may be added again to increase cell proliferation. In our own experience, establishing a cell line from such primary cultures requires great care whereas it is usually easier to establish a permanent cell line *in vitro* if the tumor cells are first transplanted *in vivo*. Moreover, addition of soluble antigen during the first weeks of cultures seems to facilitate the establishment of stable cell lines.

Once established, lymphoma cells grow rapidly, usually with doubling times between 12 and 18 Hours (Fig. 2).

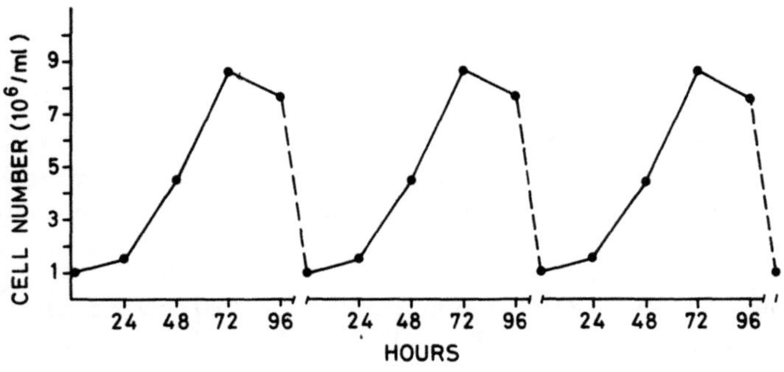

Fig. 2. Growth curve of clone LH8-105.

Routinely the medium is supplemented with 10% FCS but we have succeeded in growing cells in medium containing only 0.5% FCS.

Besides growing in stationary cultures, lymphoma cells also grow in spinner cultures at a very high cell density ($8\text{-}10\times10^6$ cells/ml). These cell lines can be cloned by limiting dilution on adherent cell feeder layers very efficiently (Fig. 3).

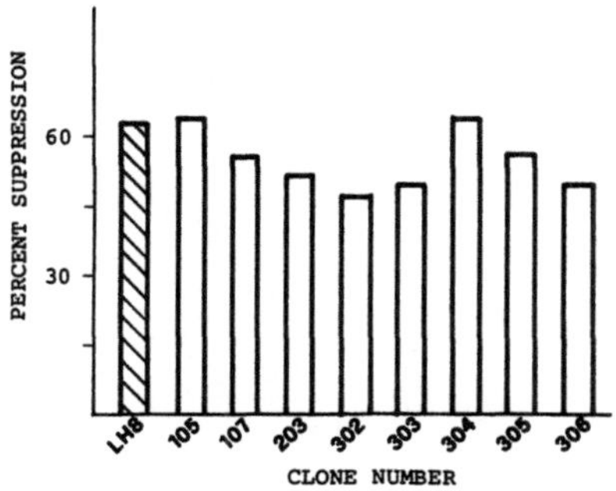

Fig. 3. Functional activity of clones derived from the HEL-specific suppressor T cell line LH8. BDF1 mice (5 mice/group) were injected i.p. with 100 µg HEL in CFA and developed anti-HEL PFC responses measured 8 days later in the parathymic lymph nodes. Culture supernatant from LH8 cell line or from clones obtained by limiting dilution from this line was injected i.v. (1 ml/ mouse) at the same time as antigen priming. Columns represent percent suppression of the control response attained by mice not injected with culture supernatant(12,579 developed anti-HEL PFC/10^6 parathymic lymph node cells).

Cells resuspended in 10% DMSO, 50% FCS and 40% medium can be stored in liquid nitrogen and are easily recovered without loss of viability and function. Metaphase analysis of established cell lines shows a normal diploid caryotype and tetraploid cells were never found. The specific biological function is maintained,without need of recloning, even after more than two years of continuous culture *in vitro*.

From a single mouse injected i.p. with 10^7 cells more than 10^9 tumor cells can be recovered in two-three weeks. These cells are easily adapted to *in vitro* culture without loss of viability. In this way several liters of homogeneous culture supernatant containing T cell-derived, antigen-specific products can be rapidly obtained.

E. Phenotypic and Functional Selection of Established T Cell Lines

Selection of established T cell lines should first be based on a genetic marker to demonstrate the donor-origin of the RadLV-induced T cell lymphoma. For this purpose, congeneic strains such as C57BL/Ka Thy 1.1 and C57BL/Ka Thy 1.2 may be used. Monoclonal anti-Thy 1.1 and anti-Thy 1.2 antibodies can be used to detect such antigens on the cell surface either by immunofluorescence or by cytotoxicity assays.

Lyt and I-region products phenotype screening may also provide useful information about the nature of a given T cell line although some authors have reported mismatching between phenotype expression and function (2). Tumor cells tend to modulate surface antigen expression and some markers may decrease when cells are continuously grown *in vitro*.

Finally, selection of antigen-specific T cell lymphomas has to be based on functional assays. Many different protocols have been described to test immunological functions of antigen-specific T cell products.

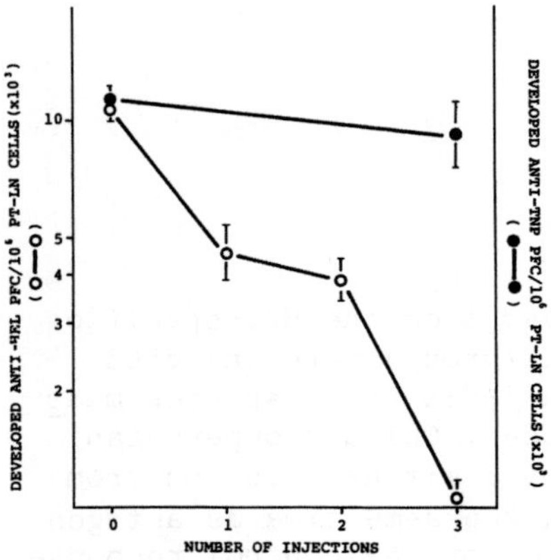

Fig. 4. HEL-specific suppressive activity of culture supernatant from clone LH8-105. BDF_1 mice (5 mice/group) were injected i.p. with 100 µg HEL or TNP-KLH in CFA and developed PFC responses measured 8 days later in the parathymic lymph nodes. Culture supernatant from clone LH8-105 (1 ml/mouse) was injected daily from the time of antigen-priming.

In the case of HEL-specific suppressor T cell lines we have been able to demonstrate antigen-specific suppressive activity both in high-speed supernatants of cell lysates and in cell culture supernatant fluids. Addition of these T cell products exerts antigen-specific suppression in different *in vitro* functional assays, as demonstrated by HEL-specific suppression of T cell proliferation (15) and by suppression of HEL-specific *in vitro* antibody response (16). Moreover, culture supernatants are able, when injected into mice, to specifically suppress the *in vivo* anti-HEL antibody response (Fig. 3-5). Suppression is observed on both primary and secondary anti-HEL antibody responses evaluated by direct and developed hemolytic plaque assays.

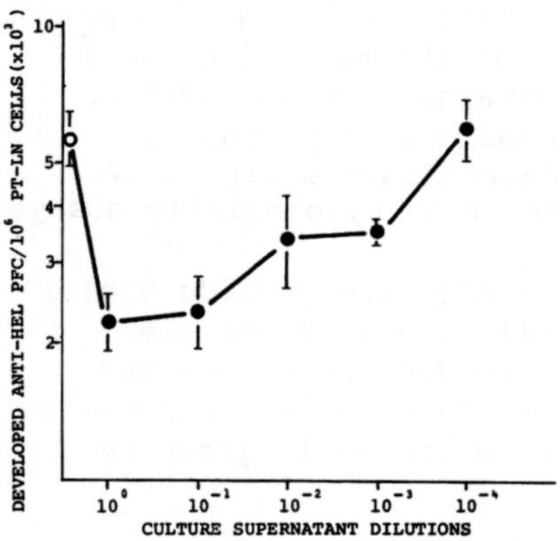

Fig. 5. Titration of suppressive factor from clone LH8-105. BDF_1 mice

(5 mice/group) were injected i.p. with 100 µg HEL and developed PFC responses measured 8 days later in the parathymic lymph nodes. Culture supernatant from clone LH8-105 (1 ml/mouse) was injected i.v., at the indicated dilutions, at the same time as antigen-priming and one day later. Open symbol represents uninjected controls.

F. T Cell Products Secreted by HEL-Specific Suppressor T Cell Lines

T cell products from a RadLV-induced, HEL-specific suppressor T cell line (LH8) have been characterized in some detail (Adorini *et al.*, submitted). Culture supernatants from this HEL-specific suppressor T cell line do not suppress the antibody response induced by a structurally related lysozyme, demonstrating the presence in the culture supernatant of a suppressive factor endowed with fine antigenic specificity. The suppressive factor is able to selectively suppress the anti-HEL antibody response induced by the N-terminal, C-terminal peptide of the HEL molecule, indicating that the fine specificity of this factor is restricted to an antigenic epitope present in this region of the HEL molecule.

The suppressive activity is restricted by genes located within the H-2 complex and analysis of the suppression induced in recombinant mice demonstrates that the interaction between HEL-specific suppressor T cell factor and its cellular target requires identity in the I-J region of the H-2 complex.

The LH8 cell line has been cloned by limiting dilution and all the clones tested displayed antigen-specific suppressive activity suggesting that this cell line, which has been maintained in continuous culture for more than 18 months, was already functionally cloned (Fig. 3).

G. Concluding Remarks

The RadLV-induced T cell transformation technique represents a very useful approach to the production of stable, antigen-specific, functional T cell clones and overcomes some of the difficulties encountered with the T cell hybridoma and IL-2-dependent long-term culture technologies. We have successfully applied this technique to establish T cell lines secreting helper or suppressive factors specific for hen egg-white lysozyme, keyhole limpet hemocyanin and azobenzenearsonate.

The possibility to obtain T cell products able to exert a precise immunoregulatory function by RadLV-transformed T lymphocytes is likely to represent a powerful tool for biochemical analysis of the antigen-specific T cell receptor. In addition, the availability of stable, functional, antigen-specific T cell clones obtained by RadLV-induced transformation may facilitate genetic studies at the DNA level of antigen-specific T cell recognition structures.

References

1. Lieberman M, Kaplan HS (1959) Leukemogenic activity of filtrates from radiation-induced lymphoid tumors in mice. Science 130: 387-388
2. Finn OJ, Boniver J, Kaplan HS (1979) Induction, establishment *in vitro* and characterization of functional, antigen-specific, carrier-primed immune T cell lymphomas. Proc Natl Acad Sci USA 76: 4033-4037
3. Kaplan HS (1967) On the natural history of the murine leukemia: presidential address. Cancer Res 27: 1325-1339
4. Lieberman M, Decleve A, Finn O, Ricciardi-Castagnoli P, Boniver J, Kaplan HS (1979) Establishment, characterization and virus expression of cell lines derived from radiation and from virus-induced lymphomas of C57BL/Ka mice. Int J Cancer 24: 168-177
5. Ricciardi-Castagnoli P, Lieberman M, Finn O, Kaplan HS (1978) T-cell lymphoma induction by radiation leukemia virus in athymic nude mice. J Exp Med 148: 1292-1310
6. Decleve A, Lieberman M, Jhle J, Rosenthal PN, Lung M, Kaplan HS (1978) Physicochemical, biological and serological properties of a leukemogenic virus isolated from cultured RadLV-induced lymphomas of C57BL/Ka mice. Virology 90: 23-35
7. Manteuil-Brutlag S, Sai-Liu, Kaplan HS (1980) Radiation leukemia virus contains two distinct viral RNAs. Cell 19: 643-652
8. Lieberman M, Kaplan HS (1976) The role of lymphoid and haematopoietic target cells in viral lymphomogenesis of C57BL/Ka mice.I. Susceptibility to viral replication. Blood Cells 2: 291-299
9. Kaplan HS, Lieberman M (1976) The role of lymphoid and haematopoietic target cells in viral lymphomogenesis of C57BL/Ka mice.II. Neoplastic transformation of bone marrow-derived cells in the thymic microenvironment. Blood Cells 2: 301-317
10. Abelson HT, Rabstein LS (1970) Lymphosarcoma: virus induced thymic independent disease in mice. Cancer Res 30: 2213-2222
11. Lieberman M, Kaplan HS (1966) Lymphoid tumor induction by mouse thymocytes infected *in vitro* with radiation leukemia virus. Nat Cancer Inst Monogr 22: 549-560
12. Meruelo D (1980) H-2 D control of leukemia susceptibility: mechanism and implications. J Immunogen 7: 81-90
13. Rowe W, Humphrey JB, Lilly F (1973) A major genetic locus affecting resistance to infection with murine leukemia viruses.III. Assignment of the Fv-1 locus to linkage group VIII of the mouse. J Exp Med 137: 850-854
14. Decleve A, Lieberman M, Jhle J, Kaplan HS (1976) Biological and serological characterization of radiation leukemia virus. Proc Natl Acad Sci USA 73: 4675-4679
15. Ricciardi-Castagnoli P, Doria G, Adorini L (1981) Production of antigen specific suppressive T cell factor by radiation leukemia virus-transformed suppressor T cell. Proc Natl Acad Sci USA 78: 3804-3808
16. Adorini L, Doria G, Ricciardi-Castagnoli P (1982) Biochemical and functional analysis of antigen-specific products obtained from virus transformed, lysozyme-specific suppressor T cell lines. Lymphokines vol 5, in press

H-2-Restricted Helper Hybridomas: One Locus or Two Control Dual Specificity?

P. Lonai, E. Arman, S. Bitton-Grossfeld, J. Grooten, G. Hämmerling

1. Introduction

T cells recognize antigen and self MHC products simultaneously. This phenomenon of *H-2* restriction was discovered some eight years ago, but it is still not understood sufficiently (1,2). The outstanding problems appear to be the following: Do T cells have one or two receptors? Is their ligand a complex or two unconnected determinants on separate molecules? Studies with T cell hybrid clones allow one to investigate the first question by means of somatic genetics: Gene complementation between the fusion partners can reveal whether dual specificity is controlled by one or by more loci.

Here we will discuss some of our recently published results and will mention new preliminary data. Special attention will be paid to the methodological problems of a more comprehensive genetic analysis.

2. Construction of Helper T Hybridomas and Characterization of Their Helper Factors (ThF)

Our hybridoma clones were derived from fusing B10.BR (H-2^k) or C57BL/6 (H-2^b) T cells specific to NP-CGG (NP-chicken gamma globulin) with BW-5147 lymphoma cells of AKR/J (H-2^k) origin. Hence, some of the lines are homologous and others heterologous in *H-2*. The resulting fusion lines (530) were screened for activity by autoradiographic antigen binding. This assay is based on the binding by IL-1 treated T cells of labeled antigen (^{125}I-CGG) released by antigen pulsed macrophages (3). Antigen binding positive fusion lines were tested for ThF production, and all were found to be active (4). ThF activity was measured with NIP-OVA (NIP-Ovalbumin) specific spleen cells, to which CGG primed T cells or hybridoma derived ThF was added in the presence of NIP-CGG, and NIP-specific plaque forming cells were counted. The active lines were cloned by limiting dilution. Details on the derivation and testing of these clones have been published previously (4,5,6).

All of the clones were found to produce carrier (CGG) specific ThF. Affinity chromatography studies showed that ThF contains V_H and I-A (and also I-E) antigenic determinants (4-6). Preliminary immunoprecipitation and SDS-PAGE separation of the reduced material supports these findings (Fig. 1). Hybridoma cells were labeled with ^{35}S-methionine, and the supernatant was fractionated either by affinity chromatography on CGG-Sepharose, HGG-Sepharose or anti V_H-Sepharose (7) or by immunoprecipitation using monoclonal anti-Ia antibodies, by a modification of the method of Shackelford and Strominger (8). It can be seen from Fig. 1 that the major component has an approximate molecular weight of 60-70,000, usually followed by two minor peaks around 25-32,000 MW. A similar pattern was obtained when the material was isolated by antigen-Sepharose or anti-V_H-Sepharose affinity separation or by immunoprecipitation with anti-Ia antibodies. Hence, V_H- and Ia-like determinants appear in one molecule or one molecular complex. It is not clear however whether both determinants are included in the major peak, and the nature of the minor components is still undefined. In non-reduced, but also in some of the reduced samples (Fig. 1D), much of the radioactivity appears at higher molecular weights (150-300,000 MW). It follows that ThF may be secreted in the form of dimers and trimers, or the polymers may become established during isolation.

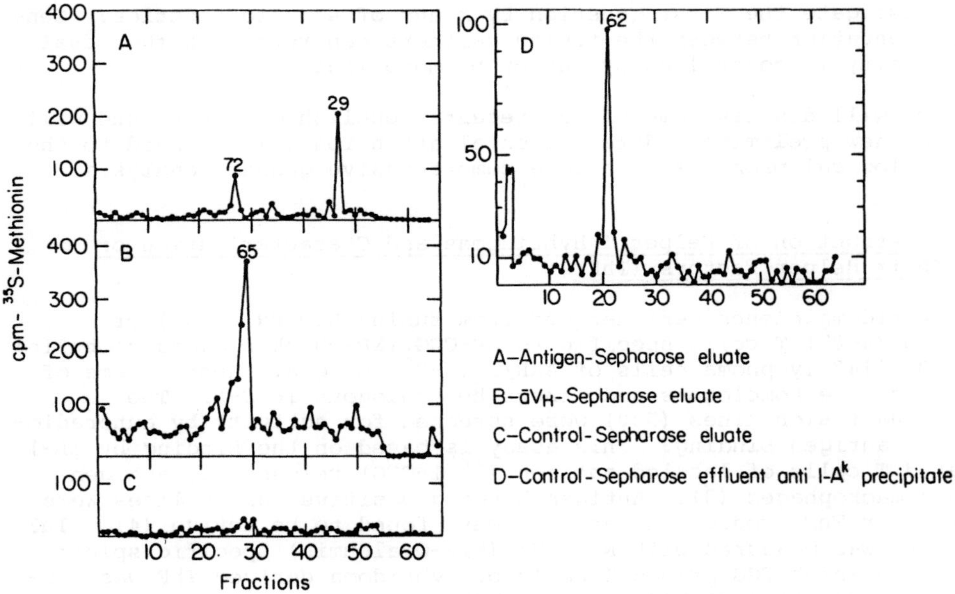

Fig. 1. Isolation of ThF *H-2* homologous hybridoma T85-109-45. Supernatant labeled with ^{35}S-methionine (900 Ci mM) 250 μCi/ml, 1% cold methionine in medium overnight. Elution: 3M KSCN. Immunoprecipitation: Ia.17 specific (anti-I-A^k) monoclonal antibody, see also ref. 8. Electrophoresis on 7-15% SDS-PAGE. Antigen-Sepharose: CGG-Sepharose. Control-Sepharose: HGG-Sepharose.

The most interesting functional characteristic of these ThRs is their *H-2* restricted activity. This was demonstrated by two experimental approaches. The ThFs activate only those B cell sources which are syngeneic in *H-2* with the hybridoma cells. Alternatively ThF can be adsorbed to normal spleen cells, and this adsorbtion is restricted to *I-A* (6). Because these hybridoma derived ThFs can replace carrier specific helper T cells, are *H-2* restricted and contain determinants which crossreact with IgV_H, it appears that they may represent a part, or all, of the *H-2* restricted T cell receptor.

3. Complementing Genes Control Dual Specificity of *H-2* Heterologous Hybridoma Cells

Because BW-5147 cells neither produce CGG specific ThF nor bind CGG, a hybridoma clone constructed from CGG specific normal T cells of C57BL/6 ($H\text{-}2^b$) origin and BW-5147 lymphoma cells of AKR ($H\text{-}2^k$) origin most likely expresses only one CGG specific gene, that of the C57BL cell. Even if the BW-5147 genome had expressed CGG specific V_H genes as a result of the cell fusion, we would not have expected to detect ThF containing only AKR-V_H by our anti-V_H affinity columns. The reason for this is that anti-$V_{H\text{-}M315}$ has much lower affinity for AKR heavy chains than for C57BL/6 heavy chains (9).

Which anti-self loci could such a hybridoma express? Because positive selection experiments with hematopoietic chimeras suggest that even early T cell precursors express anti-self receptors (10,11), it is reasonable to assume that our cells should be able to express both $H\text{-}2^b$ and $H\text{-}2^k$ specific anti-self receptors. Hence, hybridoma clones of the C57BL/6 anti-CGG plus BW-5147 type could express double specificity or may secrete two helper factors, one specific for CGG + $I\text{-}A^b$ and the other for CGG + $I\text{-}A^k$. Such a finding would demonstrate that *H-2* restricted ThF or T cell receptors are controlled by two specificity defining loci. A scheme demonstrating this reasoning is shown in Table 1.

Table 1. Complementation between specificity defining loci in *H-2*-heterologous hybridomas

Input cells	*H-2*	Possible receptor loci Anti-carrier	 Anti-self	Phenotypes	
B6 anti-CGG	b	+	+	ThF No. 1:	Anti-CGG^{B6} +Anti-I-A^{bB6}
BW-5147 (AKR)	k	-	+	ThF No. 2:	Anti-CGG^{B6} +Anti-Ia^{kAKR}

Results supporting this assumption were obtained by studying ThF secreted by *H-2* heterologous hybridoma clones. These cells produce ThF which stimulates both $H\text{-}2^b$ and $H\text{-}2^k$ B cell sources, but not allogeneic, $H\text{-}2^j$ or $H\text{-}2^q$ B cell sources. The two activities could be separated either by adsorption on normal spleen cells or by affinity chromatography on anti-$I\text{-}A^b$- or anti-$I\text{-}A^k$-Sepharose affinity columns (12). According to the above reasoning, the CGG specific, $H\text{-}2^k$ restricted helper factor was controlled by complementing genes of C57BL/6 and AKR origin, respectively. This phenomenon was observed in six out of ten hybridoma clones. It follows that dual recognition or *H-2* restriction is regulated in these ThF at least by two specificity defining loci.

Preliminary data suggest that these clones express two kinds of anti-self specificities also in their cellular antigen receptor. Antigen binding studies using CGG processed with macrophages of $H\text{-}2^b$, $H\text{-}2^k$

and $H\text{-}2^f$ origin showed that these cloned cells bound CGG processed by both syngeneic macrophages ($H\text{-}2^b$ and $H\text{-}2^k$), but that they did not bind the same antigen processed by allogeneic $H\text{-}2^k$ macrophages (5).

In the framework of the current theories for the interpretation of T cell specificity these results support dual recognition theories over altered self hypotheses (13,14). The identity of the two specificity defining loci, whether the receptor is controlled by two V_H-like or by a V_H-like and an Ia-like locus, etc., could be defined by somatic and molecular genetic methods.

A finding which is in possible contradiction with the above conclusion has been reported by Kappler et al. (15). They did not find gene complementation in 20 clones derived from fusing hybridoma cells. Antigenic specificity was assayed by antigen induced IL-2 production. Cross complementation between the genomes of hybridoma cells, however, does not have to occur in every clone, even if the relevant chromosomes have been retained. More detailed studies on the frequency of intragenomic complementation of receptor loci is necessary to resolve this problem. To investigate the possible technical differences between the results of the two groups, we have tested our hybridomas for IL-2 production. Using either Con A or antigen pulsed macrophages to induce IL-2 production, we could not detect significant IL-2 secretion among 18 hybridoma clones.

4. Technical Problems of the Progress Towards Somatic and Molecular Genetic Analysis of T Cells

It is quite common in research that when an aim which was thought to be almost unattainable is reached, progress still remains difficult as the next hurdle comes in sight. This seemed to have happened in modern immunology. By means of the different cloning techniques it is now possible to grow cloned functional T cells in almost unlimited amounts. The next problem to be solved however is the establishment of fast, accurate and simple techniques to test for phenotypic expression of T cell specificity in large numbers of clones. Solution of this problem will open the way for detailed somatic and molecular genetic analysis. For somatic genetic experiments, using cell fusion to follow gene segregation, one should be able to test up to a few hundred clones in an experiment. In the case of gene transfer studies, because of the low frequency of DNA induced transformation (16), it may be necessary to select one cell from 10^5. At present, however, there is no adequate general solution even for the first, the simpler problem.

Standard cellular immunological techniques, like chromium release assays for killer T cells and in vitro haemolytic plaque assays for suppressor and helper T cells and factors, are rather complicated. Most difficult to test are the helper T cells, but even in the easiest case, killer T cells, testing more than a few times ten clones may be cumbersome and somewhat limited by the availability of suitable target cells. Antigen induced proliferation assays for long-term T cell lines of normal cells are indeed simple, but the use of these cells is severely curtailed, because they are more difficult to grow and clone than hybridoma cells.

Two newer approaches seem to be specially worth pursuing. Some T hybridomas respond to antigenic stimulus by lymphokine production. Moreover, lymphokine producing hybridomas can be constructed from lymphokine producing fusion partners (17). Because the assay of lymphokines is relatively simple, such assays may be very useful in screening helper hybridomas. Induced IL-2 secretion has already been used for this purpose (15). Our example shows however that not all helper hybridomas produce detectable amounts of IL-2. It is possible that such clones produce other lymphokines involved in T cell-macrophage or T cell-T cell cooperation.

The other approach is antigen binding. Both Erb and Feldmann (18) and ourselves (3) have found that helper T cells recognize soluble antigen complexes containing Ia, which are released by anligen pulsed syngeneic macrophages. We have used this technique to screen fusion lines for functional helper clones (4,5). At present, however, the technique is still in a rather crude form of microscopic autoradiography. We do not yet understand why we did not obtain more than 20-40% antigen binding cells in cloned lines, and why even this figure is still subject to considerable variation. Investigation of the conditions necessary for receptor expression and preparation of the purified ligand in a suitable concentration may allow the use of simple quantitative binding assays with radiolabeled or fluoresceinated ligands, similarly to detection of hormone receptors. Such techniques, combined with use of the Fluorescence Activated Cell Sorter, could be used even to detect rare transformants and mutants in DNA transfer experiments.

References

1. Zinkernagel RM, Doherty PC (1974) Restriction of *in vitro* T cell mediated cytotoxicity in lymphocytic choriomeningitis within a syngeneic or semiallogeneic system. Nature 248: 701-702
2. Shearer GM (1974) Cell mediated cytotoxicity to trinitrophenyl-modified syngeneic lymphocytes. Eur J Immunol 4: 527-533
3. Puri J, Lonai P (1980) Mechanism of antigen binding by T cells. *H-2 (I-A)* restricted binding of antigen plus Ia by helper cells. Eur J Immunol 10: 273-281
4. Lonai P, Puri J, Hämmerling GJ (1981) *H-2* restricted antigen binding by a hybridoma clone which produces specific helper factor. Proc Natl Acad Sci USA 78: 549-553
5. Lonai P, Hämmerling GJ, Givol D, Ben-Neriah Y, Puri J (1981) Specific T-helper factor production and *H-2* restricted antigen binding by helper cells: studies with T-hybridoma clones. In: Hämmerling GJ, Hämmerling H, Kearney JF (eds) Monoclonal antibodies and T-cell hybridomas. Elsevier-North Holland, Amsterdam New York Oxford, p 457
6. Lonai P, Puri J, Bitton S, Ben-Neriah Y, Givol D, Hämmerling GJ (1981) *H-2* restricted helper factor secreted by cloned hybridoma cells. J Exp Med 154: 942-951
7. Ben-Neriah Y, Wuilmart C, Lonai P, Givol D (1978) Preparation and characterization of anti framework antibodies to the heavy chain variable region (V_H) of mouse immunoglobulins. Eur J Immunol 8: 797-801
8. Shackelford DA, Strominger JL (1980) Demonstration of structural polymorphism among HLA-DR light chains by two-dimensional gel electrophoresis. J Exp Med 151: 144-165
9. Ben-Neriah Y, Givol D, Lonai P, Simon MM, Eichmann K (1980) Allotype linked genetic control of a polymorphic V_H determinant on mouse T-helper cell receptors. Nature 285: 257-259

10. Zinkernagel RM, Callahan GN, Althage A, Cooper S, Klein PA, Klein J (1978) On the thymus in the differentiation of "H-2 self-recognition" by T cells: Evidence for dual recognition? J Exp Med 147: 882-896
11. v Boehmer H, Haas W, Jerne NK (1978) Major histocompatibility complex-linked immune-responsiveness is acquired by lymphocytes of low-responder mice differentiating in thymus of high-responder mice. Proc Natl Acad Sci USA 75: 2439-2442
12. Lonai P, Bitton S, Savelkoul HFJ, Puri J, Hämmerling GJ (1981) Two separate genes regulate self-Ia and carrier recognition in *H-2*-restricted helper factors secreted by hybridoma cells. J Exp Med 154: 1910-1919
13. Janeway CA Jr, Wigzell H, Binz H (1976) Hypothesis: two different V_H gene products make up the T cell receptors. Scand J Immunol 5: 993-1005
14. Matzinger P (1981) A one receptor view of T-cell behaviour. Nature 292: 497-501
15. Kappler JW, Skidmore B, White J, Marrack P (1981) Antigen inducible, *H-2*-restricted interleukin-2-producing T cell hybridomas. Lack of independent antigen and *H-2* recognition. J Exp Med 153: 1198-1214
16. Pellicer A, Robins D, Wold B, Sweet R, Jackson J, Lowy I, Roberts JM, Sim GK, Silverstein S, Axel R (1980) Altering genotype and phenotype by DNA-mediated gene transfer. Science 209: 1414-1422
17. Harwell L, Skidmore B, Marrack P, Kappler J (1980) Concanavalin-A-inducible interleukin-2-inducing T cell hybridoma. J Exp Med 152: 893-904
18. Erb P, Feldmann M (1975) The role of macrophages in the generation of T helper cells. III Influence of macrophage-derived factors in helper cell induction. Eur J Immunol 5: 759-766

T Cell Hybridomas Producing Antigen-Specific Factors Express Heavy-Chain Variable-Region Determinants

Z. Eshhar, T. Waks, H. Zinger, E. Mozes

A. Introduction

Amongst the approaches that have been attempted during the last years in order to immortalize functional T cells, the somatic cell hybridization was remarkably successful (1-5). Although not as abundant as continuous lines (6) nor stable as the viral transformed T cell lymphomas (7), the functional T cell hybridomas provide an adequate source for : a. somatic cell genetic studies, b. biochemical and molecular analyses of the T cell antigen recognizing unit, and c. functional analysis and characterization of soluble T cell factors that can replace the immunoregulatory activity mediated by the T cells.

We have previously described the establishment of T cell hybridomas derived from T cells that were activated to (T,G)-A--L and enriched on antigen-pulsed adherent cells (3,8,9). Two hybrid cultures were isolated which secreted antigen specific helper (R-9) and suppressor (R-11) factor. Table 1 summarizes the main properties of the (T,G)-A--L specific factors that have been reported in more detail elsewhere (8,9). Both factors exhibited antigen specific binding activity and coexpressed determinants encoded by the I-region of the major histocompatibility complex and determinants similar to those encoded by the Ig heavy chain variable genes. This intriguing observation demonstrated by the fact that both anti-H-2I^b and anti-V_H or anti-idiotypes against (T,G)-A--L specific antibodies (8-10) could absorb the activity of the factor, implies that two different and most likely separate loci are encoding for the (T,G)-A--L specific factors. Further studies have suggested to us that the active component in the R-9 helper factor is most likely composed of at least two chains: one carries the I-region and the second the V-region encoded determinants that are known to be coded by different chromosomes (17 and 11, respectively, in the mouse). The two chain recombine to yield the active factor (8,10).

In this communication we describe our recent experiments undertaken in order to establish the relationship between the V_H determinants appearing as cell bound (receptor?) or on the soluble factors produced by the T cell lines and its relevance to T cell activities. We have found a lack of correlation between the expression of V_H on the surface of some hybrid clones and the ability to produce helper or suppressor factor, and discuss the implication of this finding to the expression of V_H determinants on T cell components specific to antigen.

Table 1. Properties of (T,G)-A--L specific T helper (R-9) and suppressor (R-11) factors produced by T cell hybridomas.

1. Constitutively released into the hybridoma culture supernates.
2. Bind specifically to affinity columns of (T,G)-A--L and the activity can be recovered in the eluates.
3. R-9 derived Thf helps (T,G)-A--L primed B cells in adoptive transfer *in vivo* and hapten primed spleen cells *in vitro* in antibody responses.
4. R-11 derived Tsf suppresses antibody production of (T,G)-A--L primed spleen cells either *in vivo* or *in vitro*.
5. Do not express known determinants of the Ig constant region but possess variable region framework (V_H) and (T,G)-A--L specific idiotypic determinants.
6. Carry determinants coded by the I-A (R-9, Thf) and most likely I-J (R-11, Tsf) subregions of the $H-2^b$.
7. Glycoproteins with molecular weights of 60-70 K that may be composed of two chains.

B. Experimental Methods and Results

I. Generation and Characterization of (T,G)-A--L Specific T Cell Hybridomas

In order to increase the probability of obtaining functional and antigen specific hybridomas we potentiated the population of (T,G)-A--L reactive T cells by the restimulation *in vivo* of T cells that have been activated specifically to (T,G)-A--L in irradiated syngeneic hosts. Figure 1 illustrates schematically the procedure we employed for the generation of the (T,G)-A--L-specific T cell hybridomas (3). These activation, enrichment and fusion protocols appear to be quite efficient since out of 21 hybridoma cultures, two hybridomas secreted antigen specific potent factors into their culture fluid without any requirement for additional triggering. The activity of the soluble factors was examined in adoptive transfer experiments in which the helper factor (Thf) activity was determined using (T,G)-A--L primed B cells and the suppressor factor (Tsf) activity was evaluated using primed splenic cells (3). For larger experiments, the activity of the Thf was assessed in an *in vitro* microculture system in which the formation of NIP specific PFC was measured after coincubation of hapten primed spleen cells, (T,G)-A--L specific hybridoma supernatants and NIP-(T,G)-A--L as antigen (3,8). The suppressive effect of Tsf was assayed *in vitro* by determining the specific reduction in the formation of NIP specific PFC by NIP-(T,G)-A--L primed spleen cells (9).

The R-9 and R-11 hybridomas that were found to release (T,G)-A--L specific helper and suppressor mediators, respectively expressed on their surfaces both Thy-1 and H-2 antigens of both parental cells: Thy 1.1 and $H-2^k$ of BW5147 and Thy 1.2 and $H-2^b$ of C3H.SW. Notably, these hybridomas which did not carry any known Ig constant region determinants, could be stained weakly but specifically and repeatedly with purified antibodies prepared against the V_H of MOPC 315 which recognized framework determinants common to many mouse heavy chains (11). Likewise, the R-9 and R-11 hybrid cells could be stained with antibodies made against the major idiotypic determinants that are

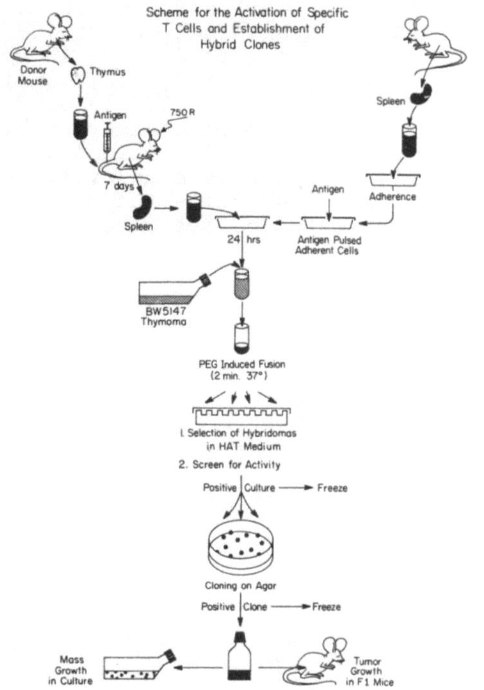

Fig. 1. Experimental protocol for the activation of antigen specific T cells and establishment of hybrid clones.

present on about 30-40% of C3H.SW anti-(T,G)-A--L antibodies (12).

II. Correlation studies between the expression of V_H determinants on the surface of hybrid clones and their ability to produce active factor.

As summarized in Table 2 , both R-9 and R-11 hybridomas expressed framework (V_H) and hypervariable (1d) determinants on their cell surfaces and on the active components of their (T,G)-A--L-specific soluble factors (8-10). Is the presence of V_H containing structures on the cell surface necessary for the hybrid cell activity? In order to approach this question and to select nonfunctional hybrid clones for biochemical and genetic analysis, we have sorted R-9 and R-11 hybrid cells, that grew for a few months in tissue culture, for the presence of V_H determinants and separated with the help of the fluorescence activated cell sorter (FACS II) between V_H positive and V_H negative clones. Following cell sorting, the hybrid cells were subcloned by limiting dilution and two-three weeks later were analysed for factor production and expression of surface V_H determinants. Fig. 2 illustrates the initial sorting profile (upper panel) where about 33% of the cells were selected as expressing V_H and about one third of the hybrids as negatives for V_H. A typical profile of positive and negative clones, analysed a few weeks after sorting and cloning, is shown in the lower panel of Fig. 2.

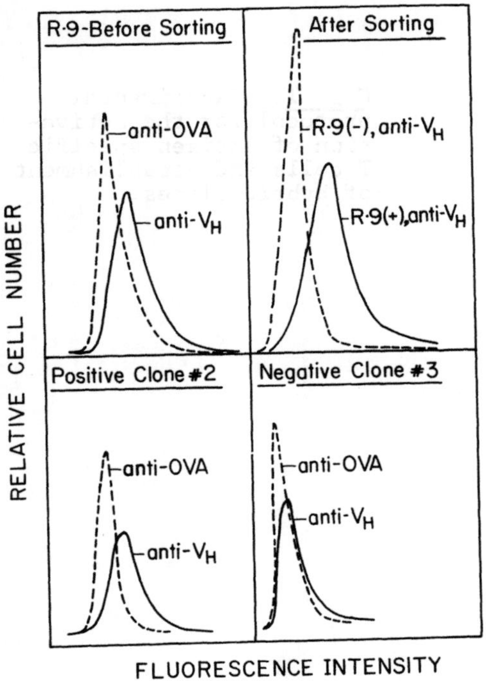

Fig. 2. Cell sorting profile of R-9 hybrid cells. 4 x 10^6 cells were reacted with 10 μg affinity purified rabbit anti-V_H antibodies. FITC labeled goat antibodies against rabbit IgG were used for staining. All reagents were absorbed with BW5147 cells and deaggregated before use.

Table 2 summarizes the results of a few experiments where we determined the ability of the V_H positive and negative subclones to produce an active factor. Comparison between surface staining for V_H and factor activity of the different subclones revealed that the majority of the subclones that were selected as positive for V_H retained the relevant determinants on their surfaces and five out of eight such clones were able to produce an active factor. One of the V_H positive (R-11·5) clones lost both surface V_H and factor activity and two clones (R.11·2, R.11·3) became inactive but retained their V_H determinants. Out of the negatively selected clones, four lost their ability to produce detectable active factor in correlation with the loss of surface V_H. The clones that were selected initially as negative for V_H but expressed V_H determinants upon subsequent determination (R-9·24·8, R-9·24·10, R-11·11) most likely represent hybrid cells that expressed low levels of V_H due to transient stage of their cell cycle. More interesting are the hybrid clones represented by R-11·18 which do not express V_H determinants but produce active factor.

Taken together, the results summarized in Table 2 suggest lack of correlation between the expression of surface molecules that bear V_H like determinants and the ability to produce antigen-specific helper and suppressor factors. It remains however to be determined whether the factors produced by the V_H negative clones carry V_H determinants and if intracellular staining with anti-V_H antibodies can be achieved using appropriate fixation procedures. Based on yet preliminary results and in analogy to the B cells surface and secreted Ig, we would like to suggest that the T cell antigen specific components - the soluble factor and the cell bound receptor, are most likely coded by the same genes and might even have a common precursor. It is also

Table 2. Relationship between V_H expression and factor activity in hybrid clones sorted with anti-V_H

Hybrid clone	Expression of V_H while sorted[1]	Factor activity[2]	Staining with anti-V_H[3]
R·9·24·3	+	+	+
R·9·24·21	+	±	±
R·9·24·6	-	-	-
R·9·24·8	-	-	+
R·9·24·10	-	-	±
R·11·1	+	±	+
R·11·6	+	+	+
R·11·8	+	+	+
R·11·2	+	-	±
R·11·3	+	-	+
R·11·5	+	-	-
R·11·11	-	-	±
R·11·13	-	-	-
R·11·14	-	+	N.D.
R·11·18	-	+	-

1. R·9·24 and R·11 hybridomas were stained with RāV_H and FITC-GāRIg, sorted in the FACS and cloned by limiting dilutions. Clones derived from the positively stained FACS fraction were designated (+), unstained clones designated (-).
2. R·9 clones were assayed for (T,G)-A--L specific helper factor in their culture supernatant. R·11 clones were examined for suppressor factor activity. Factor assay was conducted *in vitro*.
3. Two-three weeks after sorting and cloning the hybrid clones were stained again with anti-V_H.

(+), positive staining or specific activity; (±), weak staining or activity; (-), negative staining, no activity.

possible that the cellular destination of the V_H containing molecules is different - the receptor molecule is embedded into the cell surface and the functional factor is secreted outside of the lymphocyte. Although the different hybrid clones were randomly derived probably due to inherited instability of somatic cell hybrids, our ability to identify, select and analyse the different variants should contribute to our understanding of the genetic loci that code for the antigen specific molecules and the mechanism which regulates their expression and activity. Steps toward this end are the monoclonal anti-V_H antibodies (13) and the B cell hybridomas producing monoclonal anti-(T,G)-A--L antibodies that express the major anti-(T,G)-A--L idiotype (14) which share determinants with the (T,G)-A--L-specific T cell molecules expressed by the helper and suppressor hybridomas described herein. Indeed, in preliminary, unpublished experiments we have found a correlation between the ability of murine anti-(T,G)-A--L idiotypic antibody to bind to the (T,G)-A--L specific hybridoma and anti-(T,G)-A--L antibodies.

C. Concluding Remarks

Our previous experiments and the data described in this communication indicate that it is possible to immortalize antigen specific, functional helper and suppressor T cells by their hybridization to a T cell lymphoma. The hybridomas produced and secreted (T,G)-A--L specific helper and suppressor factors that coexpressed both I-region and V_H region encoded determinants. The fact that the hybrid cells expressed on their surfaces V-region like determinants as was detected by both anti-V_H and anti-idiotypic antibodies suggest that the T cells most probably use immunoglobulin like structures for their antigen-specific components. The results obtained with hybrid subclones selected for the presence of surface V_H determinants indicate lack of correlation between the ability of the cells to produce active factors and express V_H containing molecules. We believe that further comparison between such functional and non-functional hybrid clones with the help of appropriate probes and monoclonal reagents should result in the chromosome assignment of the genes coding for the T cell receptor, and hopefully in the biochemical characterization of the T cell antigen specific molecules.

Acknowledgments. We are grateful to Drs. D. Givol and Y. Ben-Neriah for providing the anti-V_H antibodies and to Ms. E. Majerowich for her cooperation in the preparation of this manuscript. Z.E. is an Incumbent of the Recanati Career Development Chair in Cancer Research.

References

1. Kontiainen S, Simpson E, Boherer E, Beverley PCL, Herzenberg LA, Fitzpatrick WC, Vogt P, Torano A, McKenzie IFC, Feldmann M (1978) T cell lines producing antigen specific suppressor factor. Nature 274: 477-480
2. Taniguchi M, Saito T, Tada T (1979) Antigen specific suppressor factor produced by a transplantable I-J bearing T cell hybridoma. Nature 278: 555-558
3. Eshhar Z, Apte RN, Löwy I, Ben-Neriah Y, Givol D, Mozes E (1980) T cell hybridoma bearing heavy chain variable region determinants producing (T,G)-A--L specific factor. Nature 286: 270-272
4. Kaufmann Y, Berke G, Eshhar Z (1981) Cytotoxic T lymphocyte hybridomas which mediate specific tumor cell lysis *in vitro*. Proc. Natl. Acad. Sci. 78: 2502-2506
5. Lonai P, Puri J. Hämmerling GJ (1981) H-2 restricted antigen binding by a hybridoma clone that produce antigen-specific helper factor. Proc Natl Acad Sci 78: 549-554
6. Möller G, ed (1981) T cell clones. Immunol Rev Vol 54
7. Finn OJ, Boniver J, Kaplan HS (1979) Induction establishment and *in vitro* characterization of functional antigen-specific, carrier primed murine T cell lymphomas. Proc Natl Acad Sci 76: 4033-4037
8. Apte RN, Eshhar Z, Löwy I, Zinger H, Mozes E (1981) Characteristics of a (T,G)-A--L-specific helper factor derived from a T cell hybridoma. Eur J Immunol 11: 931-937
9. Eshhar Z, Apte RN, Löwy I, Mozes E (1981) T cell hybridomas producing (T,G)-A--L specific helper and suppressor factors. In: Hämmerling GJ, Hämmerling U, Kearney JF (eds) Monoclonal antibodies and T cell hybridomas. Elsevier, North Amsterdam, p 469
10. Eshhar Z, Apte RN, Löwy I, Mozes E (1981) (T,G)-A--L specific immunoregulatory factors produced by T hybridomas. In: Resch K, Kirchner H (eds) Mechanism of lymphocyte activation. Elsevier, North-Holland, p 557
11. Ben-Neriah Y, Wuilmart C, Lonai P, Givol D (1978) Preparation and characterization of anti-framework antibodies to the heavy chain variable region (V_H) of mouse immunoglobulins. Eur J Immunol 8: 797-806.

12. Schwartz M, Lifshitz R, Givol D, Mozes E, Haimovich J (1978) Cross reactive idiotypic determinants on murine anti-(T,G)-A--L antibodies. J Immunol 121: 421-426
13. Eshhar Z, Gigi O, Givol D, Ben-Neriah Y (1982) Monoclonal antibodies against V_H of M 315 myeloma protein recognize a unique antigenic determinant common to different V_H regions. Submitted
14. Parhami B, Eshhar Z, Mozes E (1982) Fine specificity of monoclonal antibodies against (T,G)-A--L and analysis of their idiotypes. Submitted

Cytochrome c Specific T Cell Hybrid

S. Carel, C. Bron, G. Corradin

A. Introduction

The biochemical characterization of T cell receptor for antigen remains one of the central tasks in immunology. Progress has been hampered until recently by the lack of homogeneous cloned T cell lines specific for antigens, but the possibility now exists of obtaining a large quantity of antigen specific T cell either by somatic cell fusion or by direct cloning and should, in principle, allow one to tackle this problem with a higher probability of success. Our approach to the problem was either to obtain cloned T cells or T cell hybrids specific for small and defined cytochrome c peptides.

Here we report the characterization of one of the T cell hybrids obtained. In this case, a small cytochrome c peptide was modified by a 2,4-dinitro-5-fluoroaminophenyl group whose amino group could be modified to obtain a nitrene derivative. This can in turn be used for crosslinking experiments between the cytochrome peptide and the receptor. For this purpose, a direct interaction of the antigen with the receptor would be of paramount importance to the success of our approach. Our thought was that in spite of a vast body of literature which would indicate that unprocessed antigen cannot be recognized by a Lyt-1 positive and Lyt-2 negative proliferating T cell, a direct interaction between soluble antigen and antigen specific T cell might be measurable without previous incubation with macrophages if the peptide used were of limited size. Thus it should not differ dramatically in conformation and size from the processed peptide.

Given these assumptions, T cell hybrids were obtained by fusing long term T cells obtained from lymph nodes of SJL mice immunized with cytochrome c peptide 66-80 derivatized with a 2,4-dinitroaminophenyl and BW 5147 thymoma cells. Hybrid cells were then selected for their capacity to bind directly the homologous antigen.

B. Materials and Methods

Animals: 1-6 months old female SJL mice obtained from Bomholtgard Ltd, Denmark, were used in all experiments.

Antigens: Cytochrome c peptides 1-65, 66-80 and 81-104 prepared from horse cytochrome c type III (Sigma, St-Louis) according to Corradin and Harbury (1). These peptides were derivatized with 2,4-dinitro-5-fluoroaminophenyl (DNFA) according to Erecinska (2). In brief, 6 mg of DNFA were dissolved in 30 µl of dioxane and added to 12 mg of peptide 66-80 in 4 ml of $NaHCO_3$ 0.5 M pH 10.2. The mixture was left at 25°C for 1 hour and then dialysed overnight at 4°C against H_2O at pH 10.2 Peptide 11-25 was a generous gift of Dr. M. Juillerat, University of Geneva. Derivation of this peptide with 2-nitro-phenyl-sulphenyl chloride (NPSCl) was performed according to Fontana and Scoffone (3).

Derivatization of peptide 66-80 with citraconic acid was performed according to Atassi and Habeeb (4).

Immunization: DNAP peptide 66-80 was emulsified in complete Freund's adjuvant. Animals were injected subcutaneously at the base of the tail with a total emulsion volume of 40 µl containing 50 µg of antigen (5).

In vitro culture: LN cells were cultured according to Corradin et al. (5). Briefly, lymph node cells were homogenized and suspended in Iscove's modified DMEM (Gibco) supplemented with 2 x 10^{-5} M 2-mercaptoethanol, and 50 µg/ml of antigen; 5 % fetal calf serum FCS was added 24 h later. The cells were restimulated once a week by addition of irradiated syngeneic spleen cells and 25 µg/ml of antigen over a period of 5 weeks (6).

Proliferation test: Each week, the resultant enriched T cell populations were tested for their antigen specificity as previously described by Schrier et al. (6).

T cell hybridoma lines: These were produced according to Nabholz (7). In brief, long term T cell blasts were mixed at a ratio of 4 to 1 with AKR/J thymoma cell line BW 5147 in the presence of 40 % of PEG 1000. After fusion, the cells were distributed in 24 well Costar culture plate and cultured for 3 weeks in HAT medium and in the presence of 25 µg/ml of antigen. The resulting hybrids were screened for their capacity to bind radiolabeled antigen.

Identification of surface phenotypes: All experiments were performed with a fluorescence activated cell sorter using monoclonal anti-Thy-1.1, anti-Thy-1.2 antibodies and rabbit anti-mouse Ig.

Antigen binding test: In order to determine antigen binding by hybrid cells, 5 x 10^5 cells in 50 µl of Iscove's medium supplemented with 5 % of horse serum were placed in a round bottom microtiter plate precoated with normal horse serum. Twenty five microliters of a concentrated solution of cold antigen or medium were then added for 20 minutes at 37°C. After this incubation time, ^{125}I-labeled antigen was added in each well and again incubated for 20 minutes at 37°C. The cells were then centrifuged in 400 µl microtubes (Eppendorf microfuge) through a 50-100 % discontinuous horse serum gradient. The tubes were then frozen with dry ice and cut at the level of the pellet and counted.

C. Results and Discussion

Antigen Specific T Cell

Lymph node cells from SJL mice immunized seven days previously with DNAP-66-80 peptide were placed in culture and assessed for their capacity to respond to the homologous antigen. Proliferation was obtained only when cultures were stimulated with the homologous antigen. These cells were then restimulated each week by adding fresh irradiated spleen cells and antigen.

Characterization of Antigen Specific Hybridoma T Cell Line

Antigen specific T cells which were kept in culture for 5 weeks were fused with BW 5147 thymoma cells and were then selected for their capacity to bind directly soluble antigen.

Table 1: Antigen binding of ^{125}I-DNAP-66-80 to T cell hybrids 1/

Cell line 2/	Cold DNAP 66-80 added	Antigen bound CPM ± SE
II C3	–	7'400 ± 600
II C3	+	720 ± 40
B II A2	–	420 ± 40
B II A2	+	410 ± 29

1/ Cells were incubated with constant amount of labeled antigen; 200 fold excess of cold antigen was also added prior to addition of labeled antigen

2/ II C3 and B II A2 were clones derived from the fusion of antigen specific T cells and BW 5147 thymoma

As shown in Table 1, one of the 5 lines tested was thought to interact specifically with the peptide since radiolabeled bound antigen is displaced by preaddition of cold antigen. This line was further cloned by limiting dilution. All the clones obtained exibited a positive binding.

Surface Phenotypic Markers of Clone II C3

Surface phenotypic analysis was performed by cell flow cytometry. In addition to binding, antigen clone II C3 was positive for either Thy-1.2 and Thy-1.1 surface marker and negative for surface Ig. After about two months in culture, II C3 lost its capacity to interact specifically with the antigen. At this point, cells became negative for the Thy-1.2 marker.

Binding Analysis on Clone II C3

As shown in Table 2, clone II C3 bound specifically ^{125}I-DNAP-66-80 peptide since the binding was inhibited by preaddition of homologous cold antigen but not when cold DNAP-81-104 peptide was added. Furthermore, ^{125}I-DNAP-66-80 peptide bound very little either to the parent BW 5147 thymoma (Table 2) or to other negative hybridoma (Table 1). Similarly, other ^{125}I-DNAP peptides did not exhibit any specific binding on clone II C3 (data not shown).

Table 2. Antigen binding of ^{125}I-DNAP-66-80 on II C3 and BW 5147 cells 1/

Cold antigen added	Antigen bound CPM ± SE: II C3	Antigen bound CPM ± SE: BW 5147
–	7'400 ± 600	235 ± 40
DNAP-66-80	720 ± 40	215 ± 40
DNAP-81-104	7'450 ± 660	260 ± 45

1/ Cells were incubated with the same amount of ^{125}I-DNAP-66-80; 200 fold molar excess of cold antigen was added 20 min prior to addition of radiolabeled antigen

As shown in Table 3, the binding of labeled DNAP-66-80 peptide was completed in the period of 10 minutes and after this time a plateau level was obtained.

Table 3. Time dependency of ^{125}I-DNAP-66-80 binding on II C3 cells 1/

Incubation time in minutes	^{125}I-DNAP-66-80 bound on II C3 cells CPM	±	SE
1	2'958	±	150
3	3'608	±	200
5	4'227	±	110
10	4'780	±	60
20	5'500	±	300

1/ Cells were incubated with a constant amount of labeled antigen

Binding did not occur at 4°C in the period of 24 h tested or when cells were pretreated with trypsin for 15 minutes prior to addition of antigen, indicating that cellular interaction with the peptide was mediated by a protein moiety.

Characterization of the Segment Recognized by Clone II C3

It was found that NPS-11-25 peptide was able to inhibit the binding of DNAP-66-80 peptide and that clone II C3 bound specifically radiolabeled NPS-11-25 peptide.

Figure 1:

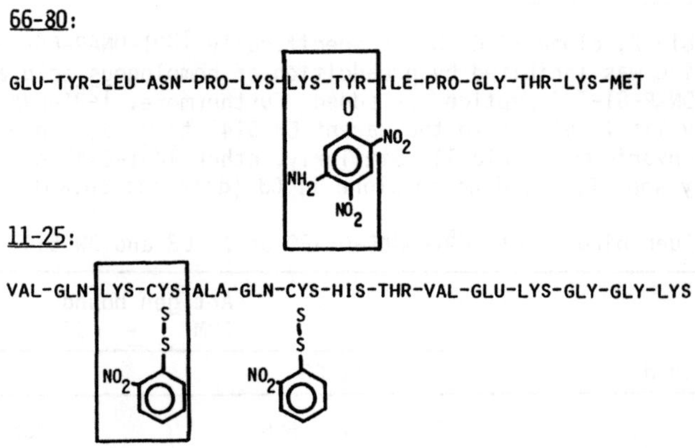

Amino acid sequence of derivatized fragments 11-25 and 66-80 from beef cytochrome c

Analysis of the amino acid sequence of peptide 11-25 and 66-80 reveals no sequence homology between the two fragments. On the contrary, introduction of DNAP group on tyrosine 74 and of NPS group on cysteine 14, as indicated in Fig. 1, brings about a structural similarity between the two derivatized peptides.

To test the hypothesis that the lysine group is part of the unit recognized by the clone II C3, lysine residues were modified by the use of the citraconic anhydride prior to modification with DNAP group.

Table 4. Effect of citraconic derivatization and deblocking on the binding of DNAP-66-80 on II C3 hybrid cells 1/

Cold antigen added	CPM
–	5'650
DNAP 66-80	645
Citraconic DNAP-66-80	5'550
Deblocked citraconic DNAP-66-80	1'210

1/ Cells were incubated with a constant amount of labeled antigen; 200 fold molar excess of cold antigen was added 20 min prior to addition of radiolabeled antigen

After the DNAP group was introduced, the citraconic was removed and the ensuing antigen was utilised to displace the ^{125}I-DNAP-66-80 peptide. As shown in Table 4, citraconic DNAP-66-80 peptide cannot inhibit the binding of ^{125}I-DNAP-66-80 while removal of citraconic groups resulted in a complete inhibition of the binding.

This is consistent with the hypothesis that the segment recognized by the hybridoma II C3 is Lys-Tyr DNAP (Fig. 1) and this modified dipeptide is necessary for the binding observed. However, a more stringent analysis is needed to confirm this hypothesis.

References

1. Corradin G, Harbury HA (1970) Cleavage of cytochrome c with cyanogen bromide. Biochim. Biophys. Acta 221: 489-496
2. Erecinska M, Vanderkooi JM, Wilson DF (1975) Cytochrome c interactions with membranes. Arch. Biochem. Biophys. 171: 108-116
3. Fontana A, Scoffone E (1972) Sulphenyl halides as modifying reagents for polypeptides and proteins. In: Hirs CHW, Timasheff SN (eds) Methods in Enzymology, Vol XXV. Academic Press, New York London, p 482-494
4. Atassi MZ, Habeeb AFSA (1972) Reaction of proteins with citraconic anhydride. In: Hirs CHW, Timasheff SN (eds) Methods in Enzymology, Vol XXV. Academic Press, New York London, p 546-553
5. Corradin G, Etlinger HM, Chiller JM (1977) Lymphocyte specificity to protein antigens. I. Characterization of the antigen-induced in vitro T cell-dependent proliferative response with lymph node cells from primed mice. J. Immunol. 119: 1048-1053

6. Schrier R, Skidmore BJ, Kurnick JT, Goldstine SN, Chiller JM (1979) Propagation of antigen-specific T cell helper function in vitro. J. Immunol. 123: 2525-2531
7. Nabholz M (1979) Hybridoma technology. In: Hybridoma technology with special reference to parasitic disease. WHO, Geneva p 23-32

The Effect of Antigen Presentation on the Fine Specificity of Anti-Cytochrome c T Cell Hybridomas

E. Heber-Katz, D. Hansburg, R.H. Schwartz

A. Introduction

An important criterion for the use of T cell hybridomas in the study of immune responses or as a source for antigen receptors or cell recognition structures is the ability to clearly demonstrate normal immune response parameters. In studying the T cell response to cytochrome c, we have been able to show that the fine specificity of cytochrome c specific T cell hybrids resembles the response of normal T cells to this antigen. Furthermore, by holding the T cell constant we have been able to study the interaction of antigen-Ia-T cell in a way which has revealed some interesting results dealing with T cell activation and Ir gene defects.

In examining the T cell responses of B10.A and B10.A(5R) mice to a family of cytochrome c peptides, we found that these two strains, differing in the K region and I-A and I-B subregions of the MHC, each had a characteristic pattern of responsiveness to a set of cytochrome c cyanogen bromide cleavage fragments (1,2,3). When T cell hybrids were derived from either of these cytochrome c immune animals, the response patterns closely resembled those of the whole lymph node populations. Remarkably, the clones from either strain responded to antigen on both B10.A and B10.A(5R) presenting cells. Furthermore, the response pattern of either set of clones resembled that of the whole B10.A lymph node population when B10.A presenting cells were used but resembled that of the B10.A(5R) lymph node population when when B10.A(5R) presenting cells were used. These data suggest that the B10.A and B10.A(5R) strains share similar T cell repertoires for certain cytochromes c and that at least part of their fine specificity differences derive from changes in the interaction of the MHC gene products with antigen alone or antigen in conjunction with the T cell receptor.

B. Methods

I. Antigen Preparation

Isolated, purified cytochrome c was cleaved with cyanogen bromide and the resulting fragments treated with acetimidyl derivatives which converts each lysine into an analog of homoarginine as previously descibed(1, 3). Tobacco horn worm moth (moth) cytochrome c 81-103 was synthesized by the Merrifield solid phase method as described elsewhere (4,5).

II. Preparation of T Lymphocytes and the Proliferation Assay

Popliteal and inguinal lymph nodes were collected 7 days after immunization with antigen in complete Freund's adjuvant in the hind footpads. T cells were obtained by passing the lymphocytes over nylon wool columns(6). The proliferative T cell assay was performed(7,8) by culturing $4x10^5$ nylon wool purified lymph node(NWPLN) cells + $1x10^5$ x-irradiated(2,000R) normal spleen cells in 0.2 ml of EHAA(9)medium plus 10% fetal bovine serum(Gibco Labs, FCS) and varying amounts of antigen. Cultures were pulsed with 1 µCi of ^{3}H-thymidine at 84 hours and harvested at 96 hours.

III. T Cell Hybridization

Bulk T cell blasts were prepared by culturing $2x10^6$ NWPLN cells and 10^6 X-irradiated spleen in 24 well Costar plates(3524,Costar,Cambridge,MA) containing 1.5 ml EHAA + 10% FCS and antigen. After 3 days of culture, cells were harvested and mixed with the HAT sensitive T cell line, BW5147, an AKR thymoma obtained from the Salk Institute Cell Distribution Center (La Jolla,CA) in a ratio of 1:2 to 1:4. Fusion was carried out according to the method of Gefter(10) employing 30%(v/v)PEG 1000 (Baker Chemical Co.,Phillipsburg, NJ) for 8 minutes. Then, $5x10^5$ cells were plated out into 96-well Costar plates(3596)containing $5x10^3$ X-irradiated(2500R) peritoneal washout cells in 0.2 ml of fusion medium which was a DMEM-based medium containing 10% NCTC 109(Micro. Assoc.,Walkerville,Md), 15%FCS, HAT and other additives as described elsewhere(11). After 6 days, hybridomas were picked and transferred to 24 well Costar plates, expanded and tested for antigen specific growth factor production.

IV. Cloning of T Cell Hybridomas

T cell hybrids were cloned by plating cells at three different concentrations: 1, 0.5, and 0.25 cells/well in the presence of $5x10^5$ thymocytes in fusing medium without HAT. The number of wells positive for growth was determined and clones were picked from the plates containing 0.25 cells/well. These clones were then tested for antigen specificity.

V. Antigen-Specific Growth Factor Assay

In the primary culture, $5x10^4$-10^5 cells from the hybridoma were added to 96 well Costar plates containing 0.2 ml fusing medium without HAT, $5x10^5$ x-irradiated spleen cells, and varying concentrations of antigen. After 2 days of culture, supernatants were collected and assayed for growth factor activity in a secondary culture in one of two ways: either using 10^5 thymocytes plus 2 µg/ml of Con A(Difco Labs,Detroit,MI)(12) or $4x10^3$ HT-2 cells, an IL-2 addicted T cell line developed by Dr. James Watson, Univ. of Auckland, Auckland, New Zealand,and provided by Dr. Phillipa Marrack, National Jewish Hospital, Denver, CO, and cultured in 0.2 ml of EHAA and 10% FCS with 25% primary culture supernatant. After 1-3 days the degree of stimulation was measured by the incorporation of ^{3}H-thymidine into DNA. Available data indicate that the supernatant activity was due to IL-2 and it will be so-called throughout. The assays were done in duplicate and the arithmetic means are reported. The cpm obtained in the HT-2 cell assay for IL-2 were shown to be sigmoidally related to the amount of IL-2 containing supernatant added.

C. Results and Discussion

B10.A(2R) mice were immunized with moth cytochrome c CnBr cleavage fragment 81-103 and their T cells tested in a proliferative assay with a panel of crossreactive cytochrome c molecules(Fig.1,upper panel). The response pattern seen we will refer to as the A pattern. If these animals are immunized with pigeon fragment 81-104,a similar pattern is seen(3,5). Immunization of B10.A(5R) mice with moth fragment 81-103 yielded a T cell response whose characteristic pattern is shown in Fig.1, lower panel, and will be referred to as the 5R pattern. B10.A(5R) mice are low responders to pigeon fragment(3).

We next examined the response of T cell clones from both immune T cell populations. Using a modification of a technique by Harwell et al. and Kappler et al.(13,14),T cell hybrids were made by fusing BW 5147 to antigen-stimulated T cells. Fusion products were tested for IL-2 production by first stimulating with x-irradiated spleen cells + antigen and then testing the supernatant for the presence of IL-2 in a thymocyte assay. In Fig.2(upper panel) the responses to the different cytochrome c fragments by a hybridoma from pigeon fragment primed B10.A animals is shown. When syngeneic B10.A spleen cells were used as antigen presenting cells(APC),the response resembled the pattern seen in the whole lymph node population. This similarity in responsiveness indicated that the hybridoma was a representative sample of the B10.A T cell repertoire for pigeon cytochrome c fragment. To examine the MHC restriction, we presented antigen in the presence of B10.A(5R) spleen cells(Fig.2,lower panel).Pigeon fragment did not stimulate but moth and fly fragments did. Surprisingly, the response profile was of the 5R pattern; not of the A pattern. All B10.A hybridomas tested showed similar responses. These findings are quite unexpected. It is generally beleieved based on chimeric experiments(15), that T cells acquire their MHC-restriction on the thymus and this in turn affects antigen specificity. It appeared from our results that the APC were affecting the fine specificity.

A reciprocal experiment was done using hybridomas made from B10.A(5R) mice primed to a synthetic analog of moth fragment 81-103. Using syngeneic B10.A(5R) spleen cells as APC(Fig.3,lower panel)we saw a response pattern that was very similar to the B10.A(5R) whole lymph node population. However,when the B10.A(5R) hybridoma was tested with B10.A spleen cells, it displayed the A response pattern,i.e. it could now respond to pigeon and tuna fragment. Here we have a case in which B10.A(5R) T cells respond better to antigen on B10.A APC even though the T cells were selected to see B10.A(5R) Ia molecules. Also,with the B10.A spleen cells, the clones respond to pigeon fragment, an antigen they cannot recognize in association with their own B10.A(5R) spleen cells. Again it appears that the APC is affecting the fine specificity of the T cell. These results strongly suggest the idea of an antigen-Ia interaction, however it is possible

Table 1. Rank order of antigen strength for stimulating the B10.A(5R) moth cytochrome c-specific T cell hybridoma

	B10.A APC	B10.A(5R) APC
Strongest	am-pigeon,am-moth	
	am-fly	
	moth,fly,pigeon	fly
	tuna	am-fly,moth
None	horse	am-pigeon,pigeon,am-moth,tuna

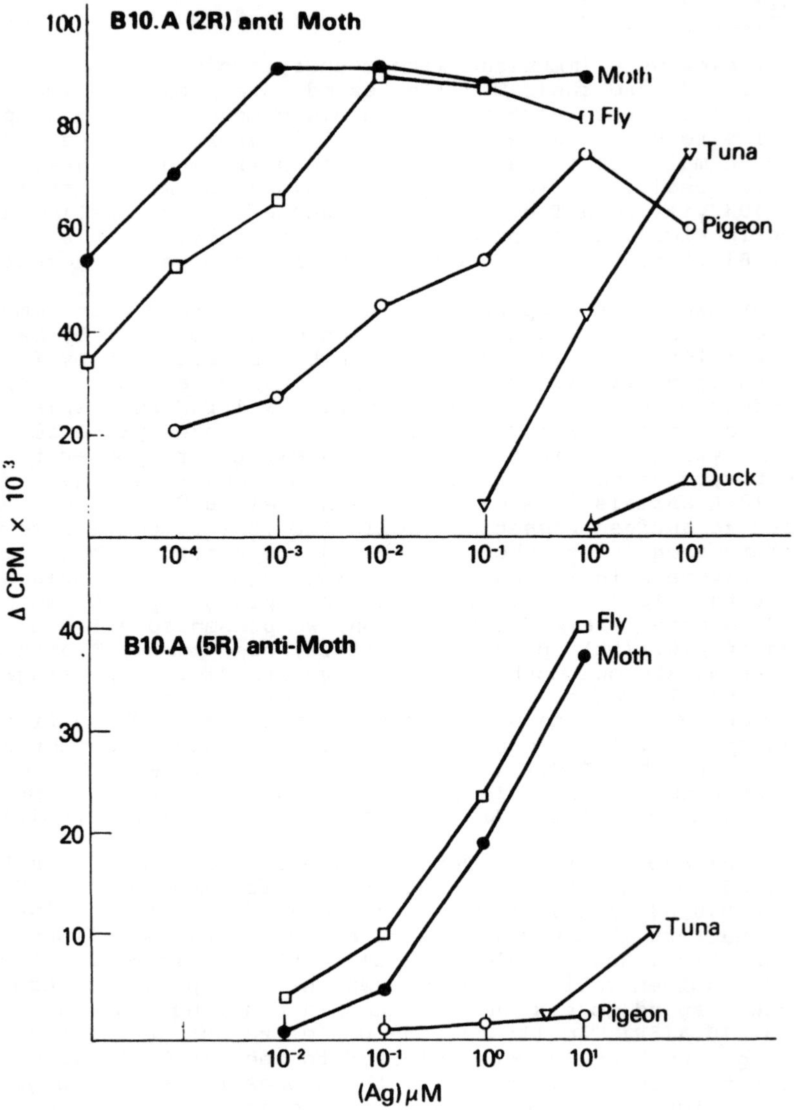

<u>Fig. 1.</u> The response pattern of lymph node T cells to various cytochrome <u>c</u> peptides after immunization with moth fragment 81-103.

B10.A(2R) mice (upper panel) were immunized with 0.5 μg of moth fragment 81-108/animal and B10.A(5R) mice (lower panel) were immunized with 50 μg of moth fragment 81-103/animal. Lymph nodes were removed and NWPLN cells were tested for their <u>in vitro</u> proliferative response to the following cytochrome <u>c</u> fragments: moth 81-103(●), fly 81-103(□), pigeon 81-104(○), tuna 81-103(▽), and duck 81-104(△). Media controls: B10.A(2R), 11,400 cpm; B10.A(5R), 330 cpm. Biological moth fragment 81-103 was used in the B10.A(5R) assay. (Adapted from Heber-Katz et al.,Contribution of APC Ia to Antigen-specific T cell Activation, J.Exp.Med., in press.)

to explain the results in terms of differences in affinity for B10.A and B10.A(5R) APC. To show an antigen-Ia interaction it was important to demonstrate a change in the order of relative strengths of the cross-reacting antigens. We utilized the observation that the 5R clone responded to both native cytochrome c fragments and their acetimidyl(am) derivatives. As seen in Table 1, the order of antigenic reactivity could be changed solely by changing the presenting cell MHC. Note, for example that am-pigeon, which is one of the strongest antigens with B10.A APC, gave no response with B10.A(5R) APC. Also, fly fragment, the strongest stimulator with B10.A(5R) APC was 25-fold less effective than am-pigeon and am-moth when used with B10.A APC.

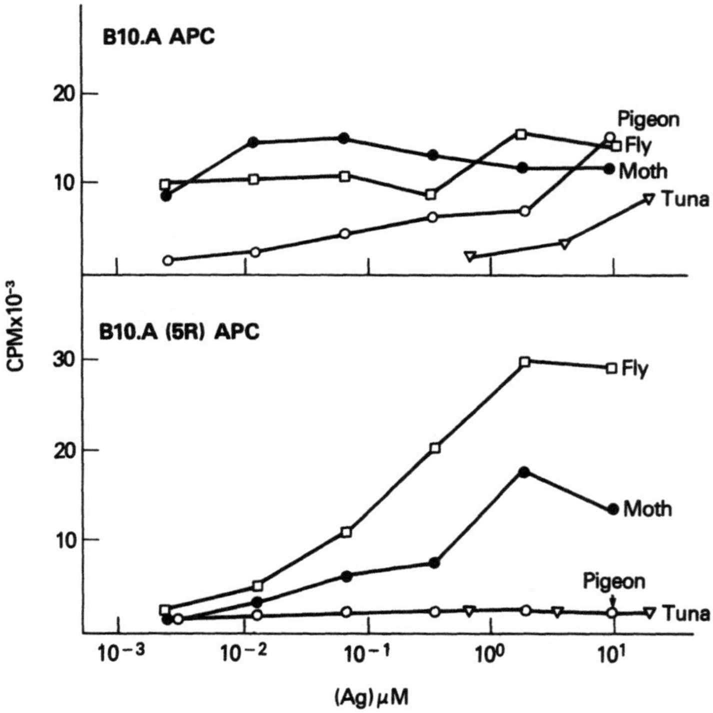

Fig. 2. The effect of different antigen-presenting cells on the cytochrome c response of a B10.A T cell hybridoma.

B10.A mice were immunized with 5μg of pigeon fragment cytochrome c 81-104 and NWPLN cells were stimulated in vitro with pigeon fragment 81-104, fused to BW5147, and selected for antigen-specific IL-2 production. The results represent a secondary assay which measures the ability of supernatants (25%v/v) from primary cultures; containing $5x10^4$ hybridoma cells, $2.5x10^5$ X-irradiated spleen, and varying concentrations of different cytochrome c fragments to stimulate the proliferation of thymocytes(10^5/well) in the presence of 2μg/ml Con A. The results are expressed in cpm of ^{3}H-thymidine incorporation. The primary cultures were stimulated with the following cytochrome c fragments: pigeon 81-104(○), fly 81-103(□), moth 81-103(●), and tuna 81-103(▽) in the presence of B10.A X-irradiated spleen cells (upper panel) or B10.A(5R) X-irradtiated spleen cells(lower panel).

Thus,by changing the presenting cell a change in the apparent antigen specificity of the T cell clone has resulted.This implies that the T cell receptor can accommodate a variety of antigen-Ia configurations but that these must be considered as pairs,i.e.the T cell specificity is affected by Ia-antigen interactions.

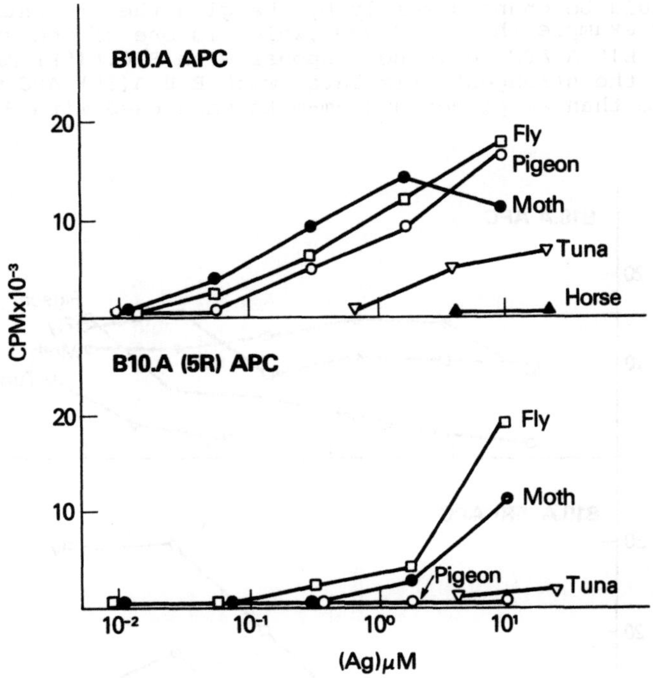

Fig. 3. The effect of different antigen-presenting cells on the cytochrome c response of a B10.A(5R) T cell hybridoma.

B10.A(5R) mice were immunized with 50μg/animal of synthetic moth fragment 81-103 and hybridomas were made and tested as in Figure 2. The primary cultures were stimulated with the following cytochrome c fragments: pigeon 81-104 (○), fly 81-103 (□), moth 81-103 (●), tuna 81-103 (▽), and horse 81-104 (▲). In the upper panel B10.A x-irradiated spleen cells and in the lower panel B10.A(5R) x-irradiated spleen cells were used to furnish APCs. (Adapted from Heber-Katz et al., J.Exp.Med.,in press.)

Table 2. Progeny test of B10.A(5R) hybridoma*

Presenting Cell:**	B10.A				B10.A(5R)			
Antigen:+	None	am-pigeon	fly	pigeon	None	am-pigeon	fly	pigeon
Subclone °:								
1	.8[++]	15.3	5.7	19.2	.5	.5	2.8	.4
2	.3	50.5	24.5	47.0	.2	.3	18.7	.3
3	.09	24.8	6.7	19.3	.1	.1	3.7	.1
4	.1	10.0	.3	1.4	.2	.3	2.9	.1
5	.3	110.0	36.9	81.1	.3	.3	12.5	.3
6	.1	61.7	11.8	43.5	.1	.1	3.4	.2
7	.07	16.7	1.9	7.9	.1	.2	8.0	.2
8	.3	27.2	7.6	15.1	.2	.1	10.0	.4
9	.1	59.2	13.4	32.0	.2	.3	22.1	.6
10	.2	19.7	3.3	19.5	.1	.2	14.3	.1
11	.2	20.3	2.6	21.4	.4	.6	13.4	.3

* The B10.A(5R) anti-synthetic moth fragment 81-103 hybridoma was subcloned at 0.25 cells/well. 11/96 wells were positive for growth.

** 2.5 x 10^5 spleen cells/well.

+ The final concentrations that were used were 8μM for pigeon and am-pigeon fragment 81-104 and 4μM for fly fragment 81-103.

° $5x10^4$ cells/well.

++ Supernatants from the primary culture were assayed for support of HT-2 cell growth as a measure of IL-2 content. The data are expressed as cpm x 10^{-3} of ^{3}H-thymidine incorporated during a 12 hour pulse 36 hours after the start of the assay.

These conclusions are based on the assumption that the response is that of a single clone. To demonstrate this unequivocally subcloning was carried out by limiting dilution. 96 wells were plated at densities of 1, 0.5, and 0.25 cells/well with positive growth of 44, 16 and 11, respectively. The 11 subclones from the lowest density plated were tested for antigen reactivity in the presence of both B10.A and B10.A(5R) irradiated spleen cells. As seen in Table 2, the fine specificity of all the subclones was similar to the original hybridoma. Although the amount of IL-2 produced by each subclone varied enormously, in all cases syngeneic B10.A(5R) APCs could only present fly fragment while allogeneic B10.A APCs could present pigeon and am-pigeon fragments in addition to fly fragment. We conclude that the lack of MHC restriction and the change in the fine specificity with a change in the APC are the properties of a single T cell and therefore that the interaction of Ia and antigen must be responsible for the observed change in the specificity of T cell activation.

It should be pointed out that even though both the B10.A and B10.A (5R) hybridomas could respond to antigen with either B10.A or B10.A(5R) APC, they were otherwise Ia-restricted. These hybridomas could not respond to any of the cytochrome c peptides in association with B10.A(4R), B10, or B10.S(9R) spleen cells. The actual involvement of Ia was demonstrated by completely blocking IL-2 release with the monoclonal antibody 17.3.3,directed against the A:E molecule, Thus, the degeneracy observed is highly selective.

References

1.Solinger,A.M.,M.E.Ultee,E.Margoliash,R.H.Schwartz(1979)The T-lymphocyte proliferative response to pigeon cytochrome c.I.Demonstration of a T-cell heteroclitic response and identification of a single antigenic determinant whose immune recognition requires two complementing MHC-linked immune-response genes.J.Exp.Med.150:830

2.Ultee,M.E.,E.Margoliash,A.Lipkowski,G.Flouret,A.M.Solinger,D.Lebwohl,L.A.Matis,C.Chen,R.H.Schwartz(1980)The T lymphocyte response to cytochrome c.II.Molecular characterization of a pigeon cytochrome c determinant recognized by proliferating T lymphocytes of the B10.A mouse. Mol.Immunol. 17:809

3.Hansburg,D.,C.Hannum,J.K.Inman,E.Appella,E.Margoliash, R.H.Schwartz (1981)Parallel cross-reactivity patterns of 2 sets of antigenically distinct cytochrome c peptides:Possible evidence for a presentational model of Ir gene function. J. Immunol. 127:1844

4.Merrifield,R.B.(1969) Solid-phase peptide synthesis.Adv.Enzm.32:221

5.Heber-Katz,E.,R.H.Schwartz,L.Matis,C.Hannum,T.Fairwell,E.Appella, D.Hansburg(1982)The Contribution of antigen presenting cell Ia to antigen-specific T cell activation.J.Exp.Med.in press

6.Julius,M.H.,E.Simpson,L.A.Herzenberg(1973)A rapid method for the isolation of functional thymus-derived murine lymphocytes.Eur.J. Immunol.3:645

7.Rosenwasser,L.J.,A.S.Rosenthal(L978)Adherent cell function in murine T lymphocyte antigen recognition.I.A macrophage dependent T cell proliferation assay in the mouse.J.Immunol.120:1991

8.Corradin,G.,H.M.Etlinger,J.M.Chiller(1977)Lymphocyte specificity to protein antigens.I.Characterization of the antigen-induced in vitro T cell-dependent proliferative response with lymph node cells from primed mice. J.Immunol.119:1048

9.Click,R.E.,L.Benck,B.J.Alter(1972)Immune responses in vitro.I.Culture conditions for antibody synthesis.Cell.Immunol.3:264

10.Gefter,M.C.,D.H.Margulis,M.D.Scharff(1977)A simple method for poly ethylene glycol-promoted hybridization of mouse myeloma cells. Somatic Cell Genet.3:231

11.Ozato,K.,N.Mayer,D.H.Sachs(1980) Hybridoma cell lines secreting monoclonal antibodies to mouse H-2 and Ia antigens.J.Immunol. 124:533.

12.Farrar,J.J.,P.L.Simon,W.J.Koopman,J.Fuller-Bonar(1978)Biochemical relationship of thymocyte mitogenic factor and factors enhancing humoral and cell-mediated immune responses. J.Immunol 121:1353

13.Harwell,L.,B.Skidmore,P.Marrack,J.Kappler(1980)Concanavalin A-inducible,interleukin-2 producing T cell hybridoma.J.Exp.Med. 152:893

14.Kappler,J.W.,B.Skidmore,J.White,P.Marrack(1981)Antigen-inducible, H-2-restricted, interleukin-2-producing T cell hybridomas: Lack of independent antigen and H-2 recognition.J.Exp.Med.153:1198

15.Longo,D.L.,L.A.Matis,R.H.Schwartz(1981) Insights into immune response gene function from experiments with chimeric animals. Crit.Rev.Immunol. 2:83

Analysis of the Anti-Self + TNP Immune Response: T Cell Lines, Clones and Hybridomas

A.M. Schmitt-Verhulst, M.A. Cooley, A. Guimezanes, C. De Préval, F. Albert, M. Buferne

A. Introduction

An immune response induced by the presentation of the hapten trinitrophenyl (TNP) on syngeneic cells can result from in vitro stimulation of T cells with syngeneic cells after trinitrobenzenesulfonate (TNBS) treatment (1) or after TNP-protein adsorption (2), or in vivo by trinitrochlorobenzene (TNCB) skin sensitization (3,4). The heterogeneity of T cell functions (5-9) is evident from the description of effector cells, including cytotoxic T cells (CTL) (1,2,4), helper T cells (TH) for CTL precursors (pCTL) (5) or B cells (6,7), as well as suppressor cells (Ts) (8-10) and delayed-type hypersensitivity (DTH) effector cells (3). In terms of the fine specificity of the response to TNBS-treated cells, it was not clear whether the apparently non-major histocompatibility complex (MHC) -restricted components of CTL responses in mice (11,12) and humans (13) could originate from T cell recognition of non-MHC and/or nonpolymorphic cell surface antigens (13) in addition to the recognition of the known class I MHC antigens. In order to analyse such a heterogenous response in terms of functional T cell subsets (14,15) and fine specificity of T cells (16,17) we established T cells lines proliferating in response to TNBS-treated syngeneic cells (14) in which we analysed interacting T cell populations (15) and from which we derived individual T cell clones (15-17). The study of the mechanism of action of TH for the development of CTL from thymic precursor cells (5,18) led us to hope that T cell hybridomas might help in the analysis of products secreted by such TH (18). Here we will briefly summarize results obtained with such procedures.

B. Materials and Methods

Mice. CBA/J (CBA), C57BL/6 (B/6), C57BL/10 (B10), B10.BR, as well as mice of other haplotypes and recombinant mice were obtained from our usual sources (15).

Media. RPMI1640, supplemented with 5% FCS, 2-mercaptoethanol ($5x10^{-5}$ M), glutamine (100 mM), Hepes (10 mM), penicillin (50 U), and streptomycin (0.5 μg/ml), was used for growth of cells (RPMI-FCS).

Sources of Interleukin-2 (Il-2). Preparation and use of concanavalin A (Con A) -stimulated rat spleen supernatants (SCA) (15) and mixed lymphocyte reaction (MLR) supernatant (16) were as previously described (15,16).

Test for Proliferative Response. Proliferative response was tested using 10^4 responding cells and $5x10^5$ stimulating spleen cells in microplates, as described (15).

Test for CTL Activity (5). CTL activity was tested during a 4-5-hr ^{51}Cr-release assay on tumor target cells (RDM4: AKR/J lymphoma; EL4: B/6 lymphoma) or on Con A- or LPS-induced splenic blast cells (15). Percent specific release = [(Exp. cpm-spontaneous cpm)/(total cpm)] x 100. Spontaneous release was for tumor target cells and for splenic blast target cells less than 10 and less than 40% of total cpm, respectively. Means and standard deviations of triplicate samples were calculated.

Establishment of Anti-Self + TNP T Cell Lines was as previously described (14,15).

Cloning by Limiting Dilution in the presence of 25% SCA and $5x10^5$ 2500-rad irradiated splenic stimulating cells was as described (15,16).

Generation of CTL in primary cultures was done by culturing $3x10^6$ splenic responding cells and $3x10^6$ splenic (2500-rad) stimulating cells in 2 ml RPMI-FCS per well of Costar 24-well cluster plates or by culturing $2.5x10^5$ splenic responding cells and $5x10^5$ (2500-rad) splenic stimulating cells in 0.2 ml RPMI-FCS per well of microtiter plates in triplicate cultures.

Fusion of T Cells with BW5147 was done in the presence of polyethylene glycol followed by selection in HAT medium as described (19).

Table 1. Comparison of the specificity of a B10 anti-B10-TNBS CTL line (BT7/SCA) and CTL clones derived from it

Strain of origin	Target cells: H-2 K I D	treatment	% specific lysis by BT7/SCA 2:1	BT7-22 2:1	BT7-20 2:1	BT7-7 2:1	BT7-2 0.3:1
B10	b b b	TNBS	57.2	55.8	55.8	53.5	26.6
		-	-0.3	-4.9	-4.7	-4.6	-5.4
B10.BR	k k k	TNBS	30.8	-2.2	1.9	-1.5	-1.5
		-	2.4	-5.4	-6.1	-5.2	-5.5
B10.MBR	b k q	TNBS	54.7	49.8	51.9	44.6	
		-	-1.7	-5.1	-8.9	-4.0	
CBA/Br	q q q	TNBS	33.5	1.7	2.3		
		-	2.4	1.6	2.2		
BALB/c	d d d	TNBS	24.4	17.7	-1.8		
		-	7.9	0.3	-1.5		
B10.RIII	r r r	TNBS	37.1	0.6	-2.4		
		-	1.2	-1.9	-4.5		
B10.WB	j j j	TNBS	26.1	-2.4	-4.4		
		-	6.2	2.1	-8.4		
C3H.NB	p p p	TNBS	16.9	2.6	3.1		
		-	4.7	3.9	2.9		
B10.S	s s s	TNBS	37.2	9.8	-5.8		
		-	4.8	0.4	-2.5		
B10.M	f f f	TNBS	45.3	0.8	11.9		
		-	6.4	0.5	5.1		
B10.SM	v v v	TNBS	8.7	-1.2	-3.8		
		-	1.9	3.4	2.0		

C. Results and Discussion

I. Fine Specificity of CTL Clones

1. Apparent Lack of MHC Restriction of Anti-B10-TNBS CTL Response Results from Clonal Heterogeneity of Crossreaction for $H\text{-}2^b$ + TNP and $H\text{-}2^k$ + TNP

As shown in Table 1, a B10 anti-B10-TNBS T cell line (BT7) propagated in the presence of B10-TNBS feeder cells and 25% SCA showed CTL activity on TNBS-treated target cells expressing any of the different mouse haplotypes tested. When CTL clones were derived from this line a much more restricted pattern of reactivity was observed: clone BT7.22 reacted with $H\text{-}2^d$ + TNP and weakly with $H\text{-}2^s$ + TNP, in addition to $H\text{-}2^b$ + TNP ($H\text{-}2K^b$), but not on TNBS-treated target cells from mice of the seven other haplotypes. Clone BT7.20 was found to react only with $H\text{-}2^f$ + TNP in addition to $H\text{-}2^b$ (K^b) + TNP, and for clone BT7.7 no crossreactivity on any of the nine different haplotypes could be detected (Table 1 and results not shown). A summary in Table 2 indicates that among 48 clones derived from three independent B10 anti-B10-TNBS T cells lines 48 were reactive to $H\text{-}2K^b$ + TNP, 25 were also reactive to TNBS-treated target cells from one or two independent haplotypes, and four were found to crossreact with untreated allogeneic target cells. Cold target blocking experiments confirmed the crossreactivities as detected by target cell lysis (17). Observed crossreactivities between self + TNP and allogeneic H-2 + TNP could (for $H\text{-}2^f$ + TNP and for $H\text{-}2^q$ + TNP) or could not (for $H\text{-}2^k$ + TNP) be correlated with known public specificities (17). This correlation might improve when other public specificities become defined by monoclonal antibodies (20,21). The low frequency of CTL clones found to react with untreated allogeneic cells contrasts with the observation of such crossreactivity for the few long-term anti-self + X clones described so far, whether CTL restricted to H-2D (22) or H-2K (23) or proliferative cells restricted to I-A (16,24). It is obvious, however, that nondetection of such crossreactvities does not preclude the existence of a crossreactive polymorphic determinant of H-2.

Table 2. Summary of reactivities of CTL clones derived from B10 anti-B10-TNBS T cell lines

Line number	H-2 haplotypes tested as target cells	CTL activity on target cells expressing: $H\text{-}2^b$ + TNP	allogeneic H-2 + TNP	allogeneic H-2
BT-5	b,k,d,q,s ± TNP	19/19 (K^b+TNP)	14/19 (K^k+TNP)	0/19
BT-6-II	b,k,d,q,s ± TNP	10/10 (K^b+TNP)	3/10 (K^k+TNP)	4/10 (K^k)
BT-6-III	b,d,d,q,s ± TNP	10/10 (K^b+TNP)	5/10	0/10
BT-7-III	b,d,d,q,s,	9/9 (K^b+TNP)	1/10 ($H\text{-}2^d$, $H\text{-}2^s$+TNP)	
	r,j,p,f,v ± TNP		1/9 ($H\text{-}2^f$+TNP) 1/9 ($H\text{-}2^q$+TNP)	

2. Reactivity of H-2^b Anti-H-2^q CTL with H-2^k + TNP

Alloreactive CTL populations have been found to crossreact with TNBS-treated target cells H-2-matched with the responding cells (25,26) (see line BT7.Q in Table 3). At the clonal level such crossreactivity is often not observed (Schmitt-Verhulst and Buferne, unpublished). For example, a particular B10 anti-H-$2D^q$ CTL clone (BT7.Q7, Table 3) derived from line BT7.Q had no reactivity on B10-TNBS target cells but showed crossreactivity on TNBS-treated, but not on untreated H-2^k target cells.

Table 3. B10 CTL clones reactive to H-$2D^q$ and to H-2^k + TNP

Effector cells	Target cells: Strain of origin:	% specific lysis of LPS blast target cells							
		B10		CBA/Br		B10.BR		B10.MBR	
	K I D	b b b		q q q		k k k		b k q	
	Treatment:	no	TNBS	no	TNBS	no	TNBS	no	TNBS
Line BT-7.Q[a)]	(2.5:1)[c)]	0	48.	41.	NT	NT	4.8	13.	23.
Clone BT-7.Q7[b)]	(2.0:1)	4.5	3.1	30.	36.	1.7	37.	32.	28.

a) Line BT7.Q: growing on CBA/Br feeder cells.
b) Clone BT-7.Q7: derived from line BT-7.Q by limiting dilution cloning on CBA/Br cells and SCA.
c) Effector to target cell ratio (E/T).

In our clonal analysis of an apparently non-H-2-restricted bulk CTL response generated against self-TNP, crossreactivities between different forms of H-2 were frequent, as exemplified by recognition of public H-2 specificities + TNP, crossreactivities between self H-2 + TNP and allogeneic H-2 determinants, as well as between an alloantigen and another alloantigen + TNP. No exception to the rule of T cell clone specificity for polymorphic H-2 determinants was observed however. We would like to suggest that the main rule in T cell recognition appears to be dictated by MHC class discrimination, which is a prerequisite for recognition of polymorphic MHC determinants and possibly neighboring cell surface antigens (17).

II. Polyfunctionality of T Cell Clones: CTL and Ts Function As An Anti-Self + TNP Clone

Starting with T cell populations obtained from CBA mice sensitized in vivo with TNCB and known to contain TH cells (5) for the development of thymic CTL, it was hoped to establish TH clones by limiting dilution cloning in the presence of 25% SCA and TNBS feeder cells 48 hrs after in vitro restimulation of the T cells with TNBS-CBA irradiated cells. One clone (3.1) which was not cytotoxic when tested on TNBS-treated RDM4 or on RDM4 in the presence of Con A (10 μg/ml) was also found negative, however, for TH activity as tested on thymic CTL precursors (5). A systematic analysis of that clone (Cooley and Schmitt-Verhulst, in preparation) revealed the following properties, summarised in Tables 4 and 5. Proliferation of the clone was both

Il-2- and stimulating cell-dependent, which allowed us to map the specificity requirement for its stimulation. It appeared that treatment with TNBS and expression of $H\text{-}2^k$ and $I\text{-}A^k$ [B10.A(4R)], but not I^k alone (B10.MBR), would allow optimal proliferation of clone 3.1. This proliferation was eliminated by treatment with anti-Thy-1 or anti-Lyt-2 monoclonal reagents followed by complement, but not by treatment with anti-Lyt-1.1 and complement. Although CTL activity was consistently negative when tested on TNBS-treated $H\text{-}2^k$ tumor target cells or Con A-induced blast cells, such activity could be detected on TNBS-treated, LPS-induced blast cells expressing $H\text{-}2^k$ and $I\text{-}A^k$ [B10.A(4R)], but not on such blasts expressing only I^k (B10.MBR). It is not clear at this point whether this clone is specific for $H\text{-}2^k$ + TNP in an $I\text{-}A^k$-restricted fashion (explaining the selective lysis of target cells expressing Ia antigens) or whether interaction with $H\text{-}2K^k$ on different types of blast cells may have different consequences for the target cells. When tested for suppressor function in a primary sensitization of CBA spleen cells to CBA-TNBS or to B/6, it was consistently found that the presence of clone 3.1 (even irradiated at 2000 rad) would significantly diminish the generation of anti-CBA-TNBS without affecting much generation of anti-B/6 CTL, unless both stimulating cells were present in the same culture (Table 5). This is consistent with the finding that a product secreted by the clone when stimulated with CBA-TNBS cells has a nonspecific suppressive activity on generation of CTL activity (results not shown). This observation would rule out elimination of stimulating cells by CTL activity as a mechanism for suppression. The nature of the suppressor activity and its mode of action are at present unknown. The possibility that immune interferon may play a role in the suppressor activity is being investigated. Evidence that CTL and suppressor activity are not necessarily linked has been obtained from analysis of subclones of 3.1: most subclones were both CTL and suppressor, but one showed CTL activity and no suppression (Cooley and Schmitt-Verhulst, in preparation).

Table 4. Summary of properties of CBA anti-CBA-TNBS clone 3.1

Stimulating or target cell: Origin	H-2 K	H-2 I-A	H-2 I-E	H-2 D	Il-2-dependent proliferation by clone 3.1 in the presence of stimulating cell[a)]: no	TNBS	Lysis of target cells by clone 3.1:[b)] Con A blasts no	Con A blasts TNBS	LPS blasts no	LPS blasts TNBS
CBA/J	k	k	k	k	-	+	-	-	-	+
B10.BR	k	k	k	k	-	+	-	-	-	+
B10	b	b	b	b	-	-	-	-	-	-
B10.A(4R)	k	k	b	b	-	+	-	-	-	+
B10.MBR	k	k	k	q	-	-	-	-	-	-

a) Measured by ^{3}H-thymidine incorporation: "-" indicates <1000.cpm and "+" indicates >8000.cpm above proliferation in the presence of Il-2, but in the absence of stimulating cells (5×10^5); no = no treatment; TNBS = 10 mM treatment of stimulating cells.
b) "-" indicates <10%; "+" indicates >30% specific lysis at E/T = 2/1.

Table 5. Suppressive action of clone 3.1 on the generation of primary CTL

Stimulating cells	clone 3.1 added	% specific lysis by CTL of[a)] CBA-TNBS Con A blasts	B6 Con A blasts
CBA-TNBS	-	45	NT
	10^5	20	NT
B/6	-	NT	40
	10^5	NT	40
B/6 + CBA-TNBS	-	45	40
	10^5	20	20

a) CTL activity generated after five days of culture of $3x10^6$ CBA responding and $3x10^6$ stimulating cells in the absence (-) or presence of 2000-rad irradiated clone 3.1 cells at E/T = 10/1.

III. Attempts at Deriving TH Hybridomas From a Fusion of BW5147 With a B10.BR Anti-B10.BR-TNBS T Cell Line

T cells obtained from TNCB-sensitized mice contain antigen-specific and H-2-restricted TH for the development of CTL responses against TNBS-syngeneic cells from otherwise unresponsive thymocyte populations (5). Such $Thy-1^+Lyt-1^+Lyt-2^-$ TH were found to secrete at least two supernatant activities upon specific antigen stimulation: the first, produced after 5-6 hrs' incubation, induced exclusively anti-allogeneic CTL; the second, produced after 15-24 hrs' incubation, induced preferentially anti-self-TNBS CTL (18). Attempts to purify the second activity led to a product similar to Il-2 in apparent molecular weight and devoid of selectivity of action (18). In all our attempts at cloning TH cells early (2-5 days after in vitro restimulation) or late (after multiple in vitro restimulations) of in vitro restimulated anti-self-TNBS T cell lines, only CTL (15,17), CTL/Ts (clone 3.1) or I-region-restricted proliferative cells without TH activity (16) could be derived on Il-2-containing SCA or MLR supernatants. Fusion with BW5147 was attempted for a T cell line (BRSF) originating from T cells from B10.BR mice pretreated with cyclophosphamide and TNCB and stimulated in vitro with B10.BR-TNBS cells as described (5,14).

After several stimulations this T cell line (BRSF) had the following characteristics: (a) it proliferated specifically in response to I^k + TNP; (b) it had lost all detectable CTL activity; (c) it had TH function for generation of anti-self + TNP CTL from thymic pCTL; (d) upon stimulation with B10.BR-TNBS cells for 24 hrs it secreted a product(s) with apparent selectivity for development of anti-self + TNP CTL, as compared to anti-allogeneic CTL. Cells from BRSF were fused with BW5147 in the presence of polyethylene glycol as described (19) and hybridomas were selected on HAT medium. Table 6 describes the screening for TH activity of 48-hr supernatants from the hybridomas growing in wells of Costar cluster plates in RPMI-10% FCS. The screening tests involved (a) measure of supernatant effect on generation of anti-CBA-TNBS CTL activity when suboptimal numbers of CBA responding cells ($2.5x10^5$) are cultured for five days in microplates in the presence of $5x10^5$ 2000-rad irradiated stimulating cells; (b) measure of supernatant effect on generation of anti-CBA-TNBS CTL activity from CBA thymocyte responding cells ($1x10^6$) cultured in microplates in the presence of $5x10^5$ stimulating cells; and (c) measure of Il-2-dependent growth induction by supernatants on 10^4 cells from CTL clone 61C2 (15) [^{3}H-thymidine pulse done after 48 hrs of culture in microplates (15)]. In all of these assays a standard preparation of SCA was taken as a positive control and supernatant from the BW5147 as a negative control.

Table 6. Test of (BRSF x BW5147) hybridoma supernatants on splenic and on thymic anti-self + TNP CTL generation and on growth of CTL clone 61C2

Supernatant added	% specific lysis of RDM4-TNBS CBA spleen cells[a)]		CBA thymocytes[a)]		Growth of CTL clone 61C2[b)]		
	+CBA	+CBA-TNBS	+CBA	+CBA-TNBS	-	+CBA	+CBA-TNBS
BW5147	0.0	15.4	-2.5	-0.5	931	1,482	845
SCA	15.0	49.4	9.1	55.3	6,474	9,275	19,732
A10	<2.0	26.0	<2.0	<2.0	<1,500	<1,500	<1,500
A12	↓	28.0	↓	↓	↓	↓	↓
A15	↓	21.0	↓	↓	↓	↓	↓
A17	↓	26.0	↓	↓	↓	↓	↓
A18	↓	29.8	↓	↓	↓	↓	↓
A20	↓	34.1	↓	↓	↓	↓	↓
B6	↓	47.3	↓	↓	↓	↓	↓
B13	↓	13.2	↓	↓	↓	↓	↓
B14	↓	24.5	↓	↓	↓	↓	↓
B15	↓	15.3	↓	↓	↓	↓	↓
B17	↓	26.6	↓	↓	↓	↓	↓
B18	↓	14.9	↓	↓	↓	↓	↓
B21	↓	30.8	↓	↓	↓	↓	↓
B23	↓	29.1	↓	↓	↓	↓	↓
B24	↓	24.9	↓	↓	↓	↓	↓
C3	↓	18.7	↓	↓	↓	↓	↓
C5	↓	17.1	↓	↓	↓	↓	↓
C17	↓	17.1	↓	↓	↓	↓	↓
C18	↓	20.2	↓	↓	↓	↓	↓
C19	↓	20.6	↓	↓	↓	↓	↓
D3	↓	48.7	↓	↓	↓	↓	↓
D4	↓	45.6	↓	↓	↓	↓	↓
D6	↓	23.7	↓	↓	↓	↓	↓
D10	↓	31.6	↓	↓	↓	↓	↓
D11	↓	13.9	↓	↓	↓	↓	↓
D13	↓	33.4	↓	↓	↓	↓	↓
D14	↓	4.8 ↘	↓	↓	↓	↓	↓
D20	↓	31.1	↓	↓	↓	↓	↓
D25	↓	26.8	↓	↓	↓	↓	↓
D31	↓	23.3	↓	↓	↓	↓	↓
E9	↓	6.8	↓	↓	↓	↓	↓
E13	↓	16.5	↓	↓	↓	↓	↓
E19	↓	10.0	↓	↓	↓	↓	↓
E25	↓	12.4	↓	↓	↓	↓	↓
2.4.F7	↓	5.6 ↘	↓	↓	↓	↓	↓
2.5.B2	↓	14.6	↓	↓	↓	↓	↓
2.5.H12	↓	8.8	↓	↓	↓	↓	↓

a) $2.5x10^5$ CBA spleen cells or $1x10^6$ CBA thymocytes as responders.
b) ^{3}H-thymidine incorporation by 10^4 61C2 clone cells.

Table 7. Test of supernatants from TB6 subclones on generation of splenic anti-self and anti-allogeneic CTL

Supernatant added	% specific lysis of				Summary of effects in response to:							
	CBA-TNBS		B10		in exp.				in exp.			
	+CBA[a]	+CBA-TNBS	+CBA	+B10	1	2	3	4	1	2	3	4
Medium	-1.9	9.7	0.0	18.2	-	-	-	-	-	-	-	-
SUB BD	7.5	26.1	-2.9	40.4	+[b]	+	+	+	+	+	+	+
TB6-1	4.5	12.8	0.9	32.6	-	-			+	-		
2	-2.9	5.6	-3.2	32.6	-	-		+	+	-	-	+
3	-3.4	4.6	-4.3	22.1	-	-			-	-		
4	-5.5	3.6	-2.4	28.9	-	-			-	-		
5	-7.2	4.0	-3.5	25.8	-	-			-	-		
6	-4.1	3.4	-2.9	18.1	-	+			-	+		
7	-0.9	3.2	-3.1	23.3	-	-			-	-		
8	-0.6	2.0	-2.2	27.5	-	-			-	-		
9	9.1	14.6	-0.9	33.1	-	-			-	-		
10	-0.4	22.5	0.0	30.9	+	-	-	+	-	-	-	+
11	-5.2	11.2	-2.5	27.9	-	-			-	-		
12	-3.4	20.2	-1/0	20.4	+	-	-	+	-	-	-	-
13	1.5	15.8	-0.9	13.0	-	+			-	-		
14	-3.7	9.8	-2.5	28.4	-	+			-	-		
15	-2.8	11.1	-1.2	17.3	-	-			-	-		
16	2.0	18.9	-2.7	20.3	+	-	+	+	-	+	+	+
17	8.9	18.5	-1.0	34.8	+	-			+	+		
18	-0.8	8.5	-2.6	40.8	-	-	-	-	+	-	+	-
19	-4.3	4.5	-1.4	34.2	-	-			+	-		
20	-0.9	15.4	-2.6	25.5	-	-			-	-		
TB6	-0.5	7.7	-7.4	28.9	-	+			-	+		

a) Indicates stimulating cells present during five days of culture.

b) + = > control ± 3 S.D.

Out of 37 hybridoma supernatants, five generated a level of splenic CTL activity at least twice (>30%) that of the negative control (15%) and two significantly decreased the CTL activity. None of the supernatants supported the generation of CTL from thymic pCTL or growth of the Il-2-dependent CTL clone (Table 6), and none of the hybridomas exhibited CTL activity when tested on RDM4-TNBS (results not shown). Hybridoma B6 (TB6) was cloned by limiting dilution at 0.5 cells per well and 48-hr supernatants from 20 clones were tested individually for their ability to enhance suboptimal generation of CTL anti-CBA-TNBS or anti-B10 when 2.5×10^5 CBA splenic responding cells were cultured in the presence of 5×10^5 stimulating cells for five days in microcultures (Table 7). MLR supernatant was taken as a positive control and RPMI-10% FCS as a negative control. Supernatants from two subclones, TB6-12 and TB6-16, increased detectable anti-self-TNP CTL activity without affecting detection of anti-B10 CTL activity, supernatants from nine subclones increased detection of anti-B10 CTL activity without affecting detection of anti-CBA-TNBS CTL activity, and supernatants from two subclones increased both CTL activities. When tested in subsequent experiments, however, these patterns did not appear to hold true for a given clone from one experiment to another (summarized in Table 7), and supernatant activity could not be increased by antigen or Con A stimulation (results not shown).

Although the transient activities which have been detected in supernatants from some of the hybridoma clones may indicate that products distinct from Il-2 (as tested by constant incapacity to induce growth of CTL clone 61C2) may be secreted by such hybridomas, the low level of the effects at low dilutions of the supernatants and the impossibility of inducing higher rates of secretion render the analysis of eventual products very difficult with the present material.

References

1. Shearer GM (1974) Cell-mediated cytotoxicity to trinitrophenyl-modified syngeneic lymphocytes. Eur J Immunol 4: 527
2. Schmitt-Verhulst AM, Pettinelli CB, Henkart PA, Lunney JK, Schearer GM (1978) H-2-restricted cytotoxic effectors generated in vitro by the addition of trinitrophenyl-conjugated soluble proteins. J Exp Med 147: 352
3. Miller JFAP, Vadas MA, Whitelaw A, Gamble J (1976) Role of the major histocompatibility complex gene products in delayed-type hypersensitivity. Proc Natl Acad Sci USA 73: 2486
4. Tagart VB, Thomas WR and Asherson GL (1978) Suppressor T cells which block the induction of cytotoxic T cells in vivo. Immunology 34: 1109
5. Cooley MA, Schmitt-Verhulst AM (1979) Specific helper T cells permit differentiation of thymic anti-self-trinitrophenyl cytotoxic precursor cells. J Immunol 123: 2328
6. Rubin B, Wigzell H (1973) Hapten-reactive helper lymphocytes. Nature 242: 467
7. Martinez-Alonso C, Coutinho A, Bernabé RR, Haas W, Pohlit H (1980) Hapten-specific helper T cells. I. Collaboration with B cells to which the hapten has been directly coupled. Eur J Immunol 10: 411
8. Asherson GL, Zembala M (1974) Suppression of contact sensitivity by T cells in the mouse. I. Demonstration that suppressor cells act on the effector stage of contact sensitivity. Proc R Soc London B 187: 329
9. Finberg R, Burakoff SJ, Benacerraf B, Greene MI (1979) The cytolytic T lymphocyte response to TNP-modified syngeneic cells. II. Evidence for antigen-specific suppressor T cells. J Immunol 123: 1210
10. Greene MI, Sugimoto M, Benacerraf B (1978) Mechanisms of regulation of cell-mediated immune responses. I. Effect of the route of immunization with TNP-coupled syngeneic cells or the induction and suppression of contact sensitivity to picryl chloride. J Immunol 120: 1604
11. Shearer GM, Rehn TG, Schmitt-Verhulst AM (1976) Role of the murine major histocompatibilty complex in the specificity of in vitro T-cell-mediated lympholysis against chemically modified autologous lympholysis. Transplant Rev 29: 223
12. Burakoff SJ, Germain RN, Benacerraff B (1976) Crossreactive lysis of trinitrophenyl (TNP) -derivatized H-2-incompatible target cells by cytolytic T lymphocytes generated against syngeneic TNP spleen cells. J Exp Med 144: 1609
13. Shaw S, Shearer GM (1978) Human cytotoxic response in vitro to trinitrophenyl-modified autologous cells. II. Diversity of self determinants recognized in association with TNP. J Immunol 121: 290

14. Guimezanes A, Schmitt-Verhulst AM (1981) I-region-restricted long-term-cultured T cells with helper function for thymic cytotoxic precursor cells. J Immunol 126: 563
15. Schmitt-Verhulst AM, Albert F, Guimezanes A, Buferne M (1981) Antigenic and genetic parameters in the stimulation and in the lytic phases of anti-hapten + self cytotoxic T cells and their derived clones: role of the T helper cells. J Supramol Struc Cell Biochem, in press
16. Guimezanes A, Albert F, Schmitt-Verhulst AM (1982) I-region-restricted T cell line stimulated with hapten-treated syngeneic cells: selection of clones with reactivity for both allogeneic Ia determinants and self-I-A + hapten. Eur J Immunol, in press
17. Guimezanes A, Davignon JL, Schmitt-Verhulst AM (1982) Clonal restriction of anti-self + hapten-polyspecific cytotoxic responses. (Submitted for publication)
18. Cooley MA, Schmitt-Verhulst AM (1982) Studies on T helper cells and the specificity of their soluble products for induction of cytotoxic T cells directed against TNP-modified self. Cellular Immunol, in press
19. Galfré G, Howe SC, Milstein C, Butcher GW, Havard JC (1977) Antibodies to major histocompatibility antigens produced by hybrid cell lines. Nature 266: 550
20. Lemke HG, Hämmerling GJ, Hämmerling U (1979) Fine specificity analysis with monoclonal antibodies of antigens controlled by the MHC and by the Qa/TL region in mice. Immunol Rev 47: 175
21. Ozato K, Sachs DH (1981) Monoclonal antibodies to mouse MHC antigens. III. Hybridoma antibodies reacting to antigens of the $H-2^b$ haplotype reveal genetic control of isotype expression. J Immunol 126: 317
22. von Boehmer H, Hengartner H, Nabholz M, Lernhardt W, Schreier MH, Haas W (1979) Fine specificity of a continuously growing killer cell clone specific for H-Y antigen. Eur J Immunol 9: 592
23. Braciale TJ, Andrew ME, Braciale VL (1981) Simultaneous expression of H-2-restricted and alloreactive recognition by a cloned lined of influenza virus-specific cytotoxic T lymphocytes. J Exp Med 153: 1371
24. Sredni B, Schwartz RH (1980) Alloreactivity of an antigen-specific T cell clone. Nature 287: 955
25. Schmitt-Verhulst AM, Shearer GM (1975) Bifunctional major histocompatibility-linked genetic regulation of cell-mediated lympholysis to trinitrophenyl-modified autologous lymphocytes. J Exp Med 142: 914
26. Lemonnier F, Burakoff SJ, Germain RN, Benacerraf B (1977) Cytolytic thymus-derived lymphocytes specific for allogeneic stimulator cells crossreact with chemically modified syngeneic cells. Proc Natl Acad Sci USA 74: 1229

Functional Characteristics of T Cell Hybridomas Obtained by Fusion of TCGF-Dependent Helper T Cell Clones with BW5147

M.H. Schreier, R. Tees, L. Nordin, A.A. Nordin

A. Introduction

Over the last three years we have established a large panel of functionally active antigen-specific T cell clones with helper T cell characteristics (1-3). Such T cells are reasonably stable and have been propagated in vitro in TCGF-supplemented medium for up to two years. Their extensive characterization in vitro and in vivo has shown that the same cell can express multiple functions and has revealed a remarkable functional heterogeneity among individual clones (4,5). Since long-term propagation of these TCGF-dependent clones is demanding and involves an expensive and time-consuming cell culture technology (4,6), it was obvious to aim at a more convenient source of monoclonal T cells with comparable functional properties. We have therefore established a large panel of T cell hybridomas by fusing the HAT-sensitive thymoma cell line BW5147 with various TCGF-dependent specific helper clones which had been shown to release biologically active factors when interacting with I-A-compatible macrophages and antigen. This allowed us to answer two questions: Are functional properties of TCGF-dependent helper clones expressed in T cell hybridomas? And, is there any correlation between TCGF production and helper activity for the in vitro antibody response?

B. Choice of Test System and TCGF-Dependent Helper Clones

Antigen-induced soluble products from specific T cell clones were tested in three experimental systems. Helper activity was evaluated in the in vitro antibody responses to a type-2 T-independent (TI) antigen, DAGG-Ficoll (7,8), and to a T cell-dependent (TD) antigen, sheep red blood cells (SRC). The same supernatants were tested in parallel for TCGF activity by measuring survival and growth of a TCGF-dependent killer clone (9).

Our recent experiments had shown that under suitable experimental conditions the in vitro antibody response to type-2 TI antigens is strictly dependent on antigen-induced soluble helper factors (4; A.A.N. and M.H.S., submitted for publication). Unlike the factor dependence of the in vitro response of *nu/nu* spleen cells to TD erythrocyte antigens, this requirement can only be reliably revealed at low cell density and in the absence of stimulatory fetal calf serum (FCS).

The generation of helper factors for an antibody response and the production of TCGF varies greatly between individual helper T cell clones when they are stimulated by specific antigen in the presence of I-A-compatible macrophages (4). Furthermore, there is no correlation between the activities that help the response of *nu/nu* spleen cells to TI and TD antigens. Table 1 demonstrates this heterogeneity with a small selected panel of TCGF-dependent helper T cell clones. The chicken egg albumin- (EA) specific clones 7 and 14, which have been studied extensively over the last year, helped to about the same extent in a secondary hapten-carrier system. Clone 14 was consistently superior to clone 7 in providing helper activity for the response to DAGG-Ficoll, while the latter always provided excellent "bystander" help for the response to sheep erythrocytes (SRC). These two clones also differed consistently in the antigen-induced release of TCGF: TCGF was almost undetectable in supernatants of clone 14, but always present in significant amounts in supernatants of clone 7. Mixing these two supernatants in various ratios revealed neither suppressive nor synergistic activities.

Table 1. Lack of correlation between helper activity in a hapten-carrier system, TCGF activity and the functional activity of antigen-induced helper factors, which promotes an antibody response of *nu/nu* spleen cells to SRC or DAGG-Ficoll

Designation of clone[1] (T'_{EA})	Hapten-carrier response[2] (IgM + IgG anti-DNP-PFC)	Antigen-induced supernatant activity[3]: TCGF activity (cpm x 10^{-3}) at 25% concentration	TCGF activity (cpm x 10^{-3}) at 50% concentration	Helper activity in cultures of *nu/nu* spleen cells containing SRC (IgM-PFC per culture)	Helper activity in cultures of *nu/nu* spleen cells containing DAGG-Ficoll (IgM-PFC per culture)
2	400	15.7	28.0	4,980	1,060
7	5,435	4.2	8.4	16,000	935
14	3,030	0.1	0.3	1,035	3,370
15	0	12.3	22.7	380	275
30	3,750	6.4	13.0	420	250

[1]Chicken egg albumin- (EA) specific T cell clones established and maintained as described in ref. 4.

[2]Ten weeks after boosting with TNP-FGG, 2×10^5 T cell-depleted spleen cells of C57BL/6J mice were cultured together with 3×10^3 cells of the indicated EA-specific clones in 0.2 ml serum-free medium containing 0.02 µg/ml TNP-EA. Anti-DNP-PFC were determined in the presence of enhancing serum after five days of culture. Figures represent the mean of four replicate cultures.

[3]Supernatants of cultures containing 5×10^4 T'_{EA}, 1×10^6 irradiated *nu/nu* spleen cells and 100 µg EA per ml serum-free medium were harvested after 48 hrs. The cell-free supernatants were passed through 0.2-µ filters and diluted 1:2 with fresh medium for the culture of 2×10^5 *nu/nu* spleen cells. These 0.2-ml cultures in Falcon II microtiter plates contained either 5×10^5 SRC or 0.01 µg/ml DAGG-Ficoll, as indicated. IgM-PFC directed against SRC or DNP were enumerated after five days of culture and are expressed as the mean of four replicate cultures. The same supernatants were diluted 1:2 (50%) or 1:4 (25%) with fresh serum-free medium and used for the culture of 10^4 cloned killer cells in a volume of 0.2 ml. After 48 hrs. of culture T cell growth was measured by ^{3}H-thymidine uptake.

C. Cell Fusion and Screening for Specific T Cell Hybrids

Due to its preferential helper activity for the in vitro antibody response to DAGG-Ficoll and to its low TCGF activity clone 14 was of special interest for our studies. By establishing a large number of T cell hybrids it was hoped that clonal T cell populations could be obtained which did not release TCGF or failed to provide help for the in vitro response to the TD antigen altogether. As is obvious from Table 2 this clone was a very good fusion partner. When $1\text{-}2x10^7$ BW cells were fused with the EA-14 clone at a ratio of 1.0 to 3.0 and distributed into 144-480 wells all cultures yielded HAT-resistant and TCGF-independent cell lines. When $5x10^5$ EA-14 cells were fused with 10^7 BW cells (ratio 1:20, Exp. 4) about half of the cultures were positive, suggesting that about one in 5000 EA-specific T cells gave rise to a hybrid. As also shown in Table 2 there is a wide variation in the fusion efficiency of individual helper T cell clones which correlates neither with their growth rates nor with the duration of their in vitro propagation.

Table 2. The fusion efficiency of individual helper T cell clones

Exp. No.	Speci- ficity	Designation of clone	BW5147 : T'	Ratio of BW5147: T'	No. of wells	No. of positive cultures
1	EA	EA-14	$6x10^7$: $6x10^7$	1.0	480	480
2	EA	EA-14	$2x10^7$: $2x10^7$	1.0	240	240
3	EA	EA-14	$1x10^7$: $3.3x10^6$	3.0	144	144
4	EA	EA-14	$1x10^7$: $5x10^5$	20.0	144	77
5	EA	EA-7	$2.5x10^7$: $1x10^7$	2.5	144	144
6	EA	EA-7	$1.25x10^7$: $1x10^7$	1.25	144	89
7	SRC	S26-14-25	$2.5x10^7$: $5x10^6$	5.0	192	2
8	SRC	S26-14-12	$2.5x10^7$: $5x10^6$	5.0	192	1
9	HRC	18-19	$1x10^7$: $3.3x10^6$	3.0	144	26
10	HRC	ATH-29	$1x10^6$: $1x10^6$	1.0	192	6
11	SRC	S26-5	$6.25x10^6$: $5x10^6$	1.25	120	42

TCGF-dependent helper T cell clones of the indicated specificities were fused at various ratios with the HAT-sensitive AKR thymoma BW5147. We used polyethyleneglycol 4000 as the fusing agent and followed essentially the method of Fazekas de St.Groth and Scheidegger (10) for the production of B cell hybrids.

In our initial screening for functionally active T cell hybridomas we harvested the cells from confluent Costar wells and cultured them for 48 hrs in the presence of congenic *nu/nu* spleen cells (10^6/ml) and antigen (50 µg EA/ml). The cell-free supernatants were then tested for their helper activity in the in vitro antibody responses to DAGG-Ficoll and SRC. The stringent requirement for antigen and I-A-compatible macrophages was established in subsequent experiments, where the number of T cell hybrids was adjusted to $2x10^5$ or $5x10^5$ cells/ml. Cells from strongly positive cultures were cloned at 0.5 cells/well, since the high fusion efficiency indicated that all wells contained multiple hybrids. The results obtained with clones derived from two positive wells (lines 363 and 397) in fusion experiment 1 (Table 2) are listed in Table 3. In striking contrast to the parental line T'_{EA}-14, three out of ten positive clones derived from line 363 produced potent helper activity for the in vitro antibody response to SRC (clones 23, 30, 34). The relative helper activity for the response to SRC and DAGG-Ficoll varied widely between individual clones. Similar results were obtained with clones derived from line 397, although on the average their helper activity for the response to the TI antigen was somewhat higher. As judged by a limited number of subcloning experiments there is also some heterogeneity in the functional activity of subclones.

Table 3. The effect of supernatants of EA-14 hybridoma clones on the in vitro antibody response of *nu/nu* spleen cells to DAGG-Ficoll and SRC[1]

	IgM-PFC directed against			IgM-PFC directed against	
Line 363	SRC	DNP	Line 397	SRC	DNP
Clone 13	345	1,400	Clone 1	480	4,320
Clone 15	30	1,880	Clone 6	1,400	6,480
Clone 23	17,200	4,220	Clone 8	1,560	760
Clone 28	530	1,820	Clone 11	65	4,160
Clone 30	24,240	2,130	Clone 15	11,760	5,640
Clone 31	2,740	445	Clone 17	16,400	4,640
Clone 34	18,160	580	Clone 19	6,400	11,040
Clone 36	4,000	1,890	Clone 21	1,370	170
Clone 44	15	1,460	Clone 26	13,920	7,520
Clone 45	2,090	1,840	Clone 28	740	1,880

$5x10^5$ cloned EA-specific hybridomas were cultured for 48 hrs in 1 ml serum-free medium in the presence of 10^6 C57BL/6J *nu/nu* spleen cells and the antigen EA (100 µg/ml). 100 µl of the cell-free membrane-filtered supernatant was added to cultures of *nu/nu* spleen cells (2x 10^5/well) which contained either SRC ($5x10^5$/well) or DAGG-Ficoll (0.02 µg/ml final concentration). After five days of culture in a total volume of 0.2 ml SRC-specific and DNP-specific PFC were enumerated. The figures represent the mean number of PFC in four replicate cultures.

Since the response to SRC may be related to the presence of TCGF in these culture supernatants, we have tested for helper activity and TCGF activity in parallel. Table 4 lists twelve selected T cell hybrids, derived from fusion experiment 4 in Table 2, according to their ability to release TCGF upon antigenic stimulation. A comparison with the in vitro antibody response promoted by the same supernatants argues against any correlation between the two activities under our experimental conditions. All supernatants have been tested in serial dilutions to exclude inhibitory effects due to excessive concentrations.

The hybrid nature of the HAT-resistant T cells which grew independently of TCGF after fusion has been proved by their expression of surface markers. All the tested cells expressed Thy-1.1 and Thy-1.2 as well as $H-2^b$ and $H-2^k$ and various amounts of Lyt-1 (11). Their H-2 restriction has been tested with a panel of congenic B10 mice. The supernatant activities were produced only in the presence of peritoneal macrophages or T cell-depleted spleen cells which expressed $I-A^b$ [C57BL/6, B10.A(5R)]. Two hybrids that released factors equally well in the presence of B10.MBR or B10.A(4R) macrophages ($I-A^k$) turned out to be constitutive producers, which lost this function over a period of three months.

Table 4. Lack of correlation between TCGF activity and helper activity in the in vitro antibody response to SRC

Designation of T cell hybrid	Growth of killer clone (cpm x 10^{-3})	IgM-PFC per culture
14/30	0.5	6,120
11	4.0	18,460
7	13.5	10
8	15.0	24,240
23	16.1	200
10	17.0	28,800
48	24.4	480
34	33.9	0
29	49.1	29,600
2	52.0	17,040
13	87.6	4,640
32	108.2	7,880
Medium	0.09	0
Partially purified Con A supernatant from rat spleen	156.0	24,440

$2x10^5$ EA-specific hybridoma cells, derived from fusion experiment 4 in Table 2, were cultured for 48 hrs in 1 ml serum-free medium in the presence of 10^6 C57BL/6J *nu/nu* spleen cells and EA (100 µg/ml). The cell-free membrane-filtered supernatants were used at 50% concentration for the culture of 10^4 TCGF-dependent cloned killer cells (9) and of $2x10^5$ nu/nu spleen cells together with $5x10^5$ SRC. The total volume of both types of culture was 0.2 ml. T cell growth was measured after 48 hrs of culture by ^{3}H-thymidine incorporation. SRC-specific direct PFC were enumerated after five days of culture. Numbers indicate the mean of four replicate cultures. Partially purified Con A supernatant from rat spleen (6), routinely used as a source of TCGF for the in vitro propagation of killer and helper T cells, was used as a positive control at optimal concentration.

D. Discussion

Our T cell fusion experiments have been carried out on the assumption that clonal T cell populations with known and stable functional properties yield T cell hybrids which express or fail to express the characteristics of the parent cell line. As judged by the TCGF production this is not the case. T cell hybridomas derived from the specific T cell clone T'_{EA}-14, which itself produced little if any TCGF, released variable and significant amounts of this factor when stimulated with antigen in the presence of I-A-compatible macrophages. This suggests that hybridization induced or enhanced a function which is not or barely expressed by either parental line. A similar observation was made with the lectin-dependent cytolytic activity of these clones (11).

The expression of the activities which are measured by the number of specific PFC directed against the TI or TD antigen is far more difficult to interpret. The helper activity, as revealed in either system, results quite certainly from the combined action of several factors, which either act directly on B cells or mediate their effect by the interaction with non-B cells in the T cell-deficient spleen cell cultures. According to our working concept all B cells, once activated from the resting state (G_o) by whatever mechanism, become susceptible to and dependent on soluble factors for proliferation and maturation to antibody secretion. The stringent requirement for helper factors in the B cell response to TI antigens has substantiated this concept. These factors are generated in an antigen-mediated and I-A-restricted interaction of T cells with macrophages and therefore their exact cellular origin is as yet obscure. The factor requirement in the B cell response to TI antigens is quite obviously less stringent, as in vitro it is readily masked at high spleen cell density and in the presence of stimulatory FCS (4; A.A.N. and M.H.S., submitted for publication).

The establishment of a reliable source of factors which meet all the requirements for the TI response, but not the TD response, would provide an ideal basis to define the additional factor requirement for the response to erythrocyte antigens. This strategy has failed so far, as individual hybrids preferentially help the response to one or the other antigen rather than expressing an all-or-nothing effect. As yet we were also unable to detect complementary activities by mixing individual supernatants. Whether the observed variation reflects qualitative differences in factor production or quantitative differences in the relative amounts of factors affecting B cell induction, proliferation and maturation remains to be established. Variability may be inherent in the test system itself, as the magnitude of the response to DAGG-Ficoll can be greatly affected by the age and the health status of the *nu/nu* mice. The critical role of macrophages or their products, either at the stage of factor induction or in the course of the B cell response, may account for at least some of the observed variation.

T cell hybridomas produce significantly less factors per cell than the TCGF-dependent helper T cell clones, whenever optimally producing clones of the two cell types are compared. On the average at least ten times more hybridoma cells are required to produce a supernatant activity comparable to that of the parent lines.

It is hoped that repeated subcloning will yield the stable factor-producing hybridomas that are required for studying the in vitro antibody response to TI and TD antigens.

Acknowledgments. The authors thank Dr. K.A. Smith for providing the CTLL cell line and for stimulating and fruitful discussions.

References

1. Schreier, MH, Tees, R (1980) Clonal induction of helper T cells: Conversion of specific signals into nonspecific signals. Int Arch Allergy Appl Immun 61: 227-237
2. Tees, R, Schreier, MH (1980) Selective reconstitution of nude mice with long-term cultured and cloned specific helper T cells. Nature 283: 780-781
3. Schreier, MH, Iscove, NN, Tees, R, Aarden, L, von Boehmer, H (1980) Clones of killer and helper T cells: Growth requirements, specificity and retained effector functions in long-term culture. Immunol. Rev. 51: 315-336
4. Schreier, MH, Tees, R, Nordin, AA (1982) Establishment and characterization of helper T cell clones. Their functional heterogeneity and effect on the B cell responses to particulate, soluble and T independent antigens. In: Feldmann, M, Schreier, MH (eds) Lymphokines, Volume 5, Monoclonal T cells and their products. Academic Press, New York, pp 443-464
5. Schreier, MH, Tees, R, Nordin, AA, Benner, R, Bianchi, ATJ, van Zwieten, MJ (1982) Functional aspects of helper T cell clones. In: Smith, KA (ed) The growth and functional aspects of T cells. Immunobiology Supplement, in press
6. Schreier, MH, Tees, R (1981) Long-term culture and cloning of specific helper T cells. In: Lefkovits, I, Pernis, B (eds) Immunological Methods, Volume II. Academic Press, New York, pp 263-275
7. Mosier, DE, Zitron, IM, Mond, JJ, Ahmed, A, Scher, I, Paul, WE (1977) Surface immunoglobulin D as a functional receptor for a subclass of B lymphocytes. Immunol Rev 37: 89-105
8. Inman, JK (1975) Thymus-independent antigens: The preparation of covalent hapten-ficoll conjugates. J Immunol 114: 704-709
9. Gillis, S, Ferm, MM, Ou, W, Smith, KA (1978) T cell growth factor: parameters of production and a quantitative microassay for activity. J Immunol 120: 2027-2032
10. Fazekas de St.Groth, S, Scheidegger, D (1980) Production of monoclonal antibodies: Strategy and tactics. J Immunol Meth 35: 1-21
11. Fischer Lindahl, K, Nordin, AA, Schreier, MH (1982) Lectin-dependent cytolytic and cytolymic T helper clones and hybridomas. (This volume.)

A Stable TCGF-Producing T Cell Hybridoma and its Thioguanine-Resistant Variant Suitable as a Tool for the Construction of New Functional T Hybridomas

K.-O. Grönvik, J. Andersson, F. Melchers, J. Zeuthen

A. Introduction

Recent studies have shown that antigens and mitogenic lectins initiate the activation of T cells. Proliferation of these activated cells, however, is dependent on T cell growth factors (TCGF or IL-2) (1-5). TCGF production requires a non-killer Ly 1^+ T cell cooperating with Ia^+ adherent cells under the influence of lectin or antigen (5-8). As most convincingly demonstrated with cloned helper T cells, TCGF production is the result of specific antigen recognition by the T cells in the context of histocompatibility antigens presented by viable adherent, macrophage-like cells (9,10).

T cell tumor cell lines (11-14) and T cell hybridomas (15-17) have been isolated which produce TCGF either spontaneously or after treatment with mitogenic lectins or phorbol esters. We report here the isolation of such a hybridoma. The antigen specificity of this line is unknown. A 6-thioguanine-resistant variant has been derived from it which is suitable for cell fusion, while retaining the property of lectin-dependent TCGF production. The line can be used as a fusion partner with normal antigen-specific T cell lines (18,19). The function of antigen recognition immortalized in the T cell hybrids can then be monitored by the antigen-specific induction of TCGF production, as originally described by Kappler et al. (20).

B. Materials and Methods

Animals: CBA/H mice were bred in the Department of Immunology, Uppsala, Sweden. B10.A(5R) and B10.MBR mice were bred at the Basel Institute for Immunology and supplied to us by Dr. H. von Boehmer.

Cell Lines: The AKR thymic lymphoma BW5147 G.1.4.Oua^r.1 ($HRPT^-$ OUR^r) (21) was supplied by Dr. R. Hyman, Salk Institute for Biological Studies, La Jolla, California, cloned anti-male-specific, TCGF-dependent killer cells from Dr. H. von Boehmer (22), and CTLL (2) from Dr. K. Smith, Dartmouth Medical School, Dartmouth, N.H., U.S.A.

Cell Cultures and Hybridization: Spleen cells from normal CBA/H mice, prepared as described (5), were cultured at $2x10^6$ cells/ml in Iscove's modified DMEM (23) supplemented with 0.5% freshly prepared homologous mouse serum and stimulated for 17 hrs. with 0.3 µg/ml concanavalin A (Con A). Harvested cells were washed once in serum-free DMEM containing 10 mg/ml of α-methylmannoside. Sixty x 10^6 washed spleen cells were fused with $15x10^6$ BW5147 lymphoma cells using 50% polyethyleneglycol (PEG) 1500 M.W. (J.T. Baker) as described

(24). The cells were thereafter resuspended in DMEM containing 5% FCS and distributed into 0.2-ml cultures in microtiter plates using per ml 10^6 2200-rad irradiated syngeneic spleen cells as "fillers". Twenty-four hrs later the medium was replaced by hypoxanthine, aminopterin and thymidine- (HAT) containing medium. For 12 days the medium was changed every third day and then exchanged by HT medium. At this point replica plating was performed on plates in which approximately 40% of all wells showed growing hybrid cells, using a FLOW multi-channel pipette. To 0.1 ml of cell culture transferred to new microtiter plates were added 0.1 ml fresh DMEM containing 5% FCS and 2 µg Con A, for a final concentration of 10 µg/ml of Con A. After 20 hrs. supernatants from these Con A-treated cultures were transferred to a new set of microtiter plates where each well was given 10^4 TCGF-dependent killer T cells in 0.1 ml of DMEM containing 5% FCS and α-methyl-mannoside at 10 mg/ml. Twenty-four hrs. later 2 µCi/well of ^{3}H-TdR (Amersham, 5 Ci/mmol) were added, the cultures incubated for 2 more hours, and incorporation into cellular DNA determined as described (5).

TCGF Production: For the production of TCGF spleen cells from mice or hybridoma cells were cultured at $5x10^6$ cells/ml or 10^6 cells/ml, respectively, in Iscove's medium supplemented with 0.03% BSA and 5 or 10 µg/ml Con A (Pharmacia Fine Chemicals, Uppsala, Sweden). Supernatant medium (1-4 liters) was harvested after 20-24 hrs. of cell culture and precipitated with ammonium sulfate at 80% saturation. The precipitate was dissolved in 0.9% NaCl containing 10 mM HEPES and dialyzed against the same buffer for 6 hrs. A small precipitate, formed during dialysis, was removed by centrifugation at 10,000 rpm for 10 min. in a Sorval SS34 head. The cleared supernatant was subjected to gel filtration on a 100 x 5-cm column of Sephacryl S200 Superfine (Pharmacia Fine Chemicals, Uppsala, Sweden) and collected fractions were tested for TCGF activity after sterilization by Millipore (0.45 µ) filtration. Fractions showing T cell blast growth-promoting activity were pooled and termed T cell growth factors (TCGF) (9).

Interferon Assay: TCGF preparations were assayed for interferon (IF) -like activity using a conventional cytopathic effect (CPE) inhibition assay.

Confluent monolayers of mouse L cells in flat-bottomed Nunclon microtiter plates (Nunc, Roskilde, Denmark) were first incubated with 0.1-ml volumes of two-fold dilutions of test samples in Dulbecco's modified MEM supplemented with penicillin (100 U/ml), streptomycin (100 µg/ml) and heat-inactivated (56°C for 30 min) fetal calf serum. After 24 hrs the culture medium was replaced with the same volumes of fresh medium containing the challenging Vesicular Stomatitis virus (Indiana strain). Complete CPE developed 24 hrs later, at which time the cultures were washed and stained with a crystal violet-formaldehyde-ethanol solution. As end-points in the titrations those dilutions protecting at least 50% of the L cells were taken. The NIH mouse interferon reference G-002-904-511 was used to calibrate our own interferon standard. The results of the interferon assays were expressed as units of interferon per ml for comparison, although strictly the antiviral activity contained in spleen cell TCGF preparations resembled immune interferon in its pH 2 sensitivity.

C. Results

I. Generation of TCGF-Producing Hybrids

Normal spleen cells treated with a mitogenic dose of Con A show optimal TCGF production after 12-20 hrs. (6,9) at which time the cultured spleen cells can be fused to BW5147, yielding TCGF-producing hybrids (20). Typically, in such fusions some 10-30% of all cultures with hybrid cells were found to produce TCGF after stimulation with Con A. So far no constitutive production of TCGF has been observed when supernatant media of more than 1000 such hybrids conditioned in the absence of Con A were screened. It remains to be elucidated whether TCGF production in the Con A-inducible T cell hybrids is a consequence of derepression of genes located on BW5147 lymphoma chromosomes or chromosomes of the normal spleen cells. The line we have selected here is derived from a fusion experiment in which after 4 weeks of only 28 of 60 cultures originally producing TCGF still had growing, Con A-inducible, TCGF-producing cells. Ten of those cell lines produced high levels of TCGF after Con A stimulation. These 10 lines are described in more detail in the following experiments.

Table 1. Various activities found in Con A supernatants of ten different T cell hybrids formed by fusion of BW5147 with Con A-activated normal CBA/H spleen cells

Hybrid line	TCGF activity (cpm x 10^{-3} per 10^4 Con A blasts) day 2	BCGF activity (cpm x 10^{-3} per 10^4 LPS blasts) day 3	TRF activity (anti-SRC PFC per 10^6 C57BL/6 nu/nu spleen cells) day 5	Natural killer cell-inducing activity	Interferon-like activity (IU/ml)
BI	2.5	9.0	6	+	<5
BII	3.4	18.0	26	+	<5
BIII	4.4	24.0	170	+	<5
BIV	5.5	12.0	74	+	<5
BV	1.6	13.0	100	+	<5
BVI	0.1	9.5	0	-	<5
BVII	2.3	12.0	56	+	<5
BVIII	0.6	12.0	8	+	<5
BIX	2.7	8.0	0	+	<5
BX	1.2	10.0	94	+	<5

II. TCGF-Producing Hybridomas Show Additional Activities

Four months after these 10 hybrid lines were established, cells were placed at 10^6/ml in serum-free medium containing 5 µg/ml Con A. The supernatant media were collected after 24 hrs and tested for various biological activities as shown in Table 1. Most lines retained the capacity to produce TCGF after this time in tissue culture. Most lines also produced B cell blast growth factor (BCGF) as measured by the ability of hybridoma supernatants to sustain the growth of 48-hr LPS blasts. However, there is no obvious correlation between TCGF and BCGF activities. Most of the hybridoma supernatants induced a sheep red blood cell (SRC) -specific antibody response with nude spleen cells of high densities (10^6/ml), i.e. they showed T cell-replacing activities (TRF) (25).

III. Hybridoma TCGF Induces Natural Killer Cells

We have previously reported that conventional TCGF preparations devoid of interferon augment the lytic activity of natural killer (NK) cells against certain teratocarcinoma cell lines lacking H-2 (9). This finding was confirmed with different hybridoma-derived TCGF preparations (Table 1). Thus, all TCGF-containing preparations induced increased lytic activity of anti-Thy 1.2-treated C57BL/6 nude spleen cells against the NK cell-sensitive target cell lines YAC-1 and Nulli-SCC 1. The hybridoma-derived TCGF preparations, which by themselves did not contain interferon, induced high levels of pH 2-sensitive interferon in serum-free nude spleen cell cultures (K.-O.G., to be published). It remains to be established whether the induced interferon causes increased NK activity.

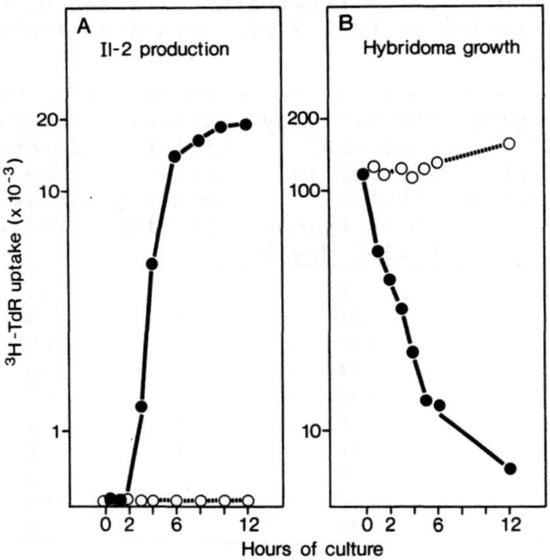

Fig. 1. Con A-induced TCGF production and concomitant inhibition of cellular proliferation of the hybridoma BIII/4. At various times after addition of Con A (10 µg/ml) to the hybridoma, supernatants were collected and assayed for TCGF content by their ability to sustain growth of killer T cells (A). Hybridomas were also pulsed with ^{3}H-TdR (2 µCi/ml) for 1 hr and uptake of the label was taken as a measure of cellular proliferation (B).

IV. TCGF Production by a T Cell Hybridoma Is Accompanied by Arrest in Cellular Proliferation

One T cell hybridoma with high levels of induced TCGF production, BIII, was selected for further analysis. As early as 3-4 hrs after addition of Con A to the hybridoma cells detectable amounts of TCGF appeared in the supernatant medium. As can be seen in Fig. 1A maximum production of TCGF was reached after 10 hrs, at which time more than

90% of the DNA synthetic capacity was inhibited (Fig. 1B). This inverse relationship between TCGF production and replicatory capacity has been observed with many TCGF-producing T cell hybridomas.

V. Properties of Hybridoma BIII-derived TCGF

Upon separation on Sephadex G100 or Sephacryl S200 hybridoma TCGF activity shows an apparent molecular weight of 30,000-40,000, identical to that produced by normal mouse spleen cells after Con A stimulation (Fig. 2). Hybridoma TCGF is stable for more than five months at pH 7.2 at 4°C. It is stable for at least 12 hrs at pH 2-8.7 at 4°C and is totally destroyed from pH 10.0 upwards. Purified hybridoma TCGF is stable at temperatures up to 70°C for 30 min, whereas all activity is destroyed at 85°C. Although preparations of conventional spleen-derived TCGF contain immune interferon (9,26,27) these physical properties make TCGF distinct from immune interferon (27-29). Hybridoma-derived TCGF does not contain any detectable levels of interferon as determined by the conventional CPE inhibition assay (Table 1).

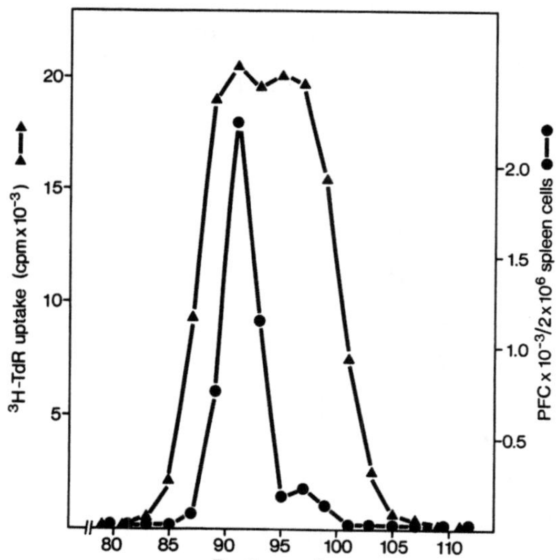

Fig. 2. TCGF and TRF activity coeluting upon gel filtration of hybridoma BIII Con A supernatants. Serum-free supernatant medium (2 liters) from 20-hr Con A-treated (10 μg/ml) hybridoma BIII cultures (10^6 cells/ml) was precipitated by 80% ammonium sulfate and subjected to gel filtration on Sephacryl S200 in a saline 10 mM HEPES buffer as described in Materials and Methods. Fractions containing 12 ml were collected and Millipore-filter sterilized. Twenty μl from collected fractions were added to 200 μl cultures in microtiter plates containing either 10^4 CTLL cells (TCGF activity, ▲—▲) or 4x10^5 C57BL/6 nu/nu spleen cells together with 5x10^5 SRC (TRF activity, ●—●). TCGF activity was measured as ^{3}H-thymidine uptake after 24 hrs of culture and TRF activity as plaque-forming cells on day 5 of culture.

Also, coeluting with the TCGF activity is the activity in BIII hybridoma supernatants which replaces T cells in the SRC-specific antibody response of nude spleen cells at high cell density (TRF activity,

Fig. 2). The TRF activity of hybridoma BIII is adherent cell-dependent and B cell haplotype-nonrestricted (data not shown).

VI. Isolation of $HPRT^-$ Variants From the BIII Hybridoma

10^3, $5x10^3$ and 10^4 BIII cells were plated in DMEM containing 10% fetal calf serum and antibiotics with 0.25% agar on a base layer of 0.5% agar, both supplemented with 20 µg/ml 6-thioguanine (6-TG), and incubated for three weeks. Seven colonies from the 10^4 cells/plate and three from the $5x10^3$ cells/plate were picked with a Pasteur pipette and transferred to 2 ml complete DMEM containing 6-TG and cultured further, after which they were transferred to normal DMEM medium and from there to Iscove's modified DMEM complete medium and their Con A supernatants (5 µg/ml Con A, 24 hrs, $5x10^6$ cells/ml) tested for TCGF activity by restimulation of 48-hr Con A blasts. This assay on three of the clones is illustrated in Fig. 3. BIII/4 and BIII/8 are clearly positive in this restimulation assay, while BIII/9 and seven other clones were completely negative for Con A-induced TCGF production. Phorbol myristate acetate (PMA) (1 ng/ml), known to be able to induce TCGF in other systems, was completely negative in induction of TCGF with all ten clones tested at various cell concentrations. The clone BIII/4 was selected for further studies and used for hybridization experiments, as it carried the $HPRT^-$ marker and was HAT-sensitive. It also was found to carry resistance to $5x10^{-4}$ M oubain (OUA). This could be expected as the BW5147 line used in the fusion to generate BIII/4 carried both the $HPRT^-$ and OUA^r markers (21), OUA being partially dominant.

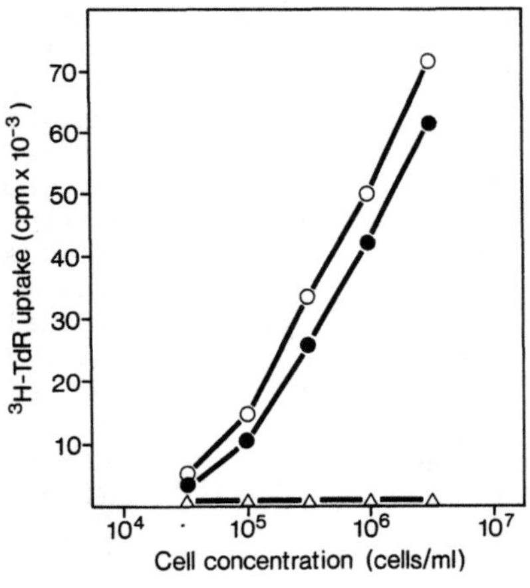

Fig. 3. Three clones of 6-thioguanine-resistant BIII hybridoma cells were treated with 5 µg/ml Con A at different cell densities for 24 hrs. Supernatants were harvested and assayed for TCGF activity on CTLL cells as described in Materials and Methods. Clone BIII/8 (o—o); clone BIII/4 (●—●); clone BIII/9 (△—△).

A large number of different fusions have successfully been carried out using the BIII/4 clone in several hybridization experiments with a variety of long-term T cell lines with specificities for different antigens (18,19). All of these original, antigen-specific T cell lines, with one exception, were inducible for TCGF production by Con A or by antigen in conjunction with the appropriate adherent cells. A large fraction of the resulting T hybridomas have been found inducible for TCGF production with Con A or with antigen plus adherent cells. Since the normal T cell parents in most cases also were capable of TCGF production after induction, it cannot be determined if it was the normal parent or BIII/4 or both genomes that were responsible for this phenotype. So far, TCGF production has been introduced into a TCGF-negative T cell line in only one case by fusion with BIII/4. From a putative alloreactive helper T cell line, noninducible to TCGF production by Con A, a cloned hybridoma was derived by fusion with EL4, which proved also to be noninducible to TCGF production by Con A (30); this hybridoma was fused with BIII/4. Hybrids were found to produce TCGF in response to Con A. In all likelihood BIII/4 had transferred the capacity to produce TCGF to the new T hybrid. This new hybrid surprisingly produced TCGF also constitutively. A further analysis of this particular experiment is presented in an accompanying paper (19,30).

Acknowledgments. The able technical assistance of Denise Richterich, Lotta Haraldson, Annick Peter, and Bente Jensen is gratefully acknowledged. The continuous propagation of T cell lines and T cell hybridomas done by Denise Richterich is especially appreciated. The Basel Institute for Immunology was founded and is supported entirely by F. Hoffmann-La Roche and Co., Ltd. J.A. is supported by the Swedish Medical Research Council.

References

1. Morgan, DA, Ruscetti, FW, Gallo, RC (1976) Selective in vitro growth of T lymphocytes from normal human bone marrows. Science 193: 1007-1008
2. Gillis, S, Smith, KA (1977) Long-term culture of tumor-specific cytotoxic T cells. Nature 268: 154-156
3. Gillis, S, Smith, KA (1977) In vitro generation of tumor-specific cytotoxic lymphocytes. Secondary allogeneic mixed tumor lymphocyte culture of normal murine spleen cells. J Exp Med 146: 468-482
4. Gillis, S, Ferm, MM, Ou, W, Smith, KA (1978) T cell growth factor: Parameters of production and a quantitative microassay for activity. J Immunol 120: 2027-2032
5. Andersson, J, Grönvik, K-O, Larsson, E-L, Coutinho, A (1979) Studies on T lymphocyte activation. I. Requirements for the mitogen-dependent production of T cell growth factors. Eur J Immunol 9: 581-586
6. Rosenstreich, DL, Farrar, JJ, Dougherty, S (1976) Absolute macrophage dependence of T lymphocyte activation by mitogens. J Immunol 116: 131-139
7. Habu, S, Raff, MC (1977) Accessory cell dependence of lectin-induced proliferation of mouse T lymphocytes. Eur J Immunol 7: 451-457

8. Wagner, H, Röllinghoff, M (1978) T-T cell interactions during in vitro cytotoxic allograft responses. I. Soluble products from activated Ly 1+ T cells trigger autonomously antigen-primed Ly 2,3+ T cells to cell proliferation and cytolytic activity. J Exp Med 148: 1523-1531
9. Grönvik, K-O, Andersson, J (1980) The role of T cell growth-stimulating factors in T cell triggering. Immunol Rev 51: 35-59
10. Schreier, MH, Iscove, NN, Tees, R, Aarden, L, von Boehmer, H (1980) Clones of killer and helper T cells: Growth requirement, specificity and retention of function in long-term cultures. Immunol Rev 51: 315-336
11. Shimizu, S, Nonaka, Y, Smith, RT (1980) Mitogen-initiated synthesis and secretion of T cell growth factor(s) by a T-lymphoma cell line. J Exp Med 152: 1436-1441
12. Gillis, S, Scheid, M, Watson, J (1980) Biochemical and biologic characterization of lymphocyte regulatory molecules. III. The isolation and phenotypic characterization of interleukin-2 producing T cell lymphomas. J Immunol 125: 2570-2578
13. Gillis, S, Watson, J (1980) Biochemical and biological characterization of lymphocyte regulatory molecules. V. Identification of an Interleukin-2-producing human leukemia T cell line. J Exp Med 152: 1709-1719
14. Farrar, JJ, Fuller-Farrar, J, Simon, PL, Hilfiker, ML, Stadler, BM, Farrar, WL (1980) Thymoma production of T cell growth factor (Interleukin 2). J Immunol 125: 2555-2558
15. Harwell, L, Skidmore, B, Marrack, P, Kappler, J (1980) Concanavalin A-inducible, Interleukin-2-producing T cell hybridoma. J Exp Med 152: 893-904
16. Schrader, JW, Clark-Lewis, I (1981) T cell hybridoma-derived regulatory factors. I. Production of T cell growth factor following stimulation by concanavalin A. J Immunol 126: 1101-1105
17. Obada, M, Yoshimura, N, Kaieda, T, Yamamura, Y, Kishimoto, T (1981) Establishment and characterization of human T hybrid cells secreting immunoregulatory molecules. Proc Natl Acad Sci USA 78: 7717-7721
18. Melchers, F, Zeuthen, J, Gerhard, W (1982) Influenza virus-specific murine T cell hybridomas which recognize virus hemagglutinin in conjunction with $H\text{-}2^d$ and display helper functions for B cells. (This volume.)
19. Andersson, J, Melchers, F, Zeuthen, J (1982) "Panreactive" T cell lines and T cell hybridomas: Their function in helping B cells. (This volume.)
20. Kappler, JW, Skidmore, B, White, J, Marrack, P (1981) Antigen-inducible H-2-restricted Interleukin-2-producing T cell hybridomas. Lack of independent antigen and H-2 recognition. J Exp Med 153: 1198-1214
21. Trowbridge, IS, Hyman, R, Mazauskas, C (1978) The synthesis and properties of T25 glycoprotein in Thy-1-negative mutant lymphoma cells. Cell 14: 21-32
22. von Boehmer, H, Hengartner, H, Nabholz, M, Lernhardt, W, Schreier, MH, Haas, W (1979) Fine specificity of a continuously growing killer cell clone specific for H-Y antigen. Eur J Immunol 9: 592-597
23. Iscove, NN, Melchers, F (1978) Complete replacement of serum by albumin, transferrin and soybean lipid in cultures of lipopolysaccharide-reactive B lymphocytes. J Exp Med 147: 923-933

24. Goldsby, RA, Osborne, BA, Simpson, E, Herzenberg, LA (1977) Hybrid cell lines with T cell characteristics. Nature 267: 707-708
25. Schimpl, A, Wecker, E (1972) Replacement of T cell function by a T cell product. Nature New Biol 237: 15-16
26. Wheelock, EF (1965) Interferon-like virus inhibitor induced in human leukocytes by phytohemagglutinin. Science 149: 310-311
27. Simon, PL, Farrar, JJ, Kind, PD (1979) Biochemical relationship between murine immune interferon and a killer cell helper factor. J Immunol 122: 127-131
28. Epstein, LB, Steven, DA, Merigan, TC (1972) Selective increase in lymphocyte interferon response in vaccinia antigen after revaccination. Proc Natl Acad Sci USA 69: 2632-2636
29. Fantes, KH (1966) Purification, concentration and physicochemical properties of interferons. In: Finter, NB (ed) Interferons. North-Holland, Amsterdam, p 119
30. Corbel, C, Andersson, J, Melchers, F, Zeuthen, J (1982) A "panre-active" T cell hybridoma which produces TCGF constitutively. (This volume.)

Influenza Virus-Specific Murine T Cell Hybridomas Which Recognize Virus Hemagglutinin in Conjunction with H-2^d and Display Helper Functions for B Cells

F. Melchers, J. Zeuthen, W. Gerhard

A. Introduction

Long-term cultures of murine T cell lines and clones are available which recognize influenza virus in conjunction with self-H-2 antigens and which help B cells to respond to influenza virus and to a "by-stander" antigen such as sheep erythrocytes (SRC) (1,2) by replication and antibody secretion (W.G. and F.M., manuscript in preparation). Continuous culture of these lines requires repeated stimulation by antigen and histocompatible spleen cells or by T cell growth factor (TCGF). Such T cell lines and clones are useful to study the molecular nature of the antigen- and H-2-recognizing T cell receptors and of the molecules which mediate specific interactions between T cells and macrophages, and T cells and B cells (3,4). However, to obtain larger quantities of T cell receptors it would be advantageous to have the expression of such molecules "immortalized" in T cell hybridomas. Furthermore, if such T cell hybridomas would also immortalize the functions of T cell lines to help B cell responses, they could then be used to study cellular cooperations of T cells with macrophages and B cells.

Fusion of antigen-specific T cells and T cell tumor lines, notably the thioguanine-resistant (HPRT-) thymoma line BW5147 (5,6), has led to functionally active T cell hybrids (7-19; see also this volume). Kappler et al. (20) have developed a method for producing T cell hybridomas that respond to stimulation with specific antigen by producing TCGF. In their case, the T cell hybrids responded to antigen only when it was presented by histocompatible irradiated spleen cells as a source of macrophages. We have used a similar approach trying to increase our chances of obtaining a functional T cell hybrid by fusing long-term cell lines in which the majority of all cells are expected to have one selected specificity and one given function.

In one case we fused an influenza virus-specific helper T cell line to the thymoma line BW5147 in the hope of immortalizing the recognition specificity of the normal T cell line as well as its function of producing TCGF upon antigen stimulation. In the other case we fused another influenza virus-specific helper T cell line to the thioguanine-resistant T cell hybridoma BIII/4 (21) which upon stimulation by Con A is induced to TCGF production. This was expected to increase our chances of being able (i) to monitor the antigen specificity of the T cell hybridoma by antigen-induced TCGF production and (ii) to also give the T cell hybridoma the capacity to produce lymphocyte growth factors needed for specific stimulation of replication and maturation to effector functions of T and B cells.

In both fusions we obtained T cell hybridomas that recognized the antigen influenza virus in conjunction with self-H-2 similarly to the normal T cell lines. Both hybridomas recognized virus hemagglutinin in a specific fashion. For one T cell line and the derived hybridoma a crossreaction with allo-H-2 was observed. Both T cell lines and the derived hybridomas responded to antigen by the production of TCGF and by the induction of specific and "bystander" help for B cells.

B. Materials and Methods

Animals. BALB/c (K^d, I-A^d, I-E/C^d, D^d), C57BL/6J (K^b, I-A^b, I-E/C^b, D^b) and C3H/HeJ (K^k, I-A^k, I-E/C^k, D^k) mice, all between four weeks and three months of age, as well as Lewis-strain rats were obtained from the Institut für Biologisch-Medizinische Forschung AG, Füllinsdorf, Switzerland. BALB/c nu/nu mice were obtained from WIGA Versuchstierzucht, Sulzfeld, F.R.G. B.10.RIII (K^r, I-A^r, I-E/C^r, D^r) and C3H/Q (K^q, I-A^q, I-E/C^q, D^q) mice were obtained from OLAC Ltd., Blackthorn, Bicester, Oxon, England.

Viruses and Antigens. The influenza strains PR8 (A/PR/8/34,H_1N_1) and HK (A/HongKong/1,H_3N_2) were grown in the allantoic cavity of 11-day-old chicken eggs and purified as described (22). The hemagglutinating activity of the virus preparations, expressed as hemagglutinating units (HAU), was determined as described by Fazekas de St.Groth and Webster (23). Due to the segmented genome of influenza viruses recombinant viruses can be produced between PR8 and HK by mixed infection and can be genotyped by RNA gel electrophoresis (24). Seven such recombinant viruses of defined genotypes (shown in Figs. 1 and 3) were kindly given to us by Drs. P. Palese and J. Schulman, Mt. Sinai School of Medicine, New York. All viruses were tested for optimal stimulatory capacity for virus-specific helper T cell lines in the presence of irradiated spleen cells as described below, using added virus in the range of 0.03 and 1000 HAU/ml. In all experiments reported here they were used at 10-100 HAU/ml in culture. Purified hemagglutinin, obtained by bromelain treatment of PR8 virus followed by lectin-affinity chromatography (25), was kindly given to us by Dr. M. Lubeck, Wistar Institute, Philadelphia, and used at its optimal stimulatory dose in culture, i.e. 0.5 μg/ml.

In Vitro Establishment and Cloning of Antigen-Specific T Cells. BALB/c mice were injected i.p. with a single dose of 100 HAU of PR8 virus in saline. Twelve days later the spleens of the immunized mice were removed. A single-cell suspension was made by gentle disruption of the spleen through a 200-mesh stainless steel screen with the aid of a rubber piston of a 10-ml plastic syringe. Two to 5 x 10^6 cells/ml were then cultured in a CO_2 incubator at 37°C at 9% CO_2 in air in Iscove's modified Dulbecco's modified Eagle's medium (26) containing human transferrin, bovine serum albumin, lecithin, and 5% freshly collected mouse serum. After five days the cells were washed and readjusted to 1-2x10^5 live cells/ml in the above medium containing 100 HAU/ml PR8 virus per ml and 2x10^6 2200 rad X-irradiated BALB/c spleen cells per ml. The cells were thereafter fed every 3-5 days with fresh medium containing virus and irradiated spleen cells as described above but without mouse serum. Two different T cell lines, vir 1 and vir 2, were established in this fashion.

Prior to in vitro assays T cells from cultures containing virus and irradiated syngeneic spleen cells were transferred at $5x10^4$ cells/ml into medium with 5% TCGF not containing virus and irradiated spleen cells and expanded for 3-5 days. TCGF was obtained by incubation of $5x10^6$/ml rat spleen cells with 5 µg Con A per ml for 24 hr at 37°C in a CO_2 incubator followed by 80% saturation with ammonium sulfate of the supernatant medium. The precipitate was redissolved, dialyzed and passed over a Sephadex G100 column (27).

Cell Fusion. Fusions were performed between the vir 2 T cell line and the BW5147 G.1.4.Oua^r.1 [6-thioguanine- and ouabain-resistant (TG^r and OUA^r)] (6) as well as between the vir 1 T cell line and a TG^r-reselected clone of a BW5147 x T cell hybridoma (designated as BIII/4) that can be induced by Con A to produce TCGF (21).

T cell blasts and HPRT- OUA^r parental cells (approximately 10^7 cells of each type) were mixed and pelleted in a 50-ml plastic tube (Falcon 2070). After removal of the supernatant the pelleted cells were fused using 0.7 ml of 50% (w/vol) polyethylene glycol (PEG) (1500 MW, British Drug Houses, England) and distributed into two 24-well Costar Trays (Costar Tissue Culture Cluster 24, Mass.) in Dulbecco's modified Eagle's medium with 10% heat-inactivated fetal bovine serum and antibiotics as well as 2% TCGF. Hypoxanthine-aminopterin-thymidine (HAT) medium was added next day. Cultures were fed every 3-4 days by exchanging 1 ml of medium. After one week in culture ouabain ($5x10^{-4}$ M) was included in the HAT medium to select against normal T cells growing in the presence of TCGF. After approximately 3 weeks growing cultures were transferred to 5-ml cultures in Falcon flasks in hypoxanthine-thymidine (HT) medium. All cultures grew in the absence of TCGF. These cultures were then expanded in Iscove's modified complete medium (26) in the absence of serum for further characterization after freezing the cells from the original cultures with serum.

Cloning. Frozen T hybridoma cells were thawed, grown for 1-2 days and then cloned by limiting dilution in liquid culture in Dulbecco's modified Eagle's medium with serum containing $5x10^6$ rat thymus cells per ml as "filler" cells (27). The clones were then grown for approximately one month and recloned after this; this was repeated once or twice.

In Vitro Assay of T Cell Proliferation. Five x 10^4 purified T cell blasts were cultured with different concentrations of viruses and $4x10^5$ 2200 rad X-irradiated spleen cells (syngeneic or allogeneic, see Results section) as 200-µl cultures in flat-bottom microculture plates. After 48-72 hrs proliferation of the T cells was measured by uptake of radioactive thymidine. One µCi ^{3}H-thymidine (2 Ci/mMol, The Radiochemical Centre, Amersham, U.K.) was added to cultures for an overnight incubation of 10 hrs and the radioactivity incorporated into DNA measured as described previously (26).

TCGF and BRMF Assay. TCGF was determined in 24-hr supernatants of T cells ($5x10^4$ cells/ml) stimulated with either Con A (5 µg/ml) or with X-irradiated (2200 rads) BALB/c spleen cells ($2x10^6$/ml) and influenza virus (10-100 HAU/ml). The assay was carried out by restimulation of CTLL cells (obtained from Dr. Kendall Smith, Dartmouth Medical School) measured by ^{3}H-thymidine uptake in a 2-hr pulse as described (29). BRMF (B cell replication and maturation factor) was measured by the ability of such supernatants to sustain the growth of 48-hr lipopolysaccharide (LPS) B blasts measured by ^{3}H-thymidine uptake in a 2-hr pulse as described (30).

In Vitro Assay of Helper Function for B Cells. Varied numbers of T cells were cultured in the presence of 10 HAU PR8 virus, 2.5x10^6 sheep erythrocytes (SRC), and 1x10^6 nu/nu BALB/c spleen cells per ml. Helper activity for B cells was assessed on day 5 or 6 by enumerating specific plaque-forming cells (PFC) against SRC added to the cultures as a "bystander" antigen (1,2,30).

B cell responses specific for influenza virus were measured by the determination of virus-specific antibodies detectable in the supernatant media of cultures after 7 days of incubation. A radioimmunoassay described previously (31) was used. Briefly, 20 HAU PR8 virus were adsorbed to each well of polyvinyl chloride microtiter "V" plates (Cooke Laboratory Products, Dynatech Div., Alexandria, Va., U.S.A.), the protein-binding capacity of the wells then saturated by incubation with 3% bovine serum albumin in 0.9% NaCl, and 20-50 µl of culture supernatant added thereafter. After 3-8 hrs the supernatant medium was removed and exchanged with radioiodinated rabbit anti-mouse Ig, detecting IgM and IgG. After 4 hrs the label was removed and the plates washed extensively with 0.9% NaCl. Individual assay wells were counted in a gamma counter after separating the wells from the plate with a heated platinum wire.

C. Results and Discussion

I. Establishment of Influenza Virus-Specific T Cell Lines

Two influenza virus-specific T cell lines with helper function for syngeneic B cells were established in vitro by injection of influenza virus strain PR8 into BALB/c mice followed by repeated restimulation of splenic T cells of the PR8 virus-stimulated mice in vitro by irradiated BALB/c spleen cells and PR8 influenza virus, as described in more detail in the Materials and Methods section. After three months of continuous culture the two T cell lines (vir 1 and vir 2) were examined for their specificity of recognition of different influenza virus antigens by assaying for their capacity to be restimulated by syngeneic or allogeneic irradiated spleen cells and different recombinant strains of influenza virus containing known assortments of the genes of the strain and the original stimulator PR8 virus. The results summarized in Fig. 1 show that vir 1 and vir 2 T cells recognize the hemagglutinin of PR8-strain virus. The detailed specificity of these lines and clones derived from them will be published elsewhere (W.G. and F.M., in preparation).

The results summarized in Fig. 2 demonstrate that the vir 1 T cell line also was restimulated by k-haplotype irradiated spleen cells even in the absence of antigen, i.e. virus. Similar crossreactions of antigen-specific, self-Ia-restricted T cells with allo-H-2 antigens have been described in other systems (32,33).

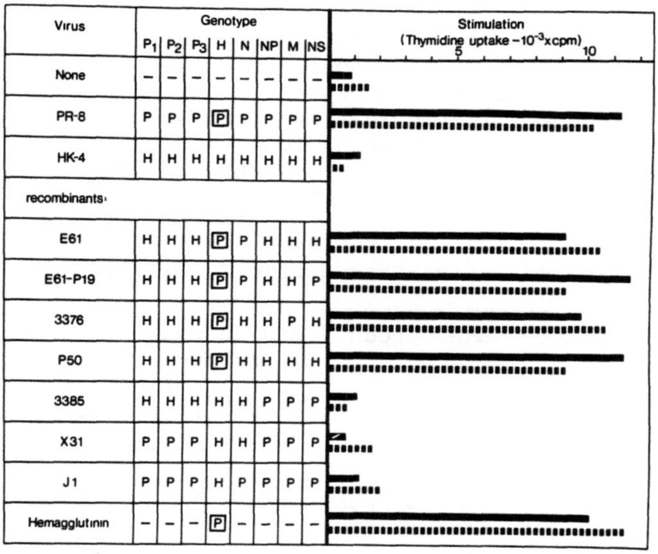

Fig. 1. Antigenic specificity of influenza-specific T cell lines, vir 1 (solid bars) and vir 2 (hatched bars), tested on PR8 [P] and HK [H] influenza virus as well as on a panel of seven recombinants between the PR8 [P] and HK [H] strains of influenza and purified PR8 virus hemagglutinin. TCGF was determined in the 24-hr supernatant media of the T cell lines, stimulated with the different viruses and X-irradiated BALB/c spleen cells, by restimulation of CTLL cells for 48 hrs followed by a 2-hr pulse of ^{3}H-thymidine.

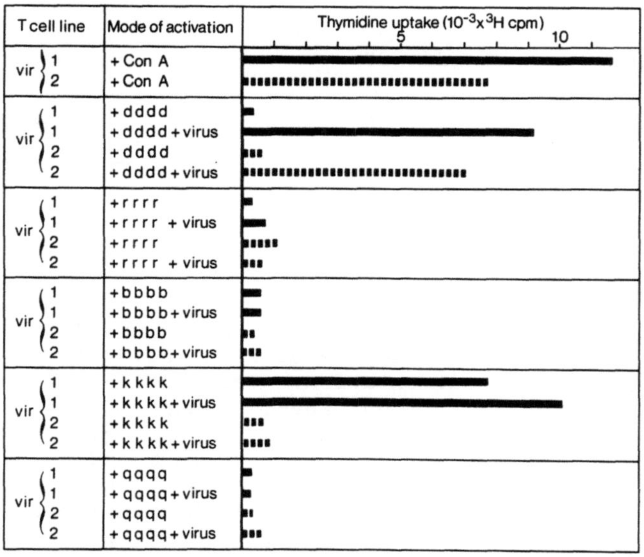

Fig. 2. H-2 restriction of influenza-specific T cell lines, vir 1 (solid bars) and vir 2 (hatched bars), tested with X-irradiated spleen cells from BALB/c (H-2^d), B.10.RIII (H-2^r), C57BL/6J (H-2^b), C3H/HeJ (H-2^k), and C3H/Q (H-2^q) mice and antigen (PR8 virus).

Both T cell lines displayed helper activity for syngeneic B cells in cooperation with syngeneic macrophages and specific antigen, i.e. PR8 virus, as is evident from an influenza virus-specific B cell response (Table 1). "Bystander" antigen-specific, SRC-specific syngeneic B cells, in the presence of syngeneic macrophages, could also be activated, provided SRC were added. Addition of Con A to purified vir 1 as well as vir 2 T cell blasts resulted in the production of TCGF and B cell replication and maturation factor (BRMF).

Table 1. Functional activities of influenza-specific T cell lines*

T cell line	Stimulation	TCGF production	BRMF production	Influenza virus-specific antibody production	"Bystander" SRC-specific antibody production
		^{3}H-TdR incorporated by CTLL cells, cpm/10^4 plated cells	^{3}H-TdR incorporated by LPS B cell blasts, cpm/10^4 plated cells	ng/10^5 plated cells	direct PFC/10^5 plated B cells
BALB/c αPR8 vir 1	+ PR8 virus, ⚡ BALB/c	11,500	9,200	320	350
	+ Con A	32,500	18,500	—	—
	without PR8 virus or without ⚡ BALB/c	<1,000	<3,000	—	—
BALB/c αPR8 vir 2	+ PR8 virus, ⚡ BALB/c	12,100	11,300	180	250
	+ Con A	28,400	26,300	—	—
	without PR8 virus or without ⚡ BALB/c	<1,000	<3,000	—	—

*For details see Materials and Methods.

II. Establishment of T Cell Hybridomas

After three months in culture, the vir 1 T cell line was expanded in TCGF-containing medium for three days and separated from residual irradiated spleen cells by velocity sedimentation. The thus enriched T cell blasts were then fused with the thioguanine-resistant T cell hybridoma BIII/4, which produces TCGF upon stimulation by Con A, but has no known specificity for antigen (21). Twenty-four independent cultures showed growth of hybrid cells, of which 20 could be induced by Con A to produce factors. Of those factor-inducible hybrid T cell cultures, four could also be induced by syngeneic irradiated spleen cells and PR8 virus to produce factors (Table 2). None of the cultures unresponsive to Con A could be induced by antigen or irradiated spleen cells. The original T cell hybridoma cultures lost their capacity to be inducible to factor production. In an attempt to stabilize this function active T hybridomas were cloned and recloned. Table 2 summarizes the results of these cloning experiments. It is evident that the capacity to produce factors upon Con A stimulation is separable from the capacity to produce factors upon stimulation by antigen and irradiated spleen cells. No clones were found which could only be induced by antigen and irradiated spleen cells but not with Con A. After the second recloning experiments T cell hybridomas could be isolated which for more than three months now have retained the function of factor production, inducible by Con A as well as by syngeneic spleen cells and antigen. Clone BIII-vir 1-5.5.2 is one of them (see below).

Table 2. Cloning of an influenza virus-specific, H-2^d-restricted T cell hybridoma* [BIII-vir 1-5.5.2] to produce TCGF upon stimulation by Con A or by antigen and H-2^d adherent cells

	Factor-producing T hybridoma clones (Number of total clones assayed [% of total])	
	Upon Con A stimulation	Upon stimulation with antigen and adherent cells
Original fusion	20 of 24 [83%]	4 of 24 [17%]
First recloning	23 of 24 [97%]	6 of 24 [25%]
Second recloning	12 of 12 [100%]	5 of 12 [45%]

*Derived from the fusion between the vir 1 T cell line and the thioguanine-resistant BIII/4 line (see the Results section).

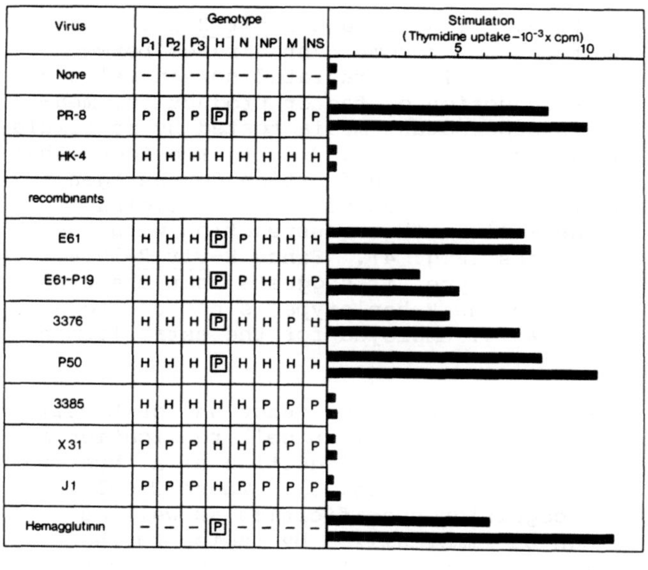

Fig. 3. Antigenic specificity of influenza-specific cloned T cell hybridomas, BIII-vir 1-5.5.2 (top line) and B-vir 2-10.4 (bottom line).

The vir 2 cell line was fused with the thioguanine-resistant thymoma BW5147. By cloning and recloning the T cell hybridoma B-vir 2-10.4 was isolated (see below). It could be induced by antigen and irradiated spleen cells as well as by Con A to produce TCGF and BRMF.

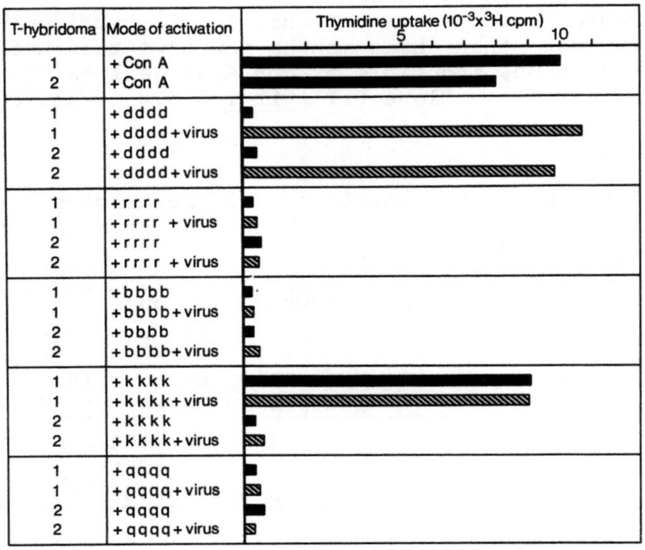

Fig. 4. H-2 restriction of influenza-specific cloned T cell hybridomas, BIII-vir 1-5.5.2 (top line) and B-vir 2-10.4 (bottom line).

III. Specificities and Function of T Cell Hybridomas

The T cell hybridoma clones BIII-vir 1-5.5.2 and B-vir 2-10.4 were then tested for recognition of various recombinant viruses between PR8 and HK (see above) in connection with irradiated spleen cells of different H-2 haplotypes. The induction of factor production, measured as the release of TCGF in supernatant media tested by restimulation of CTLL cells (29), is summarized in Fig. 3. It is evident that the two fusions immortalized the recognition function for antigen - i.e., for the hemagglutinin of the PR8 virus of the two original T cell lines vir 1 and vir 2 - as well as the crossreaction with the k-haplotype of H-2 by the vir 1 lines (Fig. 4). Since B-vir 2-10.4 does not crossreact with the k-haplotype, it suggests that the crossreactivity of B-III-vir 1-5.5.2 with the k-haplotype is not due to a reactivity of the thymoma line BW5147 employed for the establishment of both T hybridomas.

Both T cell hybridomas also retained the helper function for B cells as measured by the induction of virus-specific B cell replication and maturation to Ig secretion as well as by the induction of a "bystander" response of B cells specific for SRC (Fig. 5). These T cell hybridomas demonstrate that recognition specificities as well as functions of normal T cells can be immortalized by fusion to the thymoma line BW5147. It is interesting to note that cloning and recloning of the T cell hybridomas obtained by the fusion of BIII/4 and the vir 1 T cell line yielded clones where the abilities to produce high amounts of TCGF either by stimulation by Con A or by antigen in the presence of syngeneic spleen cells were lost, and some clones where only the capacity to produce factors upon Con A stimulation, but not upon stimulation by virus in the presence of syngeneic cells was preserved (Table 2). This may indicate that a different mechanism, maybe involvement of different receptors, is implicated in such stimulations to factor production.

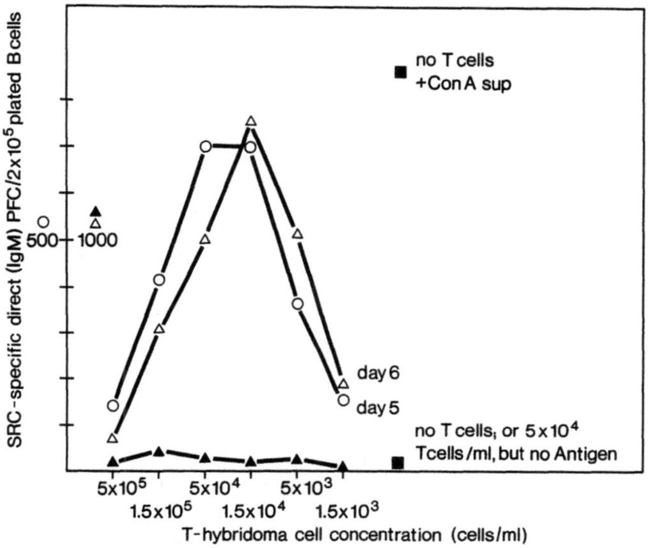

Fig. 5. Helper activity of the B-III-vir 1-5.5.2 hybridoma clone as measured by sheep erythrocyte (SRC)-specific plaque-forming cells (PFC) developing with the "bystander" antigen SRC at day 5 (o) or day 6 (Δ) in culture. Controls include T hybridoma cells in the absence of antigen (PR8 virus) (▲) and cultures without T hybridoma cells with or without Con A supernatant from rat spleen cells (■). The controls are from day 5 of culture.

These T cell hybridomas should be useful tools in studying the molecular basis of antigen and H-2 recognition by T cells and help to clarify the steps involved in cellular cooperations between antigen, T cells, macrophages and B cells.

Acknowledgments. The able technical assistance of Lotta Haraldson, Bente Jensen, Annick Peter, and Denise Richterich is gratefully acknowledged. The continuous propagation of T cell lines and T cell hybridomas done by Denise Richterich is especially appreciated. This work was in part supported by the National Institute of Allergy and Infectious Diseases, Grant No. AI13989, and National Multiple Sclerosis, Grant No. RG-851C6. The Basel Institute for Immunology was founded and is supported entirely by F. Hoffmann-La Roche and Co., Ltd., Basel, Switzerland.

References

1. Hartmann, K-U (1970) Induction of a hemolysin response in vitro. Interaction of cells of bone marrow origin and thymic origin. J Exp Med 132: 1267-1278
2. Schreier, MH, Tees, R (1979) Clonal induction of helper T cells. Conversion of specific signals into non-specific signals. Int Archs Allergy Appl Immunol 61: 227-237
3. Sprent, J (1978) Role of H-2 gene products in the function of T helper cells from normal and chimeric mice measured in vivo. Immunol Rev 42: 108-148
4. Armerding, D, Katz, DH (1974) Activation of T and B lymphocytes in vitro. II. Biological and biochemical properties of allogeneic effect factor (AEF) active in triggering specific B lymphocytes. J Exp Med 140: 19-37
5. Hyman, R, Stallings, V (1974) Complementation patterns of Thy-1 variants and evidence that loss variants "preexist" in the parental population. J Natl Cancer Inst 52: 429-435
6. Trowbridge, IS, Hyman, R, Mazauskas, C (1978) The synthesis and properties of T25 glycoprotein in Thy-1-negative mutant lymphoma cells. Cell 14: 21-32
7. Hämmerling, GJ (1977) T lymphocyte tissue culture lines produced by cell hybridization. Eur J Immunol 7: 743-746
8. Taniguchi, M, Miller, JFAP (1978) Specific suppressive factors produced by hybridomas derived from the fusion of enriched suppressor T cells and a T cell lymphoma cell line. J Exp Med 148: 373-380
9. Watanabe, T, Kimoto, M, Maruyama, S, Kishimoto, T, Yamamura, Y (1978) Regulation of antibody response in different immunoglobulin classes. V. Establishment of a T hybrid cell line secreting IgE class-specific suppressor factor. J Immunol 121: 2113-2117
10. Kontiainen, S, Simpson, E, Bohrer, E, Beverley, PCL, Herzenberg, LA, Fitzpatrick, WC, Vogt, P, Torano, A, McKenzie, IFC, Feldman, M (1978) Hybrid cell lines with T-cell characteristics. Nature 274: 477-480
11. Taniguchi, M, Saito, T, Tada, T (1979) Antigen-specific suppressor factor produced by a transplantable I-J bearing T-cell hybridoma. Nature 278: 555-558
12. Taussig, MJ, Corvalan, JRF, Binns, BR, Holliman, A (1979) Production of an H-2-related suppressor factor by a hybrid T-cell line. Nature 277: 305-308
13. Whitaker, RB, Ruddle, N (1980) Antigen-specific T-cell hybrids. I. Ovalbumin-binding T-cell hybrid. Cell Immunol 55: 56-65
14. Katz, DH, Bechtold, TE, Altman, A (1980) Construction of T cell hybridoma secreting allogeneic effect factor (AEF). J Exp Med 152: 956-968
15. Harwell, L, Skidmore, B, Marrack, P, Kappler, J (1980) Concanavalin-inducible interleukin-2 producing T cell hybridoma. J Exp Med 152: 893-904
16. Eshhar, Z, Apte, RN, Löwy, I, Ben-Neriah, Y, Givol, D, Mozes, E (1980) T-cell hybridomas bearing heavy chain variable region determinants producing (T,G)-A--L - specific factor. Nature 386: 270-272
17. Takatsu, K, Tanaka, K, Tominaga, A, Kumakura, Y, Hamaoka, T (1980) Antigen-induced T cell replacing factor (TFF). III. Establishment of a T cell hybrid clone continuously producing TRF and functional analysis of released TRF. J Immunol 125: 2646-2653

A "Panreactive" T Cell Line and T Cell Hybridoma: Their Function in Helping B Cells

J. Andersson, F. Melchers, J. Zeuthen

A. Introduction

Alloreactive T cells with specificities for Ia antigens encoded by the I-A locus of the major histocompatibility complex of the mouse are known as potent inducers of macrophages and B cells which express the corresponding Ia antigens. This induction in the presence of a "bystander" antigen, such as sheep erythrocytes (SRC), leads to the induction of "bystander" SRC-specific B cell responses (1,2). Self-antigens have also been found to act as such "bystanders", inducing autoantibody production in the corresponding B cells of the reacting host (3). Such interactions of allo-Ia-reactive T cells also lead to the production of soluble mediators which can replace these T cells in their interactions with macrophages and B cells. These mediators were originally called allogeneic effect factors (AEF) (4,5). It was shown that AEF is composed of at least two separable groups of factors: those which specifically recognize Ia antigens on macrophages and B cells (5) and others which act on activated B cells as factors stimulating successive rounds of division and effecting increased maturation to Ig secretion.

We have attempted to establish long-term lines of alloreactive T cells with helper functions for allo-Ia^b-expressing macrophages and B cells in the hope of clarifying the molecular and cellular nature of the interactions between T cells, macrophages and B cells leading to specific antibody and autoantibody production. To immortalize clones of such T cells and to thereby obtain sufficient quantities of cells for structural studies of the molecules involved in these cellular interactions, we have fused these T cells to the T cell leukemia line EL4 and the thymic lymphoma line BW5147 (6,7). Our results show that T cell hybridomas can be obtained which retain the helper function for B cells displayed by the original T cell line. Unexpectedly, however, the line originally intended to be allo-Ia^b-reactive - and all T cell hybridomas derived from it - proved to be "panreactive" with all haplotypes of Ia tested so far.

B. Materials and Methods

Animals. The following inbred strains of mice, all between six weeks and three months of age, were obtained from the Institut für Biologisch-Medizinische Forschung, A.G., Füllinsdorf, Switzerland: C57BL/6J, C57BL/6J nu/nu, BALB/c, DBA/2J, C3H/HeJ, SJL, B10.A(5R), and B10.MBR. Lewis-strain rats were from the same source. A.CA and B10.RIII-strain mice were obtained from OLAC Ltd., Blackthorn, Bicester, Oxon, England. BALB/c mice were from WIGA Versuchstierzucht,

Sulzfeld, F.R.G.

Cell cultures. All cultures were in Iscove's medium, an enriched modification of Dulbecco's modified Eagle's medium (DMEM) containing additional amino acids and vitamins (8), transferrin, albumin, and soybean lipids as replacements for serum, as well as 2-mercaptoethanol ($5x10^{-5}$ M) and Kanamycin (Bio-Cult, Irvine, Scotland). Wherever concanavalin A (Con A) was used to generate active supernatants α-methylmannoside (5 mM, Sigma Chemical Co., St. Louis, U.S.A.) was added to the cultures to neutralize possible inhibitory effects of Con A.

I. Tests for Soluble Factors

Test for T Cell Growth Factor (TCGF). Supernatant media or partially purified solutions were either tested as such (this paper) or serially diluted in culture medium in two-fold steps [see the following paper (14)]. Three cultures of each dilution were set up in Falcon Microtiter II plates (catalog no. 3040, Falcon Labware, Division of Becton-Dickinson Co., Oxnard, California, U.S.A.) or in Costar microtiter plates containing 10^4 cells per 0.2 ml culture of a TCGF-growth-dependent cytotoxic T cell line, CTLL, kindly given to us by Dr. K. Smith, Dartmouth College. TCGF-mediated stimulation of replication of the CTLL cells was assayed at 48 hrs of culture by incorporation of ^{3}H-thymidine (1 μCi per culture of ^{3}H-thymidine, 2 Ci/mmol, The Radioactive Centre, Amersham, England). Qualitative determinations of TCGF activities are expressed as [^{3}H] cpm incorporated into CTLL cells. In quantitative determinations of BRMF activities [see the following paper (14)] we define as 100 relative units of activity the concentration of TCGF giving, under the above conditions, half-maximal values for thymidine uptake with a chosen preparation of enriched rat TCGF. Such a preparation of rat TCGF was made from spleen cells of 50 Lewis-strain rats incubated for 24 hrs at 37°C in a CO_2 incubator at $5x10^6$ cells/ml in medium containing albumin, but no transferrin or lipids, and stimulated by 5 μg/ml Con A (Pharmacia A.B., Uppsala, Sweden). Rat TCGF was then enriched from the supernatant medium by ammonium sulfate precipitation at 80% saturation, dialyzed against 0.9% NaCl, 10 mM HEPES (pH 7.3), insoluble material spun out, and thereafter passed over a 6 x 90-cm Sephadex G100 column in 0.9% NaCl, 10 mM HEPES (pH 7.3). TCGF activity, eluted after albumin, was pooled to yield around 300 ml solution and finally sterilized by filtration through a 0.45-μ Millipore filter.

Test for B Cell Replication and Maturation Factors (BRMF). Supernatant media or partially purified solutions were either tested as such (this paper) or serially diluted in culture medium and cultures set up with 10^4 lipopolysaccharide- (LPS) activated B cell blasts as described above for the TCGF test with CTLL cells. LPS-activated B cell blasts were obtained in a 48-hr stimulation period by incubation of $5x10^5$ C57BL/6J nu/nu spleen cells with 50 μg LPS per ml of culture medium. LPS was a gift of Drs. C. Galanos and O. Lüderitz, Max-Planck-Institut für Immunbiologie, Freiburg, F.R.G. The LPS-activated B blasts were enriched by velocity sedimentation at 1 x g (9) before use in the assay. BRMF-mediated stimulation of replication of the B blasts was assayed at 48 hrs of culture by incorporation of radioactive thymidine as described above for TCGF activity tests. Qualitative determinations of BRMF activities are expressed as [^{3}H] cpm incorporated into CTLL cells. In quantitative determinations [see the following paper (14)] we define as 100 relative units of activity the concentration of

BRMF giving half-maximal values for thymidine uptake with a chosen preparation of enriched rat BRMF. In experiments of this paper where BRMF activity was quantitated in this way the same preparation enriched for rat TCGF (see above) was also chosen as a source for BRMF. It should be noted that absolute activities of rat TCGF and rat BRMF and their relative proportion may vary considerably from preparation to preparation.

II. Tests for Helper Activity of T Cells and T Cell Hybridomas

"Bystander" Activation of SRC-Specific B Cells. Various concentrations of nonirradiated T cells or of X-irradiated [2200-4000 rads (Philips RT305 X-ray machine, Philips Electronic Instruments Inc., Mahwah, New Jersey, U.S.A)] T cell hybridoma cells described in the Results section were incubated with various concentrations of mixtures of 2200-rad X-irradiated and nonirradiated spleen cells of various strains of mice (see Results section). 2.5×10^6 sheep erythrocytes (SRC) were added per ml of culture. Activation of B cells to a SRC-specific "bystander" (2,10) B cell response was determined by the enumeration of SRC-specific, IgM-secreting, direct PFC developing at day 5 or 6 of culture.

III. Establishment of I-Region-Antigen-Reactive Continuous T Cell Lines

B10.MBR-strain (K^b, $I\text{-}A^k$, D^q) splenic cells (5×10^6/ml) were stimulated in a one-way mixed lymphocyte reaction with 2200 rad X-irradiated spleen cells (5×10^6/ml) of B10.A(5R) (K^b, $I\text{-}A^b$, D^d) mice. After one week T cell blasts were enriched by 1 x g velocity sedimentation (9) and restimulated at 2×10^5 blasts/ml with X-irradiated C57BL/6J (K^b, $I\text{-}A^b$, D^b) spleen cells (5×10^6/ml). This restimulation of the B10.MBR-strain T blasts with X-irradiated C57BL/6J spleen cells has been continued in weekly intervals for now more than one year. The T blast concentration at the beginning of a restimulation period could be lowered to $2\text{-}5 \times 10^4$ blasts per ml after the third restimulation. This line is called MBR-anti-5R/B6-dl.

IV. Fusion of MBR-anti-5R/B6-dl with BW5147 and EL4

The parental thymic lymphoma line BW5147G.1.4.Oua^r.1 ($HPRT^-OUA^r$) (11) and the T cell leukemia line EL4.BU.1.Oua^r.1.1 (TK^-OUA^r) (12) were used. 10^7 T cell blasts were mixed with 10^7 drug-resistant tumor cells in a conical tube (Falcon 2070) and filled with GKN salt solution containing 2 mg/ml glucose, spun down and fused with 0.3 ml 50% (w/vol) polyethylene glycol (PEG) (4000 MW, Merck, Cat. No. 9727) added over 1 min. at 37^oC. After 90 seconds 10 ml GKN solution was added slowly over a period of 5 min to dilute out the PEG and incubated for 10 min at 37^oC. Cells were washed with GKN solution and plated out in 1 ml in each of 48 wells in Costar trays (Costar, Cambridge, U.S.A.) in 1 ml DMEM with 10% heat-inactivated fetal bovine serum, antibiotics and 2% TCGF. The next day an additional 1 ml of medium as above supplemented with HAT was added. Cultures were fed twice weekly with medium as above. After one week the medium was supplemented with ouabain (OUA) at 5×10^{-4} M concentration to select against normal T cells growing with TCGF. After approximately 3 weeks growing cultures were transferred to 5-ml cultures in Falcon flasks in HT medium and frozen in liquid nitrogen. All of these grew in the absence of TCGF. They were then expanded in Iscove's complete DMEM with supplements and used for further characterization.

V. Cloning

Frozen T hybridoma cells were thawed, grown for 1-2 days and then cloned by limiting dilution in liquid culture in DMEM with serum containing $5x10^6$ rat thymus cells per ml as "filler" cells. The clones were then grown for approximately one month and recloned.

Table 1. Capacity of the MBR-anti-5R/B6-dl T cell line and a hybrid, E-M-19-I-1, derived from it by hybridization with EL4 to produce TCGF and BRMF in response to X-irradiated C57BL/6J spleen cells or Con A

Concentration of X-irradiated spleen cells (cells/ml)	TCGF and BRMF activities* ^{3}H-thymidine uptake by proliferating CTLL cells or 48-hr LPS B blasts (cpm/10^4 plated cells)			
	TCGF		BRMF	
	MBR-anti-5R/B6-dl	E-M-19-I-1	MBR-anti-5R/B6-dl	E-M-19-I-1
10^6	110	80	2300	2100
$5x10^6$	2500	1400	5500	3900
10^7	2350	1200	4600	5000
none	70	n.t.	2000	n.t.
none + Con A (5µg/ml)	140	110	2100	1800

*TCGF and BRMF were assayed by restimulation of CTLL cells and 48-hr LPS B blasts as described in the Materials and Methods section.

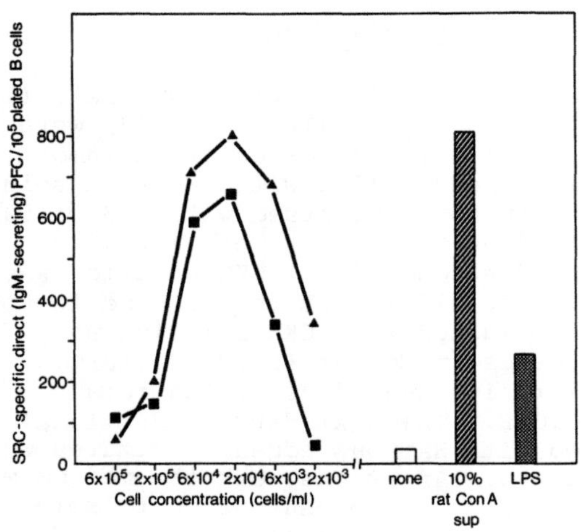

Fig. 1. Activity of the MBR anti-5R/B6-dl T cell line (■—■) and its hybrid, E-M-19-I-1 (▲—▲) in an assay for helper activity for nonirradiated "bystander" C57BL/6J B cells specific for sheep erythrocytes (SRC) with varying numbers of X-irradiated T cells. Controls with no addition of T cells (right side), with or without 10% rat Con A supernatant or LPS. Assay at day 6 in culture.

C. Results and Discussion

A T cell line intended to be specific for antigens of the b haplotype of the I-A region of the murine H-2 complex was established by stimulation of B10.MBR-strain splenic T cells with irradiated B10.A(5R) spleen cells, followed by repeated restimulation of the activated T cell blasts with irradiated C57BL/6J spleen cells. After 3-4 months in continuous culture this T cell line, called MBR anti-5R/B6-dl, was tested for its capacity to produce TCGF or BRMF in response to stimulation by Con A. No measurable activity of either factor resulted from such stimulation (Table 1). On the other hand, when high numbers of irradiated C57BL/6J spleen cells were added as stimulators these T cells induced the production of low levels of TCGF and BRMF (Table 1). Under such conditions they also were helpers for nonirradiated C57BL/6J splenic B cells specific for SRC added as bystander antigen (Figs. 1 and 2).

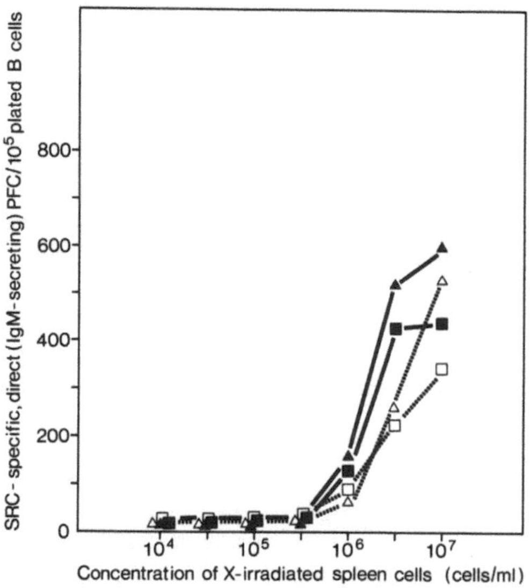

Fig. 2. Activity of the MBR anti-5R/B6-dl T cell line (■,□) and the hybridoma E-M-19-I-1 (▲,△) in an assay for helper activity for nonirradiated "bystander" C57BL/6J B cells specific for sheep erythrocytes (SRC) with varying numbers of X-irradiated C57BL/6J spleen cells. Open symbols refer to plaque-forming cells (PFC) at day 5, closed symbols at day 6.

T cell blasts of this line were then fused with the HAT-sensitive T lymphoma cell lines EL4 or BW5147. From the fusion with EL4 ten independent cultures were established, in which T cell hybrids started to grow. Five of them displayed helper activity for "bystander" B cells specific for SRC. From the fusion with BW5147 four independent cultures with growing T cell hybrids were obtained, of which one showed helper activity for B cells. This helper activity of both types of T cell hybridomas proved to be unstable and was lost after four weeks of culture. T hybridoma cells frozen away early after fusion of a helper-active culture were therefore cloned by limiting dilution. As can be seen from the cumulative data of Table 2 an active clone from the EL4 fusion had to be recloned and one of the active clones resulting from the first recloning experiment recloned again before the helper activity was stabilized, as indicated by the percentage of active clones obtained in the two recloning experiments (Table 2). Thus, T hybridoma clone E-M-19-I-1, selected in this way for helper function, has now been kept in culture for more than six months without losing activity. Stimulation of this T cell hybridoma with Con A did not yield measurable activities of either TCGF or BRMF (Table 1). On the other hand, in analogy to the original T cell line, addition of high numbers of irradiated stimulator C57BL/6J spleen cells led to the production of low levels of TCGF and BRMF induced by the T cell hybridoma cells (Table 1), as well as to helper activity for nonirradiated "bystander" C57BL/6J B cells specific for SRC (Figs. 1 and 2).

Table 2. Stabilization of the helper function for B cells of an alloreactive T cell hybridoma by recloning

	Number of functionally active clones (cultures) within total clones (cultures) assayed [% of total]
Original fusion (cultures) (MBR-anti-5R/B6-dl x EL4)	5 of 10 [50%]
First cloning*	7 of 24 [29%]
First recloning*	11 of 16 [63%]
Second recloning*	19 of 20 [95%]

*By limiting dilutions in liquid media (see Materials and Methods). A functionally active culture was cloned and functionally active clones recloned to obtain a helper-active subclone, E-M-19-I-1.

Table 3. Lack of specificity of the T cell line MBR-anti-5R/B6-dl and the T cell hybridoma, E-M-19-I-1, derived from it as measured by induced TCGF production

Mouse strain used as a source of X-irradiated spleen cells	H-2 haplotype of mouse strain			TCGF production (^{3}H-thymidine uptake, cpm x $10^{-3}/10^4$ plated CTLL cells)	
	K	I-A	D	MBR-anti-5R/B6-dl	E-M-19-I-1
C57BL/6J nu/nu	b	b	b	7.0	5.0
B10.A(5R)	b	b	d	6.5	7.5
C3H/HeJ	k	k	k	3.0	3.5
CBA/J	k	k	k	4.0	8.0
BALB/c nu/nu	d	d	d	9.0	4.0
DBA/2J	d	d	d	8.0	6.0
SJL	s	s	s	4.0	3.0
A.CA	f	f	f	2.5	2.5
B10.RIII	r	r	r	2.0	7.5
B10.MBR	b	k	q	6.0	7.0
None				0.2	0.2

The specificity of the T cell line and the T cell hybridoma, E-M-19-I-1, derived from it for the b haplotype of antigens encoded in the I-A region of the H-2 complex was tested with antigen-dependent TCGF production and "bystander" B cell activation. Unexpectedly, irradiated spleen cells from all haplotypes tested, including the parental MBR strain, were capable at high concentrations of stimulating the T cells to their functions, i.e. to effect TCGF production (Table 3) and "bystander" help for B cells (Table 4). These T cells, therefore, either recognize an antigen of the H-2 complex common to all haplotypes tested or recognize an unknown antigen present on spleen cells of all mice tested. For the latter possibility it should be mentioned that such an antigen could be similar to the antigens encoded on chromosome 17 of the mouse, for which H-2-unrestricted recognition by T cells is possible (13). The possibility that an H-2 antigenic determinant common to all H-2 haplotypes is recognized by these T cells raises the interesting question of how frequent such T cells are within the repertoire. Alternatively, these T cells may not recognize any specific structure but release nonspecific activators for TCGF and BRMF production as well as for B cell help. For practical purposes these "panreactive" T cells serve as useful tools in T cell-dependent stimulation of the production of lymphopoietic and hemopoietic factors in cells of many different H-2 haplotypes, as well as in the T cell-dependent "bystander" activation of Ig-producing B cells of different haplotypes to different "bystander" B cell antigens. One of their uses is reported in a subsequent publication (14).

Table 4. Lack of specificity of the T cell line and the T cell hybridoma derived from it as measured by induction of SRC-specific "bystander" B cells

Mouse strain used as a source of X-irradiated spleen cells	Helper activity for B cells* (SRC-specific direct PFC/10^5 plated cells)	
	MBR-anti-5R/B6-dl	E-M-19-I-1
C57BL/6J nu/nu	550	500
C3H/HeJ	350	700
BALB/c nu/nu	400	350
SJL	250	200
B10.MBR	400	500
None	<10	<10

*Determined in cultures containing $5x10^4$ T cells or 2200-rad irradiated E-M-19-I-1 cells, $2.5x10^6$ SRC and $5x10^6$ nu/nu or anti-Thy 1 + C'-treated spleen cells, all per ml as described in the Materials and Methods section.

Acknowledgments. The able technical assistance of Denise Richterich, Lotta Haraldson, Annick Peter, and Bente Jensen is gratefully acknowledged. The continuous propagation of T cell lines and T cell hybridomas done by Denise Richterich is especially appreciated. The Basel Institute for Immunology was founded and is supported entirely by F. Hoffmann-La Roche and Co., Ltd. J.A. is supported by the Swedish Medical Research Council.

References

1. Hartmann, K-U (1970) Induction of a hemolysin response in vitro.

Interaction of cells of bone-marrow origin and thymic origin. J Exp Med 132: 1267-1278
2. Schreier, MH, Tees, R (1979) Clonal induction of helper T cells. Conversion of specific signals into nonspecific signals. Int Archs Allergy Appl Immunol 61: 227-237
3. Gleichmann, E, Gleichmann, H, Wilke, W (1976) Autoimmunization and lymphomagenesis in parent → F_1 combinations differing at the major histocompatibility complex: Model for spontaneous disease caused by altered self-antigens. Transpl Rev 31: 157-224
4. Armerding, D, Katz, DH (1974) Activation of T and B lymphocytes in vitro. II. Biological and biochemical properties of allogeneic effect factor (AEF) active in triggering specific B lymphocytes. J Exp Med 140: 19-37
5. Delovitch, TL, Watson, J, Battistella, R, Harris, JF, Shaw, J, Paetkau, V (1981) In vitro analysis of allogeneic lymphocyte interaction. V. Identification and characterization of two components of allogeneic effect factor, one of which displays H-2-restricted helper activity and the other, T cell-growth factor activity. J Exp Med 153: 107-128
6. Kontiainen, S, Simpson, E, Bohrer, E, Beverley, PCL, Herzenberg, LA, Fitzpatrick, WC, Vogt, P, Torano, A, McKenzie, IFC, Feldman, M (1978) Hybrid cell lines with T-cell characteristics. Nature 274: 477-480
7. Hämmerling, GJ (1977) T lymphocyte tissue culture lines produced by cell hybridization. Eur J Immunol 7: 743-746
8. Iscove, NN, Melchers, F (1978) Complete replacement of serum by albumin, transferrin, and soybean lipid in cultures of lipopolysaccharide-reactive B lymphocytes. J Exp Med 147: 923-933
9. Miller, RG, Phillips, RA (1969) Separation of cells by velocity sedimentation. J Cell Physiol 73: 191-201
10. Andersson, J, Schreier, MH, Melchers, F (1980) T-cell-dependent B-cell stimulation is H-2-restricted and antigen-dependent only at the resting B-cell level. Proc Natl Acad Sci USA 77: 1612-1616
11. Trowbridge, IS, Hyman, R, Mazauskas, C (1978) The synthesis and properties of T25 glycoprotein in Thy-1-negative mutant lymphoma cells. Cell 14: 21-32
12. Hyman, R, Stallings, V (1976) Characterization of a TL^- variant of a homozygous TL^+ mouse lymphoma. Immunogenetics 3: 75-84
13. Fischer Lindahl, K (1979) Unrestricted killer cells recognize an antigen controlled by a gene linked to Tla. Immunogenetics 8: 71-76
14. Corbel, C, Andersson, J, Melchers, F, Zeuthen, J (1982) A "panreactive" T cell hybridoma which produces TCGF constitutively. This volume.

A "Panreactive" T Cell Hybridoma Which Produces TCGF Constitutively

C. Corbel, J. Andersson, F. Melchers, J. Zeuthen

A. Introduction

We have isolated a T cell line which shows reactivities with irradiated spleen cells of any H-2 haplotype, including self. This "panreactive" T cell line was fused to EL4 and the stable "panreactive" T hybridoma E-M-19-I-1 was shown to help in a "bystander" antibody response to sheep red blood cells (SRC) (1). This "panreactivity" is shown in this paper to be useful as it induces "constitutive" TCGF production in mixtures as well as in fusions with a TCGF-production-inducible T cell hybridoma, BIII/4 (2).

B. Materials and Methods

Fusion of E-M-19-I-1 with BIII/4. The Con A-inducible TCGF-producing hybridoma HPRT$^-$ variant BIII/4 is described elsewhere (2). It is oubain-resistant (OUA^r) as a consequence of being derived from a fusion of spleen cells with BW5147 (HPRT$^-$ OUA^r) as OUA^r is dominant. The E-M-19-I-1 reclone, derived as described in ref. 1, was selected for resistance to bromodeoxyuridine (BU^r) and was found to be HAT-sensitive (i.e., TK$^-$). Therefore, hybrids with BIII/4 could be isolated on HAT. The fusion was performed slightly differently than described in ref. 1, using normal serum-free DMEM for washing or dilution of cells fused with PEG (1500 MW, British Drug Houses, U.K.). Selection of hybrids was carried out similar to the way described previously (1), except that TCGF was not included in the medium. BRMF, TCGF and "bystander" helper activities were tested as described in the preceding articles (1,2). Cloning of T hybridomas was done as described in the preceding article (1).

C. Results and Discussion

The capacity of the T cell hybridoma E-M-19-I-1 to apparently recognize an ubiquitous structure led us to probe for the expression of such structures on the thioguanine-resistant T hybridoma BIII/4, which upon stimulation with Con A is inducible to TCGF production (2). The results presented in Table 1 show that mixed cultures of E-M-19-I-1 and BIII/4 led to the production of TCGF. This suggests that E-M-19-I-1 can induce BIII/4.

Fusion of the two T cell hybridomas with each other was consequently thought to lead to T cell hybridomas that constitutively produce

TCGF. The results of such a fusion, presented in Tables 1 and 2, indeed show that such T cell hybridomas with constitutive TCGF production could be obtained in several independent fusion events of one fusion experiment. It is this high frequency of independent occurrence of T hybrids producing TCGF constitutively (Table 2) that makes us think that a continuous self-stimulation, pictured in two alternative models in Fig. 1, may explain the apparently constitutive production of TCGF. Constitutive production of TCGF by a T cell hybridoma generated by fusion of the thymic lymphoma BW5147 with a TCGF-producing tumor cell line was reported by Stull and Gillis (3). One of the resulting T cell hybridoma cultures (number 20) cloned to yield the T cell hybridoma BIII-(E-M-19-I-1)20.21 was studied further.

Table 1. Induction of TCGF production by mixtures or hybrids of the T cell hybridoma BIII/4 and E-M-19-I-1

	TCGF production [^{3}H-thymidine uptake by proliferating CTLL cells (cpm/10^4 plated cells)]		
Cell line ($5x10^5$ cells/ml)	No additions	+ Con A	+ E-M-19-I-1
BIII/4	- (80)	+++ (100,000)	++ (26,000)
E-M-19-I-1	- (110)	- (220)	- (110)
BIII(E-M-19-I-1)20.21	+ (5,000)	++ (18,000)	not done

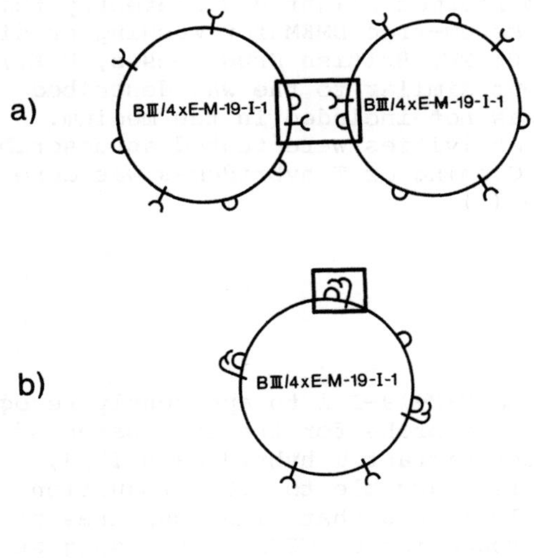

Fig. 1. Two models for constitutive factor production by a "panreactive" T hybridoma.

Table 2. Constitutive TCGF production in different original fusion events and in clones obtained from two of them by a fusion between the T hybridomas BIII/4 and E-M-19-I-1

T hybridomas[a] (BIII/4 x E-M-19-I-1)	TCGF production [^{3}H-thymidine incorporated by CTLL cells (cpm/10^4 plated cells)][b]	
Original fusion	- Con A	+ Con A
Well 5	22,500	32,000
Well 8	32,000	38,000
Well 11	650	12,000
Well 12	12,000	7,000
First cloning		
Clone 15-2	800	2,000
Clone 20-1	32,000	39,000
Clone 20-3	38,000	50,000
Clone 20-4	18,000	42,000
Clone 20-20	1,000	3,000
Clone 20-21	12,000	35,000
Controls with no T hybridoma cells		
Medium	500	500
+ 10% rat TCGF	35,000	34,000

a) Conditioned medium of $5x10^5$ T hybridoma cells/ml, irradiated with 2200 rads, incubated for 24 hrs at 37^oC in a CO_2 incubator.

b) For details see the Materials and Methods section of the preceding article (1). Data for - Con A and + Con A (5 µg/ml) obtained in separate experiments.

The constitutive TCGF production of the T cell hybridoma is dependent on cell concentration (Table 3). Thus, at $5x10^4$ cells per ml (see experiments in Fig. 2) TCGF production is not detectable. At high concentrations of T hybridoma cells high concentrations of conditioned media are inhibitory for CTLL proliferation (Table 3). This makes it impossible to simply vary the concentration of T hybridoma cells in culture and measure the quantity of TCGF produced to decide between model B of Fig. 1 in which one cell induces itself (predicting linear increases of TCGF production with increasing cell concentration) and model A of Fig. 1 in which different cells of one clone induce each other (predicting more than linear increases). At low concentrations of cells (i.e., $5x10^4$/ml) this T cell hybridoma is inducible by Con A to produce increased levels of TCGF (Table 1). It can also be stimulated by irradiated C57BL/6J spleen cells to produce increased quantities of TCGF and BRMF (Fig. 2A). The difference in the quantities of TCGF and BRMF produced by E-M-19-I-1 and by BIII-(E-M-19-I-1)20.21 should be noted.

The T cell hybridoma furthermore acts as helper for "bystander" B cells specific for SRC (Fig. 2B). Help is dependent on the concentration of T hybridoma cells and in fact inhibited a high cell concentration (1). It remains to be elucidated whether the inhibitory effects on CTLL proliferation and on "bystander" B cell responses are the same or different activities. Again, "bystander" help is detectable with spleen cells of various haplotypes, as would be expected from the "panreactivity" of the original T cell line and the T cell hybridoma, E-M-19-I-1, derived from it (1).

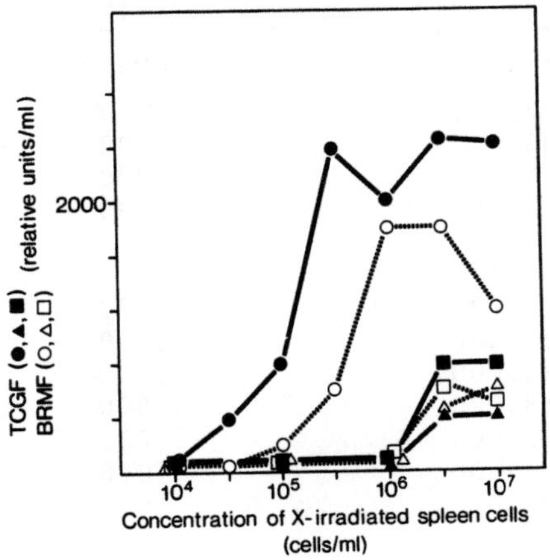

Fig. 2A. TCGF and BRMF production in response to X-irradiated C57BL/6J spleen cells. 24-hr supernatants were tested at 25% for CTLL or LPS blast-proliferative activity.

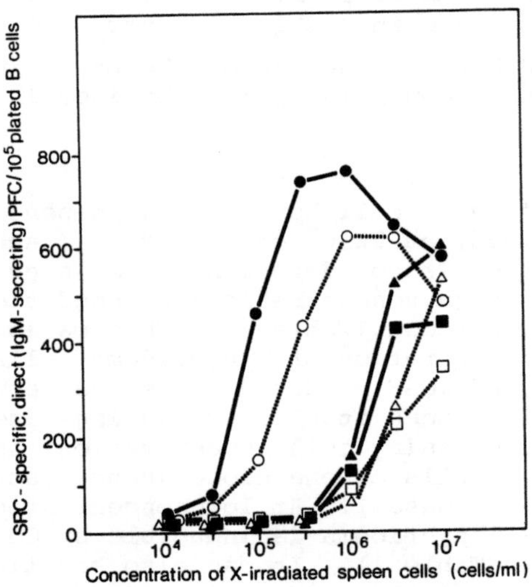

Fig. 2B. Helper activity for nonirradiated "bystander" C57BL/6J B cells specific for sheep erythrocytes (SRC) with varying numbers of X-irradiated C57BL/6J spleen cells. Open symbols refer to plaque-forming cells (PFC) at day 5, closed symbols at day 6.

Fig. 2. Activity of the MBR-anti-5R/B6-dl T cell line (■,□) (see ref. 1) and the two hybridomas, E-M-19-I-1 (▲,△) (see ref. 1) and BIII-(E-M-19-I-1)20.21 (●,○), all tested at $5x10^4$ cells/ml with the indicated numbers of irradiated spleen cells.

Table 3. TCGF inhibitory activity affecting proliferation of CTLL cells produced by different concentrations of supernatant media conditioned for 24 hrs by different concentrations of BIII-(E-M-19-I-1)20.21 T hybridoma cells

Concentration of supernatant medium (%)	Concentration of T hybridoma BIII-(E-M-19-I-1)20.21 cells		
	5×10^5 (cells/ml)	1.5×10^6 (cells/ml)	5×10^6 (cells/ml)
	^{3}H-thymidine uptake by CTLL cells (cpm per 10^4 plated cells)		
50	5,400	530	50
15	6,800	11,800	80
5	850	1,400	15,200
1.5	150	320	36,200

The added capacity of BIII-(E-M-19-I-1)20.21 T hybridoma cells to produce high levels of factors certainly increases their capacity to act as helpers for B cells. The capacity to help in a bystander B cell response appears in fact correlated with the capacity to produce high levels of factors, especially at low concentrations of irradiated stimulator spleen cells, as can be seen in a comparison of the data in Figs. 2A and B. Factor production appears to be the most important bottle-neck in the cooperation between T cells, macrophages and B cells.

Table 4. Stabilization of the capacities of T hybridoma cells by recloning to produce TCGF either by stimulation with concanavalin A or by stimulation with irradiated C57BL/6 spleen cells

	Factor-producing T hybridoma clones (Number in total clones assayed) [% of total]	
	upon stimulation with concanavalin A	upon stimulation with irradiated C57BL/6 spleen cells
Original fusion (clones)	16 of 24 [66%]	5 of 24 [21%]
First recloning	23 of 24 [97%]	7 of 24 [30%]
Second recloning	12 of 12 [100%]	9 of 12 [75%]

In this context it is interesting to note that induction of TCGF production by Con A or by lower concentrations of irradiated spleen cells ($1\text{-}5 \times 10^5$/ml, see Fig. 2A), both with low concentrations of T hybridoma cells (5×10^4/ml) of the BIII-(E-M-19-I-1) fusion are qualities which are separable in different clones and reclones of the original fusion, evident in Table 4, in the frequencies of negative, single-positive and double-positive clones. Separation of Con A-induced and antigen plus irradiated spleen cell-induced TCGF production has also been oberved in fusions of BIII/4 with influenza virus-specific T cell lines (4). All these results argue for different mechanisms, possibly different receptors, involved in such stimulations to factor production. Unfortunately, questions concerned with the molecular basis of I-region-compatible interactions between T cells and macrophages (5) and between T cells and B cells (6) will most certainly not be clarified by the T cell hybridomas reported in this paper, since our T cells do not distinguish between different haplotypes of H-2 antigens.

Acknowledgments. The able technical assistance of Denise Richterich, Lotta Haraldson, Annick Peter, and Bente Jensen is gratefully acknowledged. The continuous propagation of T cell lines and T cell hybridomas done by Denise Richterich is especially appreciated. The Basel Institute for Immunology was founded and is supported entirely by F. Hoffmann-La Roche and Co., Ltd. J.A. was supported by the Swedish Medical Research Council.

References

1. Andersson, J, Melchers, F, Zeuthen, J (1982) A "Panreactive" T cell line and T cell hybridoma: Their function in helping B cells. (This volume.)
2. Grönvik, K-O, Andersson, J, Melchers, F, Zeuthen, J (1982) A stable TCGF-producing T cell hybridoma and its thioguanine-resistant variant suitable as a tool for the construction of new functional T hybridomas. (This volume.)
3. Stull, D, Gillis, S (1981) Constitutive production of interleukin-2 activity by a T cell hybridoma. J Immunol 126: 1680-1683
4. Melchers, F, Zeuthen, J, Gerhard, W (1982) Influenza virus-specific murine T cell hybridomas which recognize virus hemagglutinin in conjunction with $H-2^d$ and display helper functions for B cells. (This volume.)
5. Rosenthal, AS, Schevach, EM (1973) Function of macrophages in antigen recognition by guinea pig T lymphocytes. I. Requirements for histocompatible macrophages and lymphocytes. J Exp Med 138: 1194-1212
6. Sprent, J (1978) Role of H-2 gene products in the function of T helper cells from normal and chimeric mice in vivo. Immunol Rev 42: 108-148

Production of Antigen-Nonspecific Immunoregulatory Lymphokines by T Cell Hybridomas

A. Altman, R.D. Schreiber, D.H. Katz

A. Introduction

A large number of non-immunoglobulin soluble mediators or lymphokines (LK) with immunoregulatory and other biological activities have been studied over the last 12 years. The majority of these LK are products of T cells activated by antigens or polyclonal mitogens. Studies on T cell-derived LK have been severely hampered by the limited amounts of LK available from conventional cultures and the heterogeneity of the cell populations producing these LK which, in turn, resulted in heterogeneous mixtures of many distinct LK. As a result of these limitations, it has been difficult to delineate the exact chemical nature of LK, the structure-function relationship among various biological activities and their biological significance.

In order to overcome these obstacles, a substantial amount of effort has been devoted in the past 4-5 years toward identifying cloned, permanent lines of LK-secreting T cells. The three methods most often used are a) screening of neoplastic T cell lines for LK secretion, particularly of the antigen-nonspecific type, b) establishment of growth factor-dependent, "normal" T cell lines, and c) construction of T cell hybridomas, almost exclusively by fusion with the 8-azaguanine-resistant variant of the murine T cell lymphoma, BW5147. The latter two methods have been used to obtain T cell lines secreting either antigen-specific or -nonspecific molecules (reviewed in 1-3).

Our laboratory has studied for the last 8 years a few LK present in supernatants of short-term secondary mixed leukocyte cultures of _in vivo_ alloantigen-activated murine T cells. An Ia^+ mediator, termed allogeneic effect factor or AEF and capable of replacing T helper cells in _in vitro_ antibody responses to T cell-dependent antigens has been identified in such supernatants (4-7). More recent studies revealed that the same culture supernatants can induce naive T lymphocytes in the absence of additional exogenous stimuli to proliferate and differentiate into Lyt-1^+23^- or Lyt-2^+ self-reactive T cells capable of proliferating in a secondary syngeneic mixed leukocyte reaction or lysing major histocompatibility complex-identical target cells, respectively (8,9). These latter activities are distinct from T cell growth factor (TCGF; IL-2), also present in AEF supernatants, which provides a proliferative stimulus for fully mature, previously-activated T cells (10,11). In addition, AEF supernatants were found to promote the differentiation of bone marrow cells in long-term culture (12) and to contain immune interferon or IFNγ (13) and macrophage-activating factor or MAF (unpublished observations).

In order to define in precise terms the relationship among various LK activities present in AEF supernatants and to obtain large quantities of biologically-active supernatants for biochemical analysis, we have decided to construct LK-secreting T cell hybridoma lines. T cells which were alloactivated according to the method used for the preparation of conventional AEF were used as a source of LK-producing cells and fused with the BW5147 T lymphoma line (14). These fusions resulted in the establishment of several cloned T cell hybridoma lines which secrete, either constitutively or following mitogen stimulation, a number of distinct LK activities (14-17). By cloning these T cell hybridoma lines and studying their soluble products it has been possible to dissociate a number of biological activities and to demonstrate non-coordinate production of distinct LK activities by different clones of the same parental hybridoma.

B. Materials and Methods

1. Mice

All inbred mice employed in these studies were obtained from the Scripps Clinic and Research Foundation mouse breeding colony, the Jackson Laboratory, Bar Harbor, ME, or from Simonsen Laboratories, Gilroy, CA.

2. Preparation of Alloactivated T Cells (ATC)

ATC were prepared as described previously (6,9,14). One x 10^8 thymocytes from young (4-6 weeks old) DBA/2 donors were mixed with an equal number of irradiated (1500 rad) spleen cells from (C3H x DBA/2) F_1 hybrid ($C3D2F_1$) mice and injected i.v. into irradiated (650 rad) DBA/2 recipient mice. The ATC recovered from the spleens of these mice 7 days later (1-1.5 x 10^7/spleen) were co-cultured with fresh irradiated $C3D2F_1$ stimulator spleen cells. Responder and stimulator cells were cultured for 18-24 hr. at 1 x 10^7/ml each in serum-free RPMI-1640 supplemented with 1 x 10^{-4} M 2-mercaptoethanol (2-ME). Cultures were maintained in a Mishell-Dutton gas mixture on a rocking platform. The cells were harvested 18 hr. later and separated on a Ficoll-Hypaque gradient. Cells recovered from the interface after centrifugation at 600 x g for 20 min. were predominantly blast-like in nature and were used for hybridization after being washed three times.

3. Construction of T Cell Hybridomas

The parental cell line used for hybridization was the 8-azaguanine-resistant derivative of the AKR-derived, BW5147 T lymphoma line obtained from Dr. Robert Hyman of The Salk Institute, La Jolla, CA. Cell fusion was performed according to Gefter *et al*. (18) with some modifications described before (14). Briefly, parental tumor cells were washed twice in Dulbecco's modified Eagle's medium (DMEM) and 2 x 10^6 tumor cells were mixed with 1 x 10^7 activated T cell blasts. The cell suspension was centrifuged and all medium was carefully removed by aspiration. The cell pellet was then resuspended in 0.2 ml of 30 percent polyethylene glycol (PEG) 1,000 in DMEM. Cells remained in this solution for a total of 8 min., during which time they were centrifuged for 6 min. at 1,000 rpm (ambient temperature). At the end of 8 min., 5 ml of DMEM were added to the tube and the cells were washed twice in this medium. They were resuspended in DMEM supplemented with 10 percent fetal calf serum (FCS) and incubated overnight at 37 C in a 10 percent CO_2-in-air atmosphere. The following day, the cells were pelleted by centrifugation and resuspended in 30 ml of FCS-supplemented DMEM containing hypoxanthine (1×10^{-4} M), aminopterin (4×10^{-6} M) and thymidine (1.6×10^{-5} M), and were distributed into individual wells of 96 well flat-bottom culture plates (Falcon, Oxnard, CA). Cultures were fed after one week by the addition of one drop of this same medium without aminopterin. Growing hybrids were transferred to 24 well tissue culture plates (Linbro, Hamden, CT) and screened for LK production when reaching confluency (see below).

4. Cloning

T cell hybridomas were cloned by limiting dilution in 200 µl flat-bottom microtiter cultures. Each well contained an average of 0.5 hybridoma cells and 2×10^5 2500 rad-irradiated BALB/c thymocytes. Growth was observed within 7-10 days and the contents of wells with growing cells were expanded by transfer into 24 well tissue culture plates and subsequently to 25 cm^2 tissue culture flasks (Falcon Plastics). One of the clones (see below) was subjected to an additional subcloning cycle.

5. Cell Surface Marker Analysis

T cell hybridoma clones were analyzed for the expression of Thy-1 and H-2 alloantigens using the fluorescence-activated cell sorter (FACS II; Becton Dickinson FACS Systems, Mountain View, CA). Five x 10^5 cells in 200 µl DMEM + 1 percent FCS were treated with a 1:200 dilution of the F7D5e7 hybridoma clone (19) -derived anti-Thy-1.2 antibodies or a 1:20 dilution of conventional alloantisera which included: AKR/J anti-AKR/Cum (anti-Thy-1.1), C57BL/6 anti-C3H (anti-$H\text{-}2^k$) and B10.BR anti-B10.D2 (anti-$H\text{-}2^d$). Following a 30 min. incubation on ice, the cells were washed three times and resuspended in 100 µl medium containing a 1:50 dilution of fluorescein-conjugated rabbit anti-mouse immunoglobulin. Controls were treated directly with the fluorescent reagent without the first-step antibodies. After additional 30 min. on ice, the cells were washed three times, resuspended in 0.5 ml medium and analyzed on the FACS II.

6. Screening for LK Production

Hybridoma cells were grown in 24 well tissue culture plates in complete DMEM + 5 percent FCS. When the cultures reached saturation density ($\sim 1 \times 10^6$/ml), the contents of each well were split equally between two new wells and each new well was supplemented with 1 ml fresh medium without or with different mitogen concentrations. The mitogens used were concanavalin A (Con A; Miles-Yeda, Rehovot, Israel) or purified phytohemagglutinin (PHA; Wellcome Research Laboratories, Beckenham, England). Alternatively, cells were harvested, counted, resuspended in fresh medium at different concentrations ($0.5\text{-}5 \times 10^6$/ml) and cultured in the absence or presence of mitogenic stimuli. Supernatants were harvested by centrifugation after 24 hr. and tested for various biological activities (see below).

7. Biological Assays

Crude or partially-purified T cell hybridoma-derived LK preparations were tested in a variety of biological assays. Supernatants which contained Con A were supplemented, where necessary, with 20 mg/ml α-methyl-D-mannoside to block the mitogenic effect of Con A.

I. In Vitro Antibody Response System

Conventional or hybridoma-derived LK preparations were tested for their ability to reconstitute primary *in vitro* antibody responses against SRBC or TNP-SRBC of DBA/2 spleen cells treated twice with monoclonal (clone F7D5e7) anti-Thy-1.2 antibody plus complement (C) or once each with a conventional AKR/J anti-C3H/St serum followed by monoclonal anti-Thy-1.2 antibody plus C. Direct plaque-forming cells (PFC) were enumerated after 5 days. The methods for establishing such cultures and determining PFC responses were described previously (4,6).

II. TCGF Assay

Serial dilutions of test supernatants were plated in 96 well flat-bottom culture plates in 100 µl volume. To each well were added 1×10^4 washed cells (in 100 µl) of the TCGF-dependent CTL line, CTLL-2 (20) (kindly obtained from Dr. J. Watson); they were maintained by two weekly passages at initial cell densities of $5\text{-}10 \times 10^3$/ml in culture medium supplemented with 25 percent v/v rat spleen-cell derived crude TCGF preparation. The medium used in this assay was RPMI-1640 supplemented with 5 percent FCS, 5×10^{-5} M 2-ME and 10 mM HEPES buffer. Cultures were pulsed for the final 4 hr. of a 24 hr. culture with 1 µCi tritiated thymidine (^{3}HTdR, 5 Ci/mmole, Amersham Corp., Arlington Heights, IL), harvested on a Titertek cell harvester (Flow Labs, Rockville, MD) and counted in a liquid scintillation counter. TCGF titers were determined in a probit analysis (21) by determining the dilution which gave 50 percent of the maximal ^{3}HTdR uptake induced by a standard rat-derived crude TCGF preparation. The standard preparation was arbitrarily assigned an activity of 10 units/ml.

III. Mitogenic Activity

Supernatants were tested either for their direct mitogenic effect on T lymphocytes or for their ability to synergize with Con A (2.5 µg/ml) in stimulation of mitogenic responses by suboptimal numbers of unfractionated thymocytes in the costimulator assay (22). In the first case, 6×10^5 unfractionated or 2×10^5 peanut agglutinin-nonagglutinable (PNA$^-$) BALB/c thymocytes were cultured for 5 days in 200 µl of Eagle-Hanks amino acid (EHAA) medium supplemented with 5 percent FCS and 5×10^{-5} M 2-ME with dilutions of test supernatants. Cultures were pulsed for the final 16 hr. with ^{3}HTdR, harvested and counted as described above. In the costimulator assay, 7.5×10^4 BALB/c thymocytes were placed in 200 µl cultures of complete RPMI-1640 medium with or without dilutions of test supernatants in the absence or presence of 2.5 µg/ml Con A. Cultures were pulsed with ^{3}HTdR for the final 16 hr. of a 3 day culture.

IV. Cytotoxic T Lymphocytes (CTL) Culture System

LK preparations were tested for their ability to serve as helper factor(s) for the induction of alloreactive CTL responses *in vitro* from thymocyte precursors or for their potential to induce self-reactive CTL in the absence of additional stimulator cells from spleen cell populations, a unique property of AEF supernatants (8,9). In the first case, 2×10^5 BALB/c thymocytes together with an equal number of 2,000 rad-irradiated C3H/St spleen cells were cultured in 200 µl complete RPMI-1640 medium in 96 well, conical-bottom Linbro culture plates in the absence or presence of test supernatants. Alloantigen-specific cytolysis was measured after 5 days by adding 5×10^3 ^{51}Cr-labeled *H-2^k* lymphoma target cells into each well and harvesting 100 µl samples of supernatants 4 hr. later for counting in a gamma counter as described before (23).

In the case of self-reactive CTL, BALB/c spleen cells (6×10^6/ 16 mm well/2 ml complete EHAA medium) were cultured in the absence or presence of various dilutions of test supernatants. No additional stimulator cells were added to these cultures. Five days later, the cells were harvested and assayed at several dilutions in a 4 hr. ^{51}Cr release against 1×10^4 ^{51}Cr-labeled *H-2*-identical, unmodified Thy-56 (*H-2^d*) lymphoma target cells (23).

V. Macrophage Activating Factor (MAF) Assay

MAF was assayed in a ^{51}Cr-release tumoricidal assay (17). In this assay, 2×10^5 adherent peptone-elicited peritoneal exudate macrophages are incubated with MAF sources and excess amounts of heat-killed *Listeria monocytogenes* as a source of second signal which has been shown to be required for macrophage activation. After 4 hr., the stimuli are washed off and 2×10^4 ^{51}Cr-labeled P815 mastocytoma cells added. After incubation for 18 hr. at 37°, the specific amount of ^{51}Cr released into the supernatant is determined. One unit of MAF is defined as the amount of MAF required to effect 50 percent of maximal ^{51}Cr release.

VI. Interferon (IFN) Assay

Interferon was assayed by the cytopathic effect assay which is based on interferon-dependent antiviral activity. Mouse L-cells (clone 929) were incubated for 24 hr. with potential sources of IFN. The resulting monolayer of cells were then challenged with purified vesicular stomatitis virus and cultured for an additional 48 hr. Viable monolayers and virus-disrupted monolayers were differentiated by light microscopy. One unit of IFN is defined as the reciprocal of the dilution which gave 50 percent protection of the L-cells from virus.

8. Partial Purification of Hybridoma Supernatants

Large volumes of LK-containing hybridoma supernatants were generated by stimulating confluent cultures of hybridoma cells in 150 cm^2 tissue culture flasks with 50 ml DMEM supplemented with 0.5 percent FCS, 5×10^{-5} M 2-ME, 10 mM HEPES buffer and 10 µg/ml Con A. Culture supernatants were harvested after 20-24 hr. by centrifugation (1400 x g, 15 min.), filtered and stored at 4°C. Volumes of 2-3 liters were fractionated at 4°C using buffers prepared with sterile, pyrogen-free water. Fractionation steps included sequential precipitation with ammonium sulfate (between 45 and 85 percent w/v), gel filtration chromatography on Sephadex G-100, ion exchange chromatography on DEAE-Sephacel and chromatofocusing. These procedures are described in detail elsewhere (16,17).

C. Results

1. Isolation and Cloning of T Cell Hybridomas (14,16)

DBA/2 T cell blasts alloactivated against $C3D2F_1$ stimulators were fused with the BW5147 T lymphoma cells using PEG. Growing hybrids were detected in 115 of 288 wells plated and were expanded in 24 well culture plates. Selected hybridoma lines were tested for the expression of Thy-1 and H-2 alloantigens and all lines examined were found to express Thy-1.2 and $H\text{-}2^d$ antigens derived from the alloactivated DBA/2 blasts as well as BW5147-derived Thy-1.1 and $H\text{-}2^k$ alloantigens. These properties as well as the production of various LK (see below) have remained stable for over a year. Cloning and subcloning were performed by seeding an average of 0.5 cells per well. All clones obtained after the second cloning cycle were positive when tested for Con A-induced production of a given LK, i.e., TCGF (16). T cell hybridoma lines were tested for either constitutive or mitogen-induced LK production.

2. AEF Production (14)

Unstimulated culture supernatants of T cell hybridomas were screened for B cell-activating properties. Approximately half of these supernatants manifested little or no positive biological effects, about one third displayed low-to-moderate biological activity, and a small number demonstrated rather potent biological effects when tested for their ability to restore the *in vitro* antibody responses of T cell-depleted B cells following stimulation with sheep red blood cells (SRBC) or TNP-SRBC. The hybridoma supernatants tested were obtained in serum-free conditions by washing the cultured hybridoma cells, re-plating them in serum-free medium and harvesting the culture supernatants after 24 hr. Culture supernatants from the unfused parental line, BW5147, failed to display any B cell-activating properties in these cultures. Two hybridomas, 27 and 34, which constitutively secreted higher levels of helper activity, were selected for further experimentation.

Analysis of these hybridoma lines has revealed the following properties of these hybridoma cells and their soluble products (14, unpublished observations):

a. Stimulation of hybridoma 34 with specific $C3D2F_1$ stimulator cells in an overnight serum-free culture resulted in an augmented B cell-stimulating activity. Furthermore, the same supernatants were mitogenic for PNA^- thymocytes and stimulated significant self-reactive CTL responses, similar to conventional AEF supernatants. The specific alloantigenic stimulation of hybridoma 34 was verified by measuring ^{3}HTdR uptake following a 24 hr. coculture with irradiated spleen cells of several H-2 haplotypes.

b. Like the conventional AEF supernatants, the B cell stimulating activity of culture supernatant from unstimulated hybridoma 34 was inhibited specifically by anti-$\underline{I-A}^d$ antibodies, consistent with the original finding that the active principle in AEF supernatants is Ia-positive and displays Ia determinants of the responder haplotype (5). These experiments are currently extended using more potent anti-$\underline{Ia}^d$ reagents, e.g., monoclonal anti-$\underline{I-A}^d$ antibody secreted by the MK-D6 hybridoma (obtained from Dr. P. Marrack, Denver, CO).

c. Upon recent recloning of hybridoma 34, many clones isolated were found to express $\underline{I-A}^d$ determinants as evidenced by their reactivity with a monoclonal anti-$\underline{I-A}^d$ antibody in a C-dependent microcytotoxicity assay. In addition, a significant proportion of the clones displayed receptors for the Fc portion of IgG, detected in a rosette assay with IgG-coated SRBC.

d. Preliminary analysis of the unstimulated culture supernatants of these clones revealed that the B and T cell activating properties segregate independently. Moreover, supernatants with potent B cell stimulating activity were completely devoid of TCGF activity. Thus, the active principle which stimulates in vitro antibody responses is distinct from TCGF.

We are currently analyzing the B cell stimulating factor(s) (BCSF) secreted constitutively by clones of hybridoma 34 with respect to their structure, mechanism of action and relationship to BCSF described by others.

3. TCGF Production (16)

Of five uncloned hybridoma lines which were tested for Con A-induced TCGF production, two (8 and 27) did not secrete detectable levels of TCGF, another two (18 and 23) secreted low amounts of TCGF and hybridoma 24 produced a relatively high level of TCGF which was, however, lower than that found in the standard TCGF preparation derived from Con A-stimulated rat spleen cells. None of these hybridomas secreted TCGF in a constitutive manner, nor did the parental tumor line BW5147 produce TCGF, either constitutively or following Con A stimulation. With the TCGF-producing hybridomas, Con A was found to be a more efficient stimulant than PHA and the optimal concentration was usually 10 µg/ml.

It is of interest to note that hybridoma 27 and another hybridoma derived from the same fusion, i.e., hybridoma 34, which were found previously to produce AEF (see above), did not produce detectable levels of TCGF. Hybridoma 24 was selected for further analysis. Eighteen out of 27 clones selected after the first round of cloning produced TCGF following Con A stimulation, including 10 clones which produced higher levels of TCGF than did mouse spleen cells cultured and stimulated under similar conditions. Clone 24A10, which secreted the highest level of TCGF, was subjected to an additional cycle of subcloning in order to verify its true clonal nature. All of twelve subclones selected randomly were found to produce TCGF upon stimulation with Con A.

In addition to the Con A-induced production of TCGF described above, the clonal derivatives of hybridoma 24 were also screened for constitutive factor secretion. Three patterns of TCGF production emerged from this analysis. About one third of the clones secreted low but significant amounts of TCGF constitutively and much higher levels upon Con A stimulation. The constitutive TCGF production was evident only when the hybridoma supernates were tested at the highest concentration (50 percent v/v) and the activity was lost upon further dilution to 12.5 percent v/v. Another half of the clones did not constitutively secrete measurable amounts of TCGF but were induced to secrete marked levels of TCGF by Con A. The remaining 15-20 percent of the clones failed to secrete TCGF even when stimulated with Con A. The fact that constitutive activity was found in the supernates of only some, and not all, of the clones strongly suggests that it reflects true TCGF production and not a nonspecific "feeder" effect. Thus, some of the T cell hybridoma clones are capable of constitutive TCGF production.

Upon biochemical analysis the biologically-active TCGF molecules were found to precipitate between 45 and 85 percent ammonium sulfate. Sephadex G-100 chromatography revealed a single peak of biological activity at a molecular weight range of 30-40,000 daltons. The biologically-active fractions obtained from the Sephadex column were pooled and subjected to ion exchange chromatography on a DEAE-Sephacel column, using a gradient of 0.05-0.5 M NaCl (pH 8.0). The biological activity was eluted as a relatively broad peak between 50-150 mM NaCl. Thus, the T cell hybridoma-derived TCGF displayed size and charge properties similar to those of murine TCGF molecules isolated from other cellular sources. This partially-purified TCGF possessed several biological activities ascribed to TCGF, in addition to its ability to support the growth of the TCGF-dependent CTL line, CTLL-2. Thus, it helped the induction of alloreactive CTL from thymocyte precursors and synergized with Con A in the costimulator assay (22). It had only a minimal BCSF activity and demonstrated weak, but significant, direct mitogenic activity on fresh thymocyte populations.

Hybridoma clone 24A10 and one of its subclones, F1, were found to grow in normal, _H-2_-compatible or in athymic nude mice. Large numbers of hybridoma cells growing both in ascitic and solid forms could be obtained following intraperitoneal inoculation and remained stable in terms of their ability to produce TCGF _in vitro_ upon Con A stimulation for an extended period of time.

4. MAF and IFN Production (17)

Hybridoma 24, which was originally selected for study as a TCGF-producer, was additionally found to produce MAF as well as IFN activities following Con A stimulation (but not constitutively). The level of MAF produced by this hybridoma represented 20-30 percent of the activity found in conventional MAF supernatants, produced by Con A stimulation of normal murine spleen cells. However, cloning of hybridoma 24 resulted in some clones producing very high levels of MAF. Of 27 clones tested, 24 produced MAF when stimulated with 10 μg/ml Con A and seven of these secreted more MAF than was present in the conventional preparation. The levels of IFN in stimulated culture supernatants of the same 27 clones were also measured. Nine clones did not produce detectable IFN and seven others produced only limited amounts of IFN. Of particular interest was clone 24G1, which produced the highest level of MAF activity (up to 55,000 units/ml compared to 2400 units/ml in the conventional MAF preparation), yet produced no detectable amounts of IFN. Levels of IL-2, MAF and IFN activities in Con A-stimulated culture supernatants of the various clones varied independently of one another (Table 1).

Table 1. Independent production of MAF, IFN and IL-2 by hybridoma 24 clones

Clone	Lymphokine Activity (Percent of Standard[1]) MAF	IFN	IL-2
A6	175	>100	2
A10	33	10	625
C7	300	>100	583
G1	2500	0	440
G4	4	50	0
G5	800	4	208

[1]Standard was a Con A stimulated normal murine spleen cell culture supernatant.

Several functional and biochemical properties of the hybridoma-derived MAF were evaluated and compared to those of conventional MAF preparations. These studies revealed great similarity between the two sources of MAF activity in that both a) required a second signal, in the form of heat-killed Listeria monocytogens, for the induction of macrophage tumoricidal activity, b) were resistant to heat treatment at 56°C for one hr. but were inactivated at 65°C, c) were sensitive to pH 4.0 but retained their activity at pH 5.0, and d) displayed molecular weight of 50-55,000 daltons following Sephadex G-100 gel chromatography. In addition, chromatofocusing of the active pool obtained by gel chromatography revealed that the major peak of activity eluted at pH 5.45 and a minor peak eluted at pH 4.90.

D. Discussion

This report documents the construction of stable T cell hybridoma lines by the fusion of alloantigen-activated T cells with the AKR/J-derived T cell lymphoma, BW5147. Various lines and clones of these T cell hybridomas produce a number of distinct LK activities, i.e., BCSF, TCGF, MAF, IFN and entity(ies) which directly activate fresh, unstimulated T cells (14-17). Some of these biological activities, namely, BCSF and TCGF are produced constitutively (but are usually augmented by mitogen or alloantigen stimulation) while others, i.e., MAF and IFN depend on mitogen stimulation. These differences may reflect distinct differentiation stages of the T cell hybrids necessary for LK production or differences in the sensitivity levels of the bioassays used. At present, the biological activity of these hybridoma lines has remained stable for about three years after fusion and up to one year after establishment in culture from a frozen stock.

These hybridomas were constructed as an extension of our earlier studies on conventional AEF supernatants (4-9,12,13). These supernatants contained multiple biological activities and, moreover, provided only limited and insufficient quantities for detailed biochemical characterization. The studies reported herein clearly demonstrate that the somatic cell hybridization technique is capable of overcoming both of the above limitations. In terms of quantities, the T cell hybridoma lines secrete much higher levels of LK activities than are found in conventional preparations. Clones of hybridoma 24 secreted up to 10-fold higher levels of TCGF and

25-fold higher levels of MAF than those found in conventional supernatants.

Second, the T cell hybridoma lines displayed a more restricted pattern of LK production than the conventional cultures of activated T cells. Cloning and subcloning were particularly useful in this respect, as exemplified by the independent production of distinct LK activities (i.e., TCGF, MAF and IFN) by clones of hybridoma 24 (Table 1). In addition, some clones of hybridoma 34 secreted constitutively BCSF without producing any detectable levels of TCGF. These results are in agreement with two recent reports which demonstrated BCSF production in the absence of TCGF by an alloreactive T cell line (24) and a T cell hybridoma (25). Such studies demonstrate that B cell-specific immunoregulatory molecules distinct from TCGF do exist and this conclusion is substantiated by the recent demonstration of direct biological effects of BCSF preparations on homogeneous B cell lines (26,27). TCGF-free preparations of BCSF may thus be extremely useful for delineating the structure and mechanism of action of these factors. The lack of TCGF in these preparations is important since it was suggested that TCGF may be directly responsible for stimulating in vitro antibody responses of T cell-depleted spleen cultures by inducing the differentiation and proliferation of functional T helper cells from pre-T cells or a few residual mature T cells (28). Parenthetically, it should be mentioned that a murine T cell lymphoma, 4-SP, which we identified as a high TCGF producer (Altman and Haas, in preparation) produces minimal or no BCSF activity. Such cell lines provide us with the ability to compare and test possible interactions between BCSF and TCGF in immune responses.

As an additional example, T cell hybridoma clone 24G1 produced very high levels of MAF but no detectable amounts of IFN (17, and Table 1). These results support the concept that MAF and IFNγ are distinct entities, in agreement with a recent report about the separation of MAF and IFNγ activities by differential adsorption to a polyinosine-Sepharose column (29). These findings are significant because in previous studies, certain types of IFN were shown to induce macrophage tumoricidal activity (30,31) and, in addition, IFNγ displays similar molecular weight, pH and heat sensitivity to those of MAF (32).

In summary, the existence of the T cell hybridoma lines described herein as well as the high levels and the restricted pattern of LK production by them provide us with the critical tools necessary to delineate the structure-function relationship among distinct LK activities, their mechanisms of production and action on cellular and molecular levels and, perhaps most importantly, their physiological role and potential use as therapeutic agents in various diseases.

Acknowledgments. We are grateful for the expert technical assistance of Robin Weiner, Anthony Sferuzza, Ph.D., Lori Hicks and Peta-Maree Stidham, and to Beverly Burgess for preparing this manuscript. This is publication number 2651 from the Immunology Departments, Scripps Clinic and Research Foundation. This work was supported by USPHS grants CA-25803, CA-24911, AI-17354, AI-07065 and a grant from Eli Lilly and Company, Indianapolis, Indiana. AA is a recipient of an American Cancer Society, Inc., Junior Faculty Research Award Number JFRA-17 and RDS is a recipient of an American Heart Association Established Investigatorship Number 77-202.

References

1. Tada T, Okumura K (1979) The role of antigen-specific cell factors in the immune response. Adv Immunol 28:1-87
2. Eshhar Z (to be published) Functional T cell hybridomas producing antigen-specific immunoregulatory factors. In: Katz DH (ed) Monoclonal antibodies and T cell products. CRC Press, Inc., Boca Raton, FL
3. Altman A, Katz DH (to be published) The biology of monoclonal lymphokines secreted by T cell lines and hybridomas. Adv Immunol
4. Armerding D, Katz DH (1974) Activation of T and B lymphocytes in vitro. II. Biological and biochemical properties of an allogeneic effect factor (AEF) active in triggering specific B lymphocytes. J Exp Med 140:19-37
5. Armerding D, Sachs DH, Katz DH (1974) Activation of T and B lymphocytes in vitro. III. Presence of Ia determinants on allogeneic effect factor (AEF). J Exp Med 140:1717-1722
6. Eshhar Z, Armerding D, Waks T, Katz DH (1977) Activation of T and B lymphocytes in vitro. V. Cellular locus, metabolism and genetics of induction and production of the allogeneic effect factor. J Immunol 119:1457-1467
7. Armerding D, Eshhar Z, Katz DH (1977) Activation of T and B lymphocytes in vitro. VI. Bochemical and physicochemical characterization of the allogeneic effect factor. J Immunol 119:1468-1474
8. Altman A, Katz DH (1980) Stimulation of autoreactive T lymphocytes by allogeneic effect factor (AEF): Relevance to normal pathways of lymphocyte differentiation. Immunol Rev 51:3-34
9. Altman A, Katz DH (1980) Existence of T cells manifesting self-reactivity indistinguishable from alloreactivity. J Immunol 125:1536-1543
10. Watson J, Mochizuki D, Gillis S (1980) T cell growth factors: interleukin 2. Immunol Today 1:113-117
11. Smith KA, Ruscetti FW (to be published) T cell growth factor and the culture of cloned functional T cells. Adv Immunol
12. Altman A, Gilmartin TD, Katz DH (1981) Bone marrow stem cell differentiation in vitro: Promotion of long term growth by a soluble lymphocyte-derived mediator. Science 211:65-67, 1981
13. Altman A, Cardenas JM, Welsh RM, Jr, Katz DH (1980) The biological effects of allogeneic effect factor on T lymphocytes. III. Interferon does not contribute to the biological activities displayed by AEF on both T and B lymphocytes. Ann Immunol (Inst Pasteur) 131C:335-347
14. Katz DH, Bechtold TE, Altman A (1980) Construction of T cell hybridomas secreting allogeneic effect factor (AEF). J Exp Med 152:956-968
15. Altman A, Schreiber RD, Katz DH (to be published) Functional T cell hybridomas producing non-specific immunoregulatory factors. In: Katz DH (ed) Monoclonal antibodies and T cell products. CRC Press, Inc., Boca Raton, FL
16. Altman A, Sferruzza A, Weiner RG, Katz DH (to be published) Constitutive and mitogen-induced production of T cell growth factor by stable T cell hybridoma lines. J Immunol
17. Schreiber RD, Altman A, Katz DH (submitted for publication) Identification of a T cell hybridoma which produces extraordinary quantities of macrophage activating factor
18. Gefter ML, Margulies DH, Scharff MD (1977) Simple method for polyethylene glycol-promoted hybridization of mouse myeloma cells. Somatic Cell Genet 3:231-236

19. Lake P, Clark EA, Khorshidi M, Sunshine GH (1979) Production and characterization of cytotoxic Thy-1 antibody-secreting hybrid cell lines. Detection of T cell subsets. Eur J Immunol 9:875-886
20. Gillis S, Smith KA (1977) Long term culture of tumour-specific cytotoxic T cells. Nature (London) 268:154-156
21. Gillis S, Ferm MM, Ou W, Smith KA (1978) T cell growth factor: Parameters of production and a quantitative microassay for activity. J Immunol 120:2027-2032
22. Paetkau V, Mills G, Gerhart G, Monticone V (1976) Proliferation of murine thymic lymphocytes in vitro mediated by the concanavalin A-induced release of a lymphokine (costimulator). J Immunol 117:1320-1324
23. Altman A, Cardenas JM, Bechtold TE, Katz DH (1980) The biological effects of allogeneic effect factor on T lymphocytes. I. The mitogenic activity and the autonomous induction of cytotoxic T lymphocytes by AEF. J Immunol 124:105-113
24. Swain SL, Dennert G, Warner JF, Dutton RW (1981) Culture supernatants of a stimulated T cell line have helper activity that synergizes with IL-2 in the response of B cells to antigen. Proc Natl Acad Sci USA 78:2517-2521
25. Takatsu K, Tanaka K, Tominaga A, Kumahara Y, Hamaoka T (1980) Antigen-induced T cell-replacing factor (TRF). III. Establishment of T cell hybrid clone continuously producing TRF and functional analysis of released TRF. J Immunol 125:2646-2653
26. Pure E, Isakson PC, Takatsu K, Hamaoka T, Swain SL, Dutton RW, Dennert G, Uhr JW, Vitetta ES (1981) Induction of B cell differentiation by T cell factors. I. Stimulation of IgM secretion by products of a T cell hybridoma and a T cell line. J Immunol 127:1953-1958
27. Howard M, Kessler S, Chused T, Paul WE (1981) Long-term culture of normal mouse B lymphocytes. Proc Natl Acad Sci USA 78:5788-5792
28. Watson J, Gillis S, Marbrook J, Mochizuki D, Smith KA (1979) Biochemical and biological characterization of lymphocyte regulatory molecules. I. Purification of a class of murine lymphokines. J Exp Med 150:849-861
29. Kniep EM, Domzig W, Lohmann-Matthes M-L, Kickhofen B (1981) Partial purification and chemical characterization of macrophage cytotoxicity factor (MCF, MAF) and its separation from migration inhibitory factor (MIF). J Immunol 127:417-422
30. Schultz RM, Chirigos MA (1978) Similarities among factors that render macrophages tumoricidal in lymphokine and interferon preparations. Cancer Res 38:1003-1007
31. Jett J, Mantovani A, Herberman RB (1980) Augmentation of monocyte-mediated cytolysis by interferon. Cell Immunol 54:425-434
32. Youngner JS, Salvin SB (1973) Production and properties of migration inhibitory factor and interferon in the circulation of mice with delayed hypersensitivity. J Immunol 111:1914-1922

Human T Cell Hybridomas with Tetanus-Toxoid-Specific Helper Activity

E.C. DeFreitas, S. Vella, A. Linnenbach, C.M. Croce, H. Koprowski

A. Introduction

Somatic cell hybridization first described by Barski *et al.* (1) initiated an era in cell biology wherein differential genetic and functional properties could be expressed and regulated in the same cell. This approach was utilized by Milstein and Kohler (2) to combine the properties of specific antibody secretion and continuous growth *in vitro* when they successfully fused normal mouse B lymphocytes with enzyme-deficient mouse myeloma cells.

Of equal interest to immunologists is the induction and activity of soluble lymphokines which regulate a variety of immune functions including delayed-type sensitivity, macrophage activation, T and B cell activation and suppression, and antibody synthesis (3). For instance, Kappler and Marrack (4), in a series of elegant studies, produced mouse T-T hybridomas by somatic cell fusion which secrete at least one such lymphokine, Interleukin-2 (Il-2) after specific antigenic stimulation. These studies and others like them (5-8) have begun to elucidate the pathways of immunological activation and regulation between lymphocytes and lymphokines.

Two major advances have made these studies feasible. The first was the discovery that normal activated mouse and human T cells could be continuously grown in culture in the presence of Il-2 (9) with maintenance of specificity and function (10, 11). The second was the ability to derive enzyme-deficient tumor cell lines of the T cell lineage (12) which can be used as fusion partners for T cell lines of defined function.

We have generated an HPRT-negative clone of a human T cell lymphoma, Jurkat, after treatment for 2 months in 6-thioguanine. Jurkat was selected for this purpose since Gillis and Watson (13) reported that this line has the genetic capability of producing Il-2 after mitogen stimulation. We reasoned that this function may improve the probability that, after fusion with a human functional T helper cell line, the hybrids would express antigen-specific helper activity known to be mediated by lymphokines (14). The results of these studies indicate that these predictions proved correct.

B. Results

I. Development of HPRT-negative human T cell lymphoma and human antigen-specific T cell lines

Jurkat T lymphoma cells were obtained from Dr. Robert Gallo, NCI, NIH, Bethesda, MD. The cultures of logarithmically growing cells were treated with 10^{-4} M 6-thioguanine (6-TG) and maintained at 37°C with periodic refeeding with fresh media plus 6-TG. After 2 weeks, the vast majority of the cells were dead. Thereafter, the cultures were examined weekly for the presence of viable clones of 6-TG-resistant cells. By 2 months post-treatment, three clones were found, picked, and expanded in 6-TG-containing media. All clones were tested for their HAT-sensitivity, surface phenotype and ability to produce Il-2 after mitogen stimulation. All clones died in HAT medium and were therefore HPRT-negative. They were of T cell origin (Table 1). One clone, Jurkat-6TG-3, produced a maximal amount (38 U/ml) of Il-2 when stimulated with 0.5 µg/ml Concanavalin A (Miles Yeda, Rehovat, Israel). PHA-P (Burroughs Wellcome, Dartford, UK) was less active and induced optimal production of Il-2 at 1 µg/ml. This clone produced less Il-2 than reported for the parental Jurkat line (13) but was the highest producer of the three mutants found. The clone Jurkat-6TG-3 was used for all fusions described.

The line of normal antigen-specific T cells used for fusions was obtained by the following protocol shown to enhance the frequency of antigen-specific helper cells in the mouse (15). Donors who had been immunized with tetanus toxoid (Tet) within the previous 2 years were bled and their peripheral blood lymphocytes (PBL) obtained after density gradient centrifugation on LSM. PBL at 2×10^6/ml were seeded in 24-well plates (Costar, Cambridge, MA) in RPMI, 10% AB human serum, 10^{-5} M 2-mercaptoethanol and Tet (Division of Biologics, Boston, MA) at 1-4 µg/ml. After 6 days, during which active cell proliferation occurs, viable cells were seeded at 1×10^5/ml in a source of lectin-free Il-2 at 10 U/ml (BRL, Bethesda, MD). Every 5 to 7 days, cultures were refed with fresh media containing antigen, Il-2, and irradiated autologous PBL or adherent cells (monocytes). This procedure was followed until the total

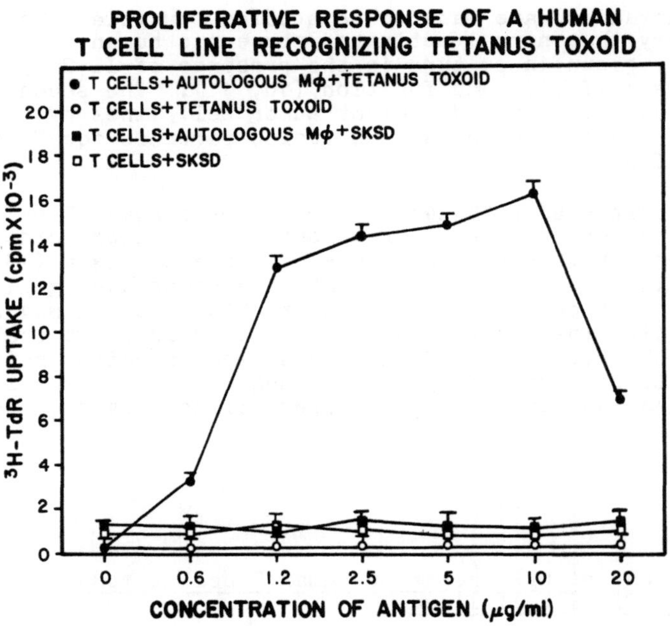

Fig. 1. Antigen-induced proliferation of human T cell line specific for tetanus toxoid.

number of viable T cells reached 10-20 x 10^6. These cells were then seeded at 1 x 10^5/ml in fresh media with Il-2 and grown for 5 to 7 days in the absence of irradiated fillers. This line was screened for antigen specificity in two methods: 1) Tet-induced proliferation and 2) Tet-induced helper activity. In the first assay, 1 x 10^4 viable T cells were seeded in 96-well microtiter plates (Costar, Cambridge, MA) with and without 2 x 10^5 irradiated autologous PBL as a source of monocytes. Wells were stimulated with and without Tet, an unrelated antigen [streptokinase-streptodornase (SKSD)] for 2 days, then pulsed for 18 hr with 1 µCi/well ^{3}H-TdR and harvested. The results given below with one of our lines indicate that this T cell line 1) specifically proliferates in response to Tet, and 2) requires autologous accessory cells to proliferate.

We also wished to determine if this T cell line would provide antigen-induced helper activity to autologous B cells. We used a modification of the Mishell-Dutton culture for in vitro antibody induction (16) suggested by Michael Hoffman (17). Cultures of whole PBL were seeded in RPMI plus 10% FCS, tetanus toxoid and 10% monocyte-conditioned media (MCM). After 24 hr of culture 10% human serum was added from which anti-tetanus antibody had been adsorbed. After several days in culture, the cells were harvested, the media containing antigen replaced with fresh media and the cells recultured for an additional 2 days. The supernatants were then assayed for anti-tetanus antibody in a solid-phase RIA (18). To obtain optimal antibody responses to Tet in vitro, we found several parameters appear to vary with the donor: 1) number of PBL per/well, 2) Tet concentration, and 3) day of peak antibody production. These were determined empirically.

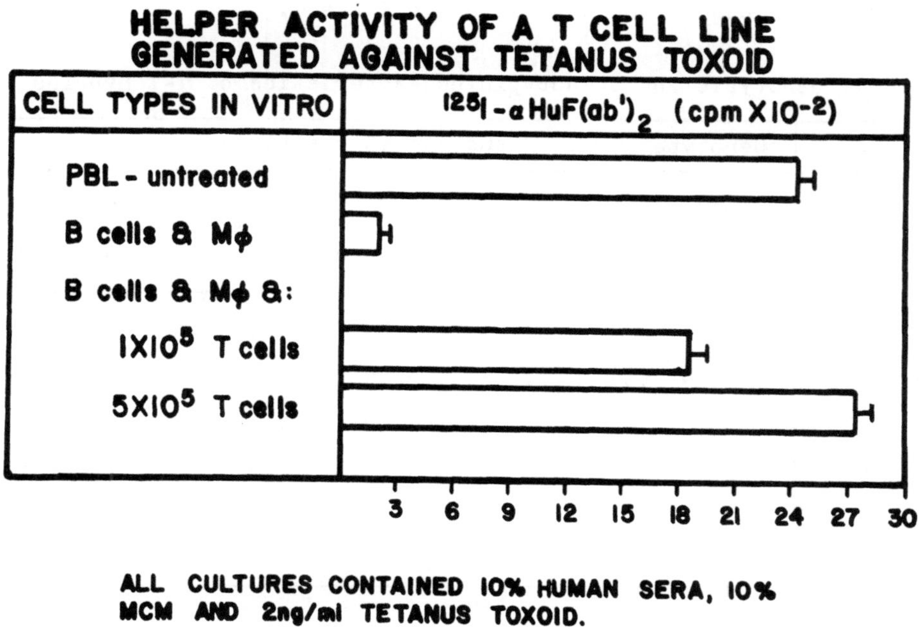

Fig. 2. Helper activity of human T cell line specific for tetanus toxoid.

In order to test the helper activity of the Tet-specific T cell line, we removed the normal T cells from the PBL of the autologous donor by both SE_N rosetting and OKT3 + C' treatment. The combined treatments yielded a population of non-PHA responsive cells when tested at 2×10^6/ml. These B cells and monocytes were seeded at 1×10^5/well and various numbers of the Il-2-propagated Tet-reactive T cell line added. Cultures were processed simultaneously with the whole PBL. The results below indicate that the T cell line provided dose-dependent help to autologous B cells, resulting in the production of specific anti-tetanus antibody. Cultures containing no antigen or heterologous antigens (SKSD or ovalbumin) induced no detectable antibody.

C. Production of Human T-T Hybrids

Fusions between Jurkat-6TG-3 and the Tet-specific helper T cell line were performed as described by Kontiainen *et al*. (19). Il-2 propagated T cells were mixed with Jurkat-6TG-3 at a ratio of 5:1 (usually 10^7 and 5×10^7, respectively) and fused with a 50% solution of polyethylene glycol (PEG) 1600 in serum-free media. The cells were resuspended in HAT selective medium containing 10% FCS, seeded into 3 microtiter plates, gassed with 5% CO_2 and taped closed. They were fed with fresh HAT medium weekly by replacing half the volume of each well. Although the unfused T cell blasts and the Jurkat-6TG-3 died within 1 to 2 weeks, the earliest growing T cell hybrids were observed at 7 weeks post-fusion. Each fusion yielded 5 to 10 viable clones which grew exponentially thereafter. The genotypic and phenotypic characteristics of representative clones from one such fusion are listed in Table 1.

Table 1. Genotypic and phenotypic characteristics of T-T hybrids

	Genotype	Phenotype (% positive)					
Cell type	Number of chromosomes mean (range)	SE receptor	FC receptor	Surface DR	OKT 3	OKT 4	OKT 8
SH1		> 95	> 99	< 1	86	28	46
SH2	72 (44-106)	> 95	> 99	< 1	58	46	67
SH3		> 95	> 99	< 1	50	55	59
SH5		> 95	> 99	< 1	72	83	83
Jurkat-6TG-3 (parent)	48	> 95	> 99	< 1	98	28	13

All clones expressed the T cell markers for SE receptor and Fc receptor as did the parental Jurkat-6TG-3. No clones expressed Dr antigen as detected by flow cytometry with a panel of mouse monoclonal antibodies directed against framework Dr determinants, despite the fact that Dr antigens were reportedly found on some activated human T cells (20). The OKT monoclonal antibody-defined human T cell antigens described by Kung et al. (21) were present on all hybrids to varying degrees. None of the hybrids uniformly expressed the pan T cell OKT3 marker as did the parental Jurkat-6TG-3. OKT4- and 8-defined antigens, which are correlated with mutually exclusive populations of helper and suppressor/cytotoxic T cells, were also expressed in varying proportions of each clone. Clearly, at least one clone (SH5) contained cells that expressed both the OKT4- and 8-defined antigens simultaneously. Perhaps the expression of these antigens is not regulated in hybridomas as it is in normal T cells. Therefore, we could not select a potential helper T-T hybrid based on these criteria.

Two approaches were used to evaluate the antigen specificity of these hybrids. The first was the ability to produce Il-2 after antigenic stimulation. We initially evaluated the ability of the hybrids to produce Il-2 in the presence of varying concentrations of the mitogens PHA-P and Con A using the protocol of Gillis et al. (22) which measured Il-2 activity in supernatants by their ability to support the growth of an Il-2-dependent mouse T cell line, CTLL, obtained from Dr. Kendall Smith, Darthmouth Medical School, Dartmouth, NH. The results are presented in Table 2 below.

Table 2. Ability of T-T hybrids to produce interleukin-2 after mitogen stimulation

Hybrids tested[a]	Induction of Il-2 with tetanus toxoid (μg/ml)[b]					PHA-P[c]	ConA
	0	0.2	1	5	10		
SH1	1454	224	1387	959	819	11,054	4279
SH2	943	1390	512	821	420	12,923	7267
SH3	514	867	397	485	415	16,646	16,724
SH5	624	614	795	567	364	9026	3609
Jurkat-6TG-3 (parent)	313	591	463	524	391	3842	8632

[a] T hybrids were washed free of HAT media, resuspended in RPMI and 10% FCS to 2 x 10^6/ml and seeded in 96-well microtiter plates with and without antigen or mitogen. Supernatants were collected after 24 hours of culture and tested in varying dilutions on Il-2-dependent mouse CTLL cell line.

[b] Results expressed are the cpm of ^{3}H-TdR incorporated by mouse CTLL cells after a 24-hour incubation in the presence of supernatants from the sources indicated.

[c] Results expressed for mitogen-stimulated supernatants represent the amount of Il-2 produced in the presence of optimal mitogen concentrations.

They indicate that all the hybrids produced Il-2 when stimulated with either Con A or PHA-P although each hybrid showed optimal Il-2 production at different concentrations of each mitogen ranging from 0.1 μg/ml to 10 μg/ml. In addition, some hybrids appeared to respond significantly better to PHA-P than to ConA (eg., SH1 and SH2) despite the fact that the parental Jurkat-6TG-3 shows the opposite pattern. Clone SH3 appears to respond equally well to both mitogens. This suggests that the normal T cell parent of these clones may come from the mutually exclusive populations of PHA-P and ConA-reactive T cells described in the mouse (23). Nevertheless, each clone clearly has the genetic and metabolic requirements to produce Il-2 after mitogenic stimulation. In contrast, neither the hybrids nor the parental Jurkat-6TG-3 responded to Tet at any concentration. This was expected since even the antigen-specific normal T cell line used for fusion did not respond to Tet except in the presence of autologous irradiated accessory cells (see Fig. 1).

We therefore tested the ability of the hybrids to produce Il-2 after antigenic stimulation in the presence of autologous irradiated adherent cells from peripheral blood (monocytes). The results are presented in Table 3 below.

Table 3. Recognition of tetanus toxoid by human T-T hybrids by interleukin-2 production in the presence of autologous monocytes

Hybrids tested[a]	Induction of Il-2 with: Tetanus toxoid (µg/ml)[b]						
	0	0.2	1	5	10	PHA-P[b]	ConA
SH1	1365	1328	3105	2929	1806	14,527	20,327
SH2	2614	3916	4380	4873	11,277	22,112	6028
SH3	2201	1636	7373	1517	1181	7476	12,616
SH5	719	443	588	1077	461	15,879	15,538
Jurkat-6TG-3 (parent)	887	328	316	416	402	13,911	35,189
Mϕ alone	1107	804	785	333	361	1128	2724

[a] T hybrids, washed free of HAT media, were seeded at 2×10^6/ml in RPMI and 10% FCS on 3-hour adherent monolayers from peripheral blood cells which had been rosetted with SE_N to remove T cells and irradiated with 5000 rads from a cesium source. Microcultures were then stimulated with antigen or mitogen for 24 hours. Supernatants from each group were collected and tested for their ability to support growth of a continuous line of Il-2-dependent mouse CTLL as described.

[b] Optimal concentrations of each mitogen were used to obtain maximal Il-2 production.

Clones SH1 and SH3 responded minimally to Tet by exhibiting only a three-fold increase in Il-2 production when exposed to antigen. In contrast, SH2 showed a five-fold increase in Il-2 production at 10 µg/ml Tet. SH5, in contrast, was totally unresponsive as was the parental line Jurkat-6TG-3. Irradiated peripheral blood monocytes alone were also nonproductive. The presence of irradiated monocytes appeared to augment the ability of all hybrids and Jurkat-6TG-3 to respond to mitogens. This effect may be due to the production of Il-1 by the monocytes which has been shown to be the second signal in the induction of T cell-derived Il-2 by mitogens (24, 25).

A second approach to determine antigen-recognition by these T hybrids utilized the observations that antigen-specific T lymphocytes from guinea pigs bound specifically to homologous antigen-pulsed autologous macrophages *in vitro* (26). By prelabeling the hybrids with ^{3}H-TdR, we could determine the percentage of hybrids bound to monocyte cultures in the presence and absence of specific antigen. The results in Table 4 show that twice as many SH1 and SH3 cells attached to the Tet-pulsed monocytes as to the SKSD-pulsed monocytes. The majority (57%) of SH2 cells bound to Tet-pulsed monocytes. These findings correlate with the Il-2 induction experiments which also show that SH1 and SH3 are minimally recognitive of Tet on autologous monocytes, while SH2 demonstrates significant antigen-specific recognition. These experiments do not distinguish between the possibilities that a greater proportion of SH2 expresses the receptor for monocyte-associated Tet or that the SH2 receptor has a higher affinity for antigen. As was seen in the induction of Il-2, SH5 and Jurkat-6TG-3 were totally negative for binding. Less than 1% of all hybrids attached to wells without monocytes or to empty wells prepulsed with 20 µg/ml Tet (not shown).

Table 4. Recognition of tetanus toxoid by human T-T hybrids: binding of hybrids to antigen-pulsed autologous monocytes

Hybrids[a] tested	Binding to autologous monocytes prepulsed with[b]:		
	Media	Tetanus toxoid	SKSD
SH1	564 (13%)	1097 (26%)	604 (14%)
SH2	500 (12%)	2386 (57%)	469 (11%)
SH3	429 (13%)	814 (26%)	405 (13%)
SH5	325 (7%)	298 (7%)	216 (5%)
Jurkat-6TG-3 (parent)	108 (1%)	292 (2%)	146 (1%)

[a] T hybrids and Jurkat-6TG-3 cells were prelabeled with ^{3}H-TdR, washed, counted, and equal numbers of all clones added to a 24-hr culture of autologous monocytes prepulsed with nothing (media), Tet (20 µg/ml) or SKSD (20 µg/ml). After 3 hr at 37°C, the non-adherent cells were washed away and the adherent cells harvested on a Titertek cell harvester after detachment with trypsin-EDTA.

[b] The results are expressed as the mean cpm of ^{3}H-TdR collected on filter strips from the cells binding to the adherent cell population. SEM <10% mean. The number in parentheses is the percentage of T cells which adhered to the attached monocytes out of the total number added to the adherent cell monolayer.

Our goal of obtaining antigen-specific helper T hybrids seemed best accomplished by using clone SH2. Initial experiments demonstrated small but significant helper activity provided by SH2 for autologous B cells when stimulated with tetanus toxoid in a Mishell-Dutton culture for *in vitro* antibody production (data not shown). We therefore subcloned SH2 in an effort to obtain lines with greater activity. SH2 was plated at 1 cell per well in 96-well microtiter plates. Thirteen subclones were obtained from 96 wells and these were evaluated for their ability to induce autologous B cells to generate anti-Tet antibody *in vitro*. The results of one such experiment with three subclones are presented in Table 5.

Table 5. Antigen-specific helper activity of human T-T hybrids

Cells in culture	Anti-tetanus antibody[b] (^{125}I-anti-Hu F(ab)'$_2$; cpm x 10^2)[a]		
	Tetanus toxoid *in vitro*		
	0	2 µg/ml	2 ng/ml
PBL	3.4	16.4	45.0
PBL + Il-2	3.8	11.6	39.4
B + Mϕ	9.9	14.0	9.3
B + Mϕ + Il-2	10.0	18.0	7.0
B + Mϕ + T cells	2.8	13.9	16.8
B + Mϕ + T cells + Il-2	2.9	19.0	41.2
B + Mϕ + T hybrids:			
SH2E5	16.9	<u>168.2</u>	<u>136.3</u>
SH2E5 + Il-2	14.8	<u>132.7</u>	77.8
SH2E6	13.6	<u>98.6</u>	<u>127.4</u>
SH2E6 + Il-2	12.9	<u>158.0</u>	55.0
SH2E12	19.9	26.8	10.7
SH2E12 + Il-2	24.7	64.1	<u>224.5</u>

[a] All data are expressed as cpm of ^{125}I-anti-human F(ab)'$_2$ bound to plates pretreated with 20 µg/ml tetanus toxoid and then incubated with the cell-free supernatants from the *in vitro* cultured cells. The SEM was <10% of the mean. The nonspecific binding of the cell-free supernatants to plates without tetanus toxoid was always less than 600 cpm.

Whole PBL showed maximal specific antibody production with 2 ng/ml Tet as previously found. Addition of 2 U/ml partially purified lectin-free human Il-2 (Associated Biomedics, Buffalo, NY) did not appreciably affect this response. After depletion of T cells, no antibody production could be detected. Supplementation of the B cells with 10^4 normal peripheral blood T lymphocytes (obtained by rosetting with SE_N and lysis of the erythrocytes by hypotonic shock) increased the antibody response of the B cells in the presence of Tet. However, this level of antibody was only one-third that seen with whole PBL. Addition of Il-2 to the B cells and 10^4 T cells resulted in specific antibody production equal to that of whole PBL. These results indicate that in this system: 1) nanogram amounts of antigen stimulate normal T and B cells to produce optimal levels of antibody; 2) signaling of B cells by T cells is required for specific antibody production for which Il-2 cannot substitute; 3) 10^4 normal peripheral blood T cells provide suboptimal levels of helper activity for B cells, but in the presence of Il-2 this number of T cells becomes optimal probably due to T cell proliferation.

In the same experiment, the subclones of SH2 (SH2E5, SH2E6, SH2E12) were added to the B cells and monocytes as a source of helper activity. The same number of hybrids (10^4) induced a dramatic increase in the amount of specific antibody induced. Interestingly, SH2E5 and SH2E6 provided significant help at both microgram and nanogram concentrations of antigen. SH2E5 and SH2E6 were also similar in that both responded in the absence of exogenous Il-2. In fact, the presence of exogenous Il-2 suppressed the antibody response at nanogram concentrations of antigen. In contrast, SH2E12 demonstrated no helper activity unless exogenous Il-2 was added. At 2 ng/ml antigen, this subclone was clearly the best helper. All clones induced higher antibody backgrounds in the absence of antigen for unknown reasons.

The functional heterogeneity of these subclones suggests that these hybrids may be losing chromosomes which regulate their functional properties. Extensive karyotype analyses are in progress to answer this question.

The role of Il-2 in the augmentation of helper function is also of interest. Unlike normal T cells, exogenous Il-2 has no effect on the replication of T hybrids, and therefore its effect on SH2E12 is not as a growth factor *per se*. Il-2 appears to be working in conjunction with SH2E12 as a differentiation factor for antibody synthesis. Experiments are in progress to determine whether the effect is at the level of the T cell or the B cell.

C. Conclusions

We have generated human T cell hybrids with antigen-specific helper activity for B cells. The antigen-induced lymphokines produced by these cells will serve to characterize the functional and biochemical nature of the signals required for human antibody responses *in vitro*.

Acknowledgements. The authors thank Dr. Robert Gallo and Dr. Kendall Smith for their gifts of cell lines. The excellent and steadfast technical assistance of Ms. Kathleen Haines and Mr. Joseph DiCarlo is gratefully acknowledged. We thank the Editorial Department for typing the manuscript. Finally, we are indebted to Drs. Michael Hoffman and Philippa Marrack for their helpful suggestions in the course of this work.

References

1. Barski G, Sorieul S, Cornefert F (1961) "Hybrid" type cells in combined cultures in two different mammalian cell strains. J Nat Cancer Inst 26: 1296-1291

2. Köhler G, Milstein C (1975) Continuous culture of fused cells secreting antibody of predefined specificity. Nature 256: 495-497
3. Cohen S, Pick E, Oppenheim J (eds) (1979) Biology of the Lymphokines. Academic Press, New York.
4. Kappler JW, Skidmore B, White J, Marrack P (1981) Antigen-inducible, H-2-restricted, interleukin-2-producing T cell hybridomas. Lack of independent antigen and H-2 recognition. J Exp Med 153: 1198-1214
5. Katz DH, Bechtold TE, Altman A (1980) Construction of a T cell hybridoma secreting allogeneic effect factor (AEF). J Exp Med 152: 956-968
6. Taussig MJ, Holliman A (1979) Structure of an antigen-specific suppressor factor produced by a hybrid T-cell line. Nature 277: 308-310
7. Taniguchi M, Miller JFAP (1978) Specific suppressive factors produced by hybridomas derived from the fusion of enriched suppressor T cells and a T lymphoma cell line. J Exp Med 148: 373-382
8. Nabholz M, Cianfriglia M, Acuto O, Conzelmann A, Haas W, von Boehmer H, MacDonald HR, Johnson JP (1980) Cytolytically active murine T cell hybridomas. Nature 287: 437-440
9. Morgan D, Ruscetti F, Gallo R (1976) Selective in vitro growth of T lymphocytes from normal human bone marrow. Science 193: 1007-1008
10. Schreier MH, Tees R (1980) Clonal induction of helper T cells: Conversion of specific signals into nonspecific signals. Int Arch Allergy Appl Immun 61: 227-237
11. Kurnick JT, K-O Grönvik, Kimura AK, Lindblom JB, Skoog VT, Sjöberg O, Wigzell H (1979) Long term growth in vitro of human T cell blasts with maintenance of specificity and function. J Immunol 122: 1255-1260
12. Grillot-Courvalin C, Brouet J-C, Berger R, Bernheim A (1981) Establishment of a human T-cell hybrid line with suppressive activity. Nature 292: 844-845
13. Gillis S, Watson J (1980) Biochemical and biological characterization of lymphocyte regulatory molecules. V. Identification of an Interleukin-2-producing human leukemia T cell line. J Exp Med 152: 1709-1719
14. Schimpl A, Wecker E (1973) Stimulation of IgG antibody response in vitro by T cell-replacing factor. J Exp Med 137: 547-552
15. Schrier R, Skidmore B, Kurnick J, Goldstein S, Chiller JM (1979) Propagation of antigen-specific T cell helper function in vitro. J Immunol 123: 2525-2531
16. Mishell RI, Dutton RW (1967) Immunization of dissociated spleen cell cultures from normal mice. J Exp Med 126: 423-442
17. Hoffmann MK (1980) Antigen-specific induction and regulation of antibody synthesis in cultures of human peripheral blood mononuclear cells. Proc Natl Acad Sci USA 77: 1139-1143
18. Zollinger WD, Dalrymple JM, Artenstein MS (1976) Analysis of parameters affecting the solid phase radioimmunoassay quantitation of antibody to meningococcal antigens. J Immunol 117: 1788-1798
19. Kontiainen S, Simpson E, Bohrer E, Beverley PCL, Herzenberg LA, Fitzpatrick WC, Vogt P, Torano A, McKenzie I, Feldmann M (1978) Hybrid cell lines with T-cell characteristics. Nature 274: 477-480

20. Kornbluth J, Dupont B (1980) Cloning and functional characterization of primary alloreactive human T lymphocytes. J Exp Med 152s: 164-181
21. Kung PC, Talle MA, DeMaria ME, Butler MS, Lifter J, Goldstein G (1980) Strategies for generating monoclonal antibodies defining human T-lymphocyte differentiation antigens. Transpl Proc 12: 141-146
22. Gillis S, Ferm MM, Ou W, Smith KA (1978) T cell growth factor: Parameters of production and a quantitative microassay for activity. J Immunol 120: 2027-2032
23. Stobo JD, Paul WE (1973) Functional heterogeneity of murine lymphoid cells. III. Differential responsiveness of T cells to phytohemagglutinin and concanavalin A as a probe for T cell subsets. J Immunol 110: 362-375
24. Larsson E-L, Iscove NN, Coutinho A (1980) Two distinct factors are required for the induction of T cell growth. Nature 283: 664-666
25. Smith KA, Lachman LB, Oppenheim JJ, Favata MF (1980) The functional relationship of the interleukins. J Exp Med 151: 1551-1556
26. Lipscomb MF, Ben-Sasson SZ, Uhr JW (1977) Specific binding of T lymphocytes to macrophages. J Immunol 118: 1748-1754

Functional Analysis of Il-2 Produced by T-Cell Hybridomas: Il-2 Promotes T-Cell Growth But Does Not Mediate T-Cell or B-Cell Maturation and Differentiation

K. Pfizenmaier, M. Röllinghoff, H. Wagner

A. Introduction

In vitro stimulation of mixtures of T-cells and macrophages/-dendritic cells with either antigen or mitogen results in the production of a variety of mediators involved in lymphocyte activation. Supernatants (Sn) of such cultures have been used for many years as a rich source of nonantigen-specific helper factors for both B-cell- and T-cell- responses (1 - 5). Interleukin-2 (Il-2), a factor with a mol. wt. of approximately 30 000, has been identified as a major constituent of such Sn (6 - 9). The characteristic feature of Il-2 is its capacity to promote T-cell growth. In addition, other biological activities were also found in Il-2-rich preparations of ConA-stimulated mouse spleen cells including B-cell stimulatory activity as well as maturation and differentiation of both pre-T-cells and thymocytes into cytotoxic effector T-cells (9 - 14). Because these activities copurified under conventional separation procedures, it was suggested that Il-2 is critically involved in all these processes. However, whether or not Il-2 is the only mediator remained unclear. Thus, it is conceivable that other, so far unidentified factor(s) of a similar biochemical nature contaminate the Il-2 preparations even when vigorous separation procedures are applied. These putative factor(s) may, either by themselves or synergistic with Il-2, be responsible for the distinct biological activities detected in different assay systems.

In order to approach this question we sought for a homogeneous source of Il-2 producing cells. Accordingly, we have fused splenic T-cells with the BW 5147 thymoma and screened for hybridomas which constitutively produced Il-2. We have recently succeeded in the establishment of such hybridomas. Sn of cloned hybridoma TH-14.83 contained a 34 000 mol. wt. factor that promoted growth of Il-2 dependent cytolytic T-cells (15 - 16). We therefore concluded that the hybridoma produced IL-2.

We then compared the biological activity of the hybridoma derived Il-2 with that of Il-2 preparations obtained from ConA-stimulated spleen cells. The test systems employed here were (i) induction of H-2 restricted CTL-responses from thymocytes and (ii) induction of plaque forming cell responses from T cell deficient nu/nu spleen cells. The data obtained revealed that Il-2 was apparently the only lymphokine produced by the hybridomas; its functional activity was restricted to the clonal expansion of mature, antigen primed T cells. Our data therefore imply that for maturation and differentiation of T cells and of B cells to occur, other factors must be involved.

B. Methods

I. Origin of hybridomas and production of Il-2

The establishment and characterization of the Il-2 producing hybridomas has been described (15). They were obtained by fusion of alloimmunized C3H splenic T cells with BW 5147.

Hybridoma cells were screened for constitutive Il-2 production in the CTL growth assay as described below. Positive clones were isolated and subsequently tested for enhanced Il-2 production by mitogenic stimulation as described in detail elsewhere (16). The functional activity of the Il-2-producing hybridomas was found to be variable and not very long-lived; all of the clones and subclones isolated lost their capacity to produce Il-2 after only a few months of culture. During this time period the following protocols were found to yield reasonable Il-2 activity: a) 24 h culture of 1 x 10^6 cells/ml in the presence of 1-2 µg/ml of ConA plus 10 ng/ml PMA in normal tissue culture medium supplemented with 1-2% fetal calf serum; b) 24 h culture of hybridomas obtained from the peritoneal cavity of tumor bearing syngeneic mice, which were irradiated (450 R) before injection of cells. Hybridomas were cultured in normal tissue culture medium supplemented with 5 % fetal calf serum. Il-2 preparations used in this study were obtained from clone TH 14.83 according to protocol b. Culture supernatants were either used directly, or fractionated by gel filtration after they were 20 x concentrated by ultrafiltration using Amicon YM-10 membranes. 2 ml of the concentrated material was then passed over a calibrated 2.5 x 55 cm AcA 54 column (LKB Produkter, Sweden). The column was equilibrated with PBS (pH 7.4, conductivity 15 mS) and the samples eluted at a constant flow rate of approximately 7 ml/h. Individual fractions (2 ml) were tested for Il-2 activity in the CTL growth assay, and the positive fractions were pooled and concentrated 10 x before use.

II. Preparation of Il-2 from normal spleen cells

This was done as described previously (5). Briefly, 10^7 spleen cells/ml were incubated for 20 hours in serum-free culture medium supplemented with 1 µg/ml ConA (Pharmacia, Uppsala, Sweden). Partially purified Il-2 preparations were obtained by gel filtration on Sephadex G-100 (5).

III. Biological assays for Il-2 activities

1. Growth of CTL : 2 x 10^4 C3H or CBA derived CTL obtained after repeated in vitro stimulation with BALB/c spleen cells were seeded in a total volume of 1 ml culture medium containing

various concentrations of the supernatants to be tested. Growth of CTL was assessed both microscopically and by determining the cytolytic activity generated at day 2-3 of culture in a 3 hours ^{51}Cr-release assay as described (11) using 10^4 P815 (H-2^d) tumor cells as target. For each determination, positive control cultures were set up with a standard preparation of Il-2. CTL cultured in the absence of Il-2 were usually dead within 24 to 48 h, indicating the strict Il-2 dependence of the CTL.

2. Generation of primary CTL responses from thymocytes: This was performed as previously described (11). Briefly, 4 x 10^6 thymocytes, obtained from either CBA/J or BALB/c mice, were stimulated with 1 x 10^6 irradiated (2000 R) allogeneic or TNP-conjugated, syngeneic splenic stimulator cells. The enhancement of a CTL-response in the presence or absence of Il-2 containing supernatants was assessed at day 6 of culture. Serial dilutions of the effector cells generated were prepared and tested in a standard ^{51}Cr-release assay against P815 (H-2^d), and LS (H-2^k) tumor target cells, which, for assaying TNP specific CTL, were conjugated with TNP as described (11).

3. Induction of a plaque forming cell response in nu/nu spleen cells: Culture supernatants of ConA stimulated spleen cells and of hybridomas or the Il-2 containing fractions after gel chromatography were assayed for their ability to induce a sheep red blood cell (SRBC)-specific IgM plaque forming cell response of 2 x 10^6 C3H-nu/nu spleen cells as previously described (10). The supernatants to be assayed for TRF activity were added two days after onset of the culture. IgM plaques were determined at day 5.

C. Results and Discussion

I. Comparison of Il-2 preparations from hybridomas and spleen cells for functional activity in primary CTL responses.

The Il-2 content of a 24 h Sn of TH 14.83 cells was estimated in the CTL growth assay in comparison with a ConA-stimulated spleen cell Sn, fractionated by Sephadex G-100 gel filtration. Table 1 indicates that the maximum response was obtained at about 5 % concentration of hybridoma Sn, whereas with the spleen cell derived Il-2 preparation plateau levels were already reached at a concentration of 1 - 2 %.

Table I. Comparison of Il-2 activity in Sn of TH 14.83 and ConA-stimulated spleen cells

Source of Il-2	Concentration of supernatant (%) 0.5	1.0	2.5	5.0	10
spleen	65	85	90	90	87*
TH 14.83	40	52	90	97	89

+Il-2 activity was determined in the CTL-growth assay as described in Methods.

*Percent specific lysis of P815 (H-2^d) target cells

Various concentrations of each preparation were then tested for their capacity to induce a primary CTL response using thymocyte responder cells. Because we knew already from the initial screening experiments that TH-14 Sn enhanced allospecific CTL-responses in a nonantigen-specific manner (15), we now were interested in whether or not Il-2 derived from TH-14 cells can also induce H-2 restricted CTL responses. Accordingly, in the experiment depicted in Table II A where we used hybridoma Sn as a source of Il-2, we set up MLC's with CBA and BALB/c thymocytes responding against syngeneic, TNP conjugated, splenic stimulator cells. In addition, the enhancement of an allospecific CTL response (CBA anti BALB/c) as well as the CTL growth activity was assessed. The results in Table II are expressed as percent specific lysis obtained with the optimal dose of each of the tested Il-2 preparations. Comparable Il-2 activity of spleen cell and hybridoma Sn was revealed in the CTL-growth assay and in the allospecific primary CTL response. Surprisingly, we noted that in contrast to the standard Il-2 preparation, the hybridoma Sn was ineffective for the induction of primary TNP specific CTL responses from both CBA and BALB/c thymocytes (Table IIA).

Table II. Il-2 is necessary, but not sufficient, for the induction of CTL responses from thymocytes

	Assay system	Source of supernatant: spleen	TH 14.83	none
A	CTL growth	35*	41	Ø
	1° MLC: k anti d	51	38	3
	k anti k-TNP	32	5	3
	d anti d-TNP	39	8	5
B	CTL growth	87	76	2
	1° MLC: k anti d	42	40	Ø
	k anti (kxd)	31	Ø	Ø
	k anti k-TNP	52	8	3

* % specific lysis of target cells syngeneic to the stimulator cell type at 20:1 effector to target cell ratio.

A similar result was obtained using gel chromatography fractionated Il-2 from TH 14.83 (Table IIB). Again, TH 14.83 derived Il-2 was equally effective as the spleen cell Il-2 preparation in the CTL growth assay, yet it failed to induce a TNP-specific CTL response. As shown before, the same Il-2 preparations induced allospecific CTL provided fully allogeneic stimulator cells were used. However, when semiallogeneic stimulator cells were used in order to avoid positive allogeneic effects, no CTL response could be generated. The data therefore show that the hybridoma Sn lack factor(s) which are required for the induction of both alloreactive and H-2 restricted CTL when thymocytes are used as responder cells. As these supernatants contain high amounts of Il-2, we conclude that Il-2 by itself cannot initiate the whole cascade of events leading to effector function. Apparently, other factor(s) are involved. They are clearly contained in Sn of ConA stimulated spleen cell cultures and can also be elicited during an allogeneic MLC, in the experiments described here most likely from radioresistant cells within the stimulator cell population. From the data available it is unclear, at which time during the activation process of thymocytes, the putative molecules act. Thus, it is conceivable that a maturation factor precedes Il-2 function. One can further speculate that clonal expansion and differentiation of precommitted T cells into effector cells are separate events which are driven by distinct molecules, the first being Il-2 and the latter a putative differentiation factor.

II. Hybridoma derived Il-2 preparations do not contain TRF activity.

We have addressed the question of whether or not hybridoma derived Il-2 also exerts T cell replacing factor (TRF) activity in B cell responses. We followed the original protocol (2), where TRF was described as the "late acting signal" for B cell differentiation Accordingly we added, at day two of culture, various concentrations of the above described Il-2 preparations to nu/nu spleen cells responding against sheep red blood cells (SRBC). IgM plaques were determined at day 5. A representative experiment (out of 4) is outlined in Table III.

Table III. Lack of TRF activity TH 14.83 culture supernatants

Concentration of Sn (%)	SRBC specific PFC/culture obtained with ConA spleen Sn	TH 14.83 Sn
Ø	50	50
5	500	70
10	1000	110
20	2200	160

For this experiment, unfractionated Sn of hybridoma 14.83 was used. Similar data were also obtained with the 30 - 40 000 mol. wt. fractions of this Sn. Clearly, hybridoma Sn did not effectively mediate differentiation of B cells into plasma cells and thus did not contain TRF activity, whereas the Il-2 positive fractions obtained after gel chromatography of a ConA stimulated spleen cell Sn were also rich in TRF (Table III). Our data therefore support the finding that Il-2 and TRF are different molecules with distinct functional activity: The biochemical separation of these two activities from ConA Sn of spleen cells has recently been achieved (17). Although we can conclude that Il-2 per se does not stimulate B cell responses, the data presented here do not exclude the possibility that Il-2 is, in conjunction with other factor(s), involved in B cell stimulation (18).

In conclusion, this study evidenced that hybridoma technology can be used as a tool to dissect the heterogeneous group of lymphokines and to define their function during the process of lymphocyte activation. The data presented here on the biological activity of Il-2 produced by a hybridoma suggest that the function of Il-2 is more restricted than it was envisaged in earlier studies. From this restricted functional activity of Il-2, which is T cell growth, it is implied that other factors are involved in lymphocyte maturation and differentiation.

References

1. Waldmann H, Munro AJ (1973) T cell-dependent mediator in the immune response. II. Physical and biological properties. Immunology 27: 53-64

2. Schimpl A, Wecker E (1975) A third signal in B cell activation given by TRF. Transplant Rev 23: 176-188

3. Altman A, Cohen IR (1975) Cell-free media of mixed lymphocyte cultures augmenting sensitization in vitro of mouse T lymphocytes against allogeneic fibroblasts. Eur J Immunol 5: 437-444

4. Plate JMD (1976) Soluble factors substitute for T-T cell collaboration in generation of T-killer lymphocytes. Nature 260: 329-331

5. Wagner H and Röllinghoff M (1978) T-T cell interactions during in vitro cytotoxic allograft responses. I. Soluble products from activated Ly 1^+ T cells trigger autonomously antigen primed Ly 23^+ T cells to cell proliferation and cytolytic activity. J Exp Med 148: 1523-1538

6. Watson J, Gillis S, Marbrook J, Mochizuki D, Smith KA (1979) Biochemical and biological characterization of lymphocyte regulatory molecules. I. Purification of a class of murine lymphokines. J Exp Med 150: 849-861

7. Wagner H, Hardt C, Heeg K, Pfizenmaier K, Solbach W, Bartlett R, Stockinger H, Röllinghoff M (1980) T-T cell interactions during cytotoxic T lymphocyte (CTL) responses: T cell derived helper factor (interleukin 2) as a probe to analyse CTL responsiveness and thymic maturation of CTL progenitors. Immunol Rev 51: 215-256

8. Smith KA (1980) T cell growth factor. Immunol Rev 51: 337-357

9. Watson J, Mochizuki D (1980) Interleukin 2: A class of T cell growth factors. Immunol Rev 51: 257-278

10. Stötter H, Rüde E, Wagner H (1980) T cell factor (Interleukin-2) allows in vivo induction of T helper cells against heterologous erythrocytes in athymic (nu/nu) mice. Eur J Immunol 10: 719-722

11. Wagner H, Hardt C, Bartlett R, Röllinghoff M, Pfizenmaier K (1980) The CTL immunocompetence of peanut agglutinin-positive (cortical) and negative (medullary) Lyt 123 thymocytes. J Immunol 125: 2532-2538

12. Gillis S, Union N, Baker P, Smith K (1979) The in vitro

generation and sustained culture of nude mouse cytolytic T-lymphocytes. J Exp Med 149: 1460-1476

13. Hünig T, Bevan MJ (1980) Specificity of cytotoxic T cells from athymic mice. J Exp Med 152: 688-702

14. Wagner H, Hardt C, Bartlett R, Pfizenmaier K, Röllinghoff M, Heeg, K (1980) T-lymphocyte progenitors from thymus deficient (nu/nu) mice differentiate in vitro into H-2 restricted, hapten-specific cytotoxic effector cells. Behring Inst Mitt 67: 105-109

15. Pfizenmaier K, Röllinghoff M, Wagner H (1980) Establishment of hybridomas producing Il-2. Behring Inst Mitt 67: 68-74

16. Pfizenmaier K (1982) Interleukin-producing T-cell hybridomas: A biological characterization of mediators controlling lymphocyte activation. Lymphokines, 5: 323-352

17. Schimpl A, Hübner L, Wong C, Wecker W (1980) Distinction between T helper cell replacing factor (TRF) and T cell growth factor (TCGF). Behring Inst Mitt 67: 221-235

18. Leibson H, Marrack P, Kappler IW (1981) B cell helper factors. I. Requirement for both Interleukin 2 and another 40 000 mol. wt. factor. J exp Med 154: 1681-1693

Molecular Characterization of Interleukin 2 Produced from Tumor Cell Lines and T Cell Hybridomas

S. Gillis, D. Mochizuki

I. Introduction

As a laboratory which has been keenly interested in the molecular characterization and mechanism of action of T-cell growth factor (now referred to as Interleukin 2, IL-2) we initially turned to malignant cell lines and T-cell hybridomas as sources of crude conditioned medium from which to purify IL-2. Our hope from such an approach was to uncover cellular sources of IL-2 which would generate a higher titer conditioned medium than could be obtained by conventional protocols involving mitogen stimulation of either splenocytes or peripheral blood leukocytes. We chose to investigate IL-2 production both in cases in which putative producer cells were tested for constitutive lymphokine production and in cultures containing mitogens (PHA, Con-A, pokeweed, lipopoly-saccharide) and/or phorbol esters. Because other investigators had previously shown that some murine lymphomas produced lymphokines only after lectin-triggering (1); we felt that a search solely for constitutive IL-2 producer cells might be too limited in terms of its chances for success. Abstracted below are the results of our initial studies together with experimental evidence detailing our recent progress on the biochemical characterization of IL-2. Such characterization has been further enhanced by our development of another hybridoma (of B-cell origin) whose monoclonal immunoglobulin reacts with a determinant present on the IL-2 molecule.

II. Improved Cellular Reagents for IL-2 Production and Characterization

Following culture in either the presence or absence of mitogen/phorbol ester, human and murine tumor cell line supernates were tested for the presence of IL-2 activity in a standard T-cell proliferation induction microassay (2).

Of the cell lines tested only two (one each in both human and mouse systems) was found to produce high titer IL-2 upon mitogen-stimulation. Of the murine T cell lines tested, 1% phytohemaglutinin (PHA) stimulation of cloned LBRM-33 cell lines resulted in culture supernates which contained between 100 and 500 times the amount of biologically active IL-2 which was routinely generated by identical numbers (10^6 cells/ml) of optimally stimulated rat or mouse splenocytes (3). Similarly, only PHA and phorbol ester stimulation of the human leukemia T cell line Jurkat-FHCRC produced between 100 and 300 times the amount of human IL-2/ml normally generated by lectin-stimulated human PBL or spleen cells (4).

Although the identification of high titer tumor line sources for both murine and human IL-2 have allowed extension of preparative purification protocols to include recovery of biological activity from single protein bands following detergent polyacrylamide gel electrophoresis (see below) (5, 6), two methodological problems continued to plague tumor cell line IL-2 production. The first was a problem only from the standpoint of using tumor cell line supernates as a source of lectin free IL-2. Unfortunately, tumor line IL-2 production in either murine or human model systems was not constitutive. Both Jurkat-FHCRC and LBRM-33 cell lines required mitogen and/or phorbol ester stimulation to yield supernates containing appreciable IL-2 activity. Thus, resultant supernates contained contaminating amounts of mitogen in addition to IL-2. This residual mitogen placed some limitation on the use of such crude tumor line supernates for initiation of IL-2 dependent cultures. Such conditioned medium (regardless of the dilution which is made) invariably contains small but meaningful concentrations of T cell mitogen. Mitogen may serve to activate other T cells in the starting seed population which, when proliferating in the presence of exogenous IL-2, may evolve into subpopulations with innumerable functions and antigen specificities. Although ammonium sulfate concentration and Sephadex gel filtration chromatography removes 90-95% of the Con-A present in conventional conditioned medium (2), (similar effective removal of PHA may be achieved by passage of crude IL-2 supernates over porcine-thyroglobulin-Sepharose), even minute concentrations of residual mitogen will activate a small percentage of cells to become IL-2 responsive.

The second constraint involved in using IL-2 producer tumor lines was the result of a fundamental observation regarding mitogen induced IL-2 production. As intimated above, tumor cell IL-2 biosynthesis and secretion was not constitutive and required simulation with mitogenic concentrations of lectin. Interestingly, such mitogen stimulation in addition to triggering IL-2 production also led to tumor cell death (3, 4). Therefore as a matter of practicality, IL-2 producer tumor cells must be maintained in multiple large volume cultures in order to obtain sufficient quantities of supernate for preparative purification of molecularly homogeneous IL-2.

Based on these two exigencies (the necessity for mitogen to stimulate Il-2 production and concomitant tumor cell line death) we recently attempted to create a constitutive producer of murine IL-2 activity. Using mitogen induced mortality as an inherent selection vector, we reasoned that fusion of PHA stimulated LBRM-33 cells with a drug marked T-cell lymphoma, followed by hybrid cloning in selective media, might result in initiation of a hybridoma cell line which constitutively synthesized and secreted IL-2 activity (7).

LBRM-33 lymphoma cells were stimulated for 12 hrs with 1% PHA (Grand Island Biological Co., Grand Island, NY). Following several washings with RPMI 1640-5% FCS, 2×10^7 PHA-stimulated LBRM-33 cells were mixed with 5×10^6 hypoxanthine-aminopterin-thymidine (HAT)- sensitive BW5147 cells and pelleted (5 min 160xg centrifugation). Fusion was initiated by the drop-wise addition of 1 ml of 40% polyethylene glycol (PEG) in RPMI 1640 (w/v) pH 7.2. PEG-treated cells were then pelleted again by a 12 min centrifugation (300xg). The pellet was gently resuspended by careful addition of 10 ml of RPMI 1640 medium supplemented with 10% FCS and the cell solution pelleted for a final time by a 5 min centrifugation at 160xg. The resultant cell pellet was then resuspended in 100 ml of HAT-containing Click's medium supplemented with 10% FCS, 1 mM sodium pyruvate, 25 mM HEPES buffer, 16 mM $NaHCO_3$, 50 U/ml penicillin, 50 µg/ml streptomycin and 300 µg/ml fresh L-glutamine. 2×10^8 BALB/c thymocytes (as a source of feeder cells) were added to the cell suspension and the entire cell population dispensed in 200 µl aliquots to 480 flat-bottom microplate wells (#3596, Costar Inc., Cambridge, MA). After 10 days of culture, supernates from microculture wells positive for hybrid cell growth were screened for IL-2 activity as assessed by the capacity of the culture supernate (1:10 dilution) to stimulate ^{3}H-Tdr incorporation of an IL-2 dependent T cell line.

Supernates harvested from 10 day old BW5147/LBRM fusion products stimulated significant T cell line ^{3}H-Tdr incorporation. Of the 500 microcultures tested, approximately 27 were selected as potentially containing constitutive IL-2 producer hybridoma cells. Medium control supernates or culture medium harvested from parent BW5147 cells stimulated no CTLL cell replication. Repetitive cloning of such hybrid cells led to the elaboration of cloned cell lines whose supernates stimulated extremely high levels of T cell line proliferation. In fact, supernates conditioned by several hybrid clones stimulated as much T cell line ^{3}H-Tdr incorporation as identical CTLL cultures conducted in the presence of a 1:10 dilution of a standard 3U/ml rat IL-2 preparation. Unfortunately, although IL-2 produced by the hybrid line appeared biochemically indistinguishable from LBRM-33 or murine spleen cell derived IL-2, the titer of the hybrid cell supernate (even under optimal conditions) never approached that produced by mitogen stimulated parent IL-2 producer cells (7).

Furthermore, although production of murine IL-2 producer cell lines was relatively straightforward, attempts to fuse Jurkat-FHCRC human IL-2 producer tumor cells with either the drug marked CEM or HSB-2 human T-cell leukemias have been repeatedly unsuccessful.

III. Application of Producer Tumor/Hybridoma Cell Lines for Biochemical Characterization of IL-2

Even though fusion strategies for isolation of constitutive IL-2 producer cells were only marginally successful, mitogen-induced IL-2 production by LBRM-33 and Jurkat-FHCRC cells as well as hybridoma IL-2 production has had a dramatic effect on further molecular characterization of IL-2. By being able to begin preparative protocols with a markedly higher titered CM we were able to (i) run more biologically active material through standard fractionation protocols (successive gel filtration, ion exchange chromatography and isoelectric focussing, IEF); and (ii) finish preparative IEF, retaining sufficient activity and protein to warrant further molecular characterization.

A. Development of SDS-Polyacrylamide Gel (PAGE) System for Further Fractionation of IL-2

Based in part on the relative insusceptibility of IL-2 activity to treatment with SDS and reducing agents, we chose as our next step in IL-2 purification to develop preparative PAGE systems for further separation of IEF-processed IL-2. Based in part on the application of a discontinuous, mono-tris bicene buffer system developed by Wiley and Wallace (8), we have been successful in using PAGE to recover biologically active IL-2 from electrophoresed, Coomassie blue stained and destained gels. Technical aspects of the gel system are addressed below:

1. Sample Preparation for SDS-Page

IEF pure IL-2 was dialyzed against 50 mM ammonium bicarbonate, aliquoted into appropriate volumes and lyophilized. Freeze-dried IL-2 was maintained at -70°C. Lyophilized fractions were rehydrated with sample buffer (10mM mono-tris, 10% glycerol, 0.005% bromophenol blue, 0.1-1% SDS, 0-1% 2-mercapterethanol) and heated to 70°C for 10 minutes immediately prior to electrophoresis.

2. Gel Systems

Twelve x twelve cm, 1 to 2 cm thick slab gels were 13-15% acrylamide and were electrophoresed using a 13% acrylamide, 1 cm separation gel and a 5% acrylamide, 2 cm stacking gel. Stock acrylamide solution was 30% by weight acrylamide and 0.8% by weight N, N^1-bis methylene acrylamide (bis). Separation gels were 0.112 M mono-tris, 0.029 M HCl, 0.05% TEMED, and 0.13% ammonium persulfate. Stacking gels were 0.0526 M mono-tris, 0.052 M HCl, 0.1% TEMED, and 0.5% ammonium persulfate. Electrode buffer used was 50 mM tris, 50 mM bicine, 0.03% SDS.

3. Recovery of Biological Activity

Following electrophoresis, the gel was fixed and stained with 0.25% Coomassie Blue R (Sigma, St. Louis, Missouri) in 50% methanol, 7.5% acetic acid (prefiltered through Whatman number 1). The gels were stained at room temperature while rocking for 15-30 min. The gels were destained in a solution of 20% ethanol, 7.5% acetic acid with multiple changes of the destaining solution over 2-3 hr. The individual protein bands (visualized with a light box) were cut out of the gel using a scalpel. Coomassie blue staining proteins were eluted from gel slices by electrophoresis using an apparatus obtained from CBS Scientific (Del Mar, California). The gel slices were first equilibrated in the electrode buffer (50 mM tris-acetate pH 2.8, 0.1% SDS) and then placed in the sample unit. The sample unit consisted of two vertical chambers with bore diameters of 0.1 cm and 0.25 cm, connected by a horizontal channel to allow buffer flow between the two chambers. A piece of dialysis membrane was held in place on the end of each chamber with rubber washers and screw caps. The gel slices were placed in the large bore chamber (anode). The two chambers were then connected with buffer. The sample was electrophoresed at constant voltage with 60-70 V ($\leq$0.25 watts) for 15-20 hr with a buffer change every 7-10 hr. The proteins were eluted from the gel into the smaller bore buffer (cathode) chamber. At the end of the elution procedure it was possible to see the Coomassie stain adjacent to the dialysis membrane in the smaller bore chamber. Most of the buffer was then removed from the smaller elution chamber and the protein was resuspended in a small volume (100-200 µl) of buffer by gentle pipetting. The protein sample was then assayed immediately in IL-2 microassay.

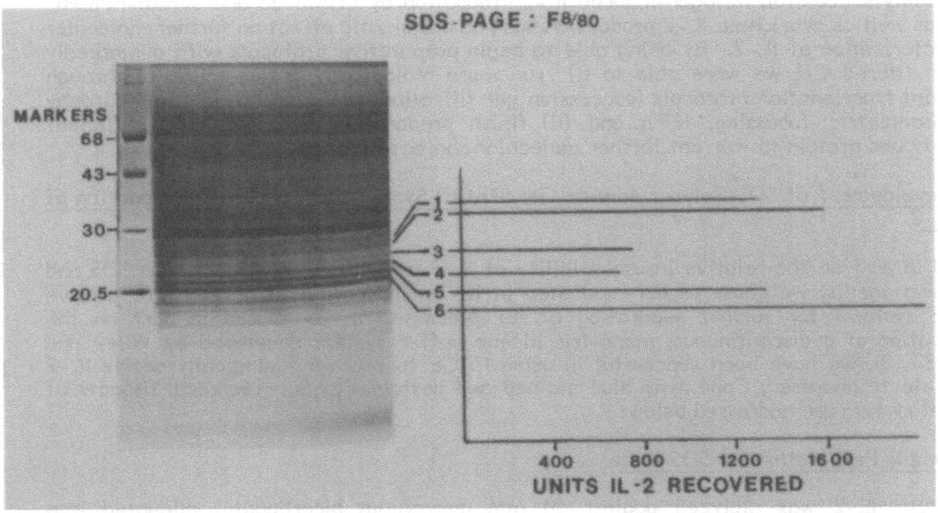

Fig. 1. The Coomassie blue protein staining pattern of LBRM IL-2 preparation F8/80 following 13% SDS-PAGE on 2 mm thick gels. The gel was loaded with 2 X 10^5 units of IL-2 which had been sequentially purified by AcA54 gel filtration, DEAE-Sephacel, and preparative IEF. The protein bands which migrated between the carbonic anhydrase (30,000 daltons) and SBTI (21,500 daltons) markers were cut out, and the protein in a portion of the gel was eluted and titrated in an IL-2 microassay. The recovery of IL-2 activity from each band is also shown on the right. Recovery of biological activity after electrophoresis, staining/destaining of the slab gel and subsequent elution was 21%. The apparent molecular weights of bands 2 and 6 were determined to be 25,000 and 21,000 daltons, respectively.

B. Page Analysis of Murine IL-2

Figure 1 shows the Coomassie blue staining pattern of LBRM-335A4 conditioned medium preparation F8/80. Prior to SDS-PAGE this CM preparation had been exposed to sequential gel filtration and ion exchange chromatography and preparative flatbed IEF. PAGE analysis diagrammed in Figure 1 is of the pI 4.9 peak. Much to our surprise, after gels were electrophoresed, stained and destained, 7-9 protein bands were visible between the markers of soybean trypsin inhibitor (20,500 m.w.) and carbonic anhydrase (30,000 m.w.). Two cm from each stained band was sliced from the gel and the protein present was electrophoretically eluted. Following elution, each band was assayed for IL-2 activity. Interestingly, two bands corresponding to proteins of 25,000 m.w. and 21,500 m.w. contained large amounts of IL-2 activity. The recovery of biological activity was approximately 20% of the total number of IL-2 units applied to the slab gel. To re-affirm molecular homogeneity, both 25 K and 21.5 K species were subjected to a second round of SDS-PAGE. The Commassie blue staining pattern of second cycle gels confirmed that both proteins migrated as well resolved bands (Figure 2). Each major IL-2 protein band had been successfully separated from minor contaminating species which had migrated slightly ahead or behind the major activity peaks. Subsequent electroelution of second set gels confirmed that both 21.5 K and 25 K proteins possessed significant IL-2 activity.

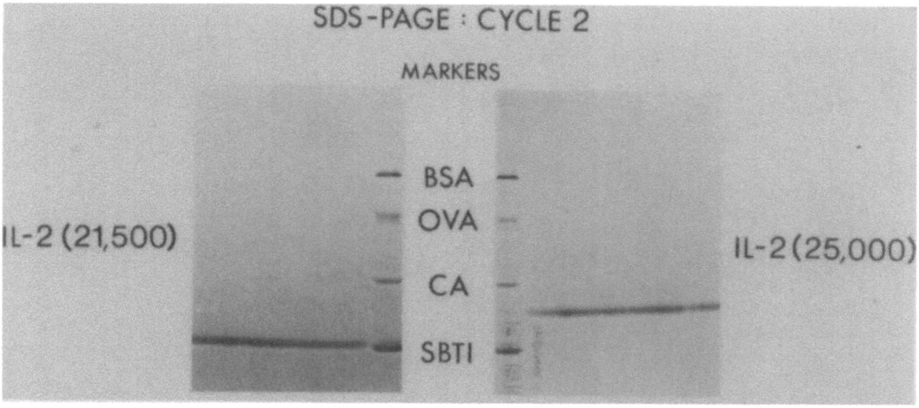

Fig. 2. The Coomassie blue protein staining pattern of LBRM IL-2 preparation F8/80 following a second cycle of SDS-PAGE on 11.6% acrylamide, 2 mm thick slab gels. The 25,000 dalton and 21,000 dalton species recovered from 2 mm thick slab gels (Fig. 1) were separately pooled and re-electrophoresed. The major 21,000 dalton and 25,000 dalton protein bands were cut out and the protein in a portion (8%) of the band was eluted and titrated in an IL-2 microassay. Each apparently homogeneous protein band was found to contain IL-2 activity.

IV. Use of Monoclonal Anti-IL-2 Antibody for Biochemical and Molecular Characterization of Human IL-2.

Recent experimentation conducted in our laboratory described the generation of B-cell hybridomas whose antibody product neutralized IL-2 dependent T-cell replication (10). Spleen cells harvested from BALB/c mice previously immunized with rat IL-2 were fused with the drug marked myeloma SP-2. Several of the resultant HAT resistant hybrid cell clones secreted a soluble product which significantly inhibited (greater than 50%) IL-2 dependent T-cell proliferation when tested in conventional IL-2 microassays. Of the potential anti-IL-2 secreting hybridomas, the clone designated 4E12B2H5 was selected for further characterization based on the capacity of this cell line's supernate to totally inhibit IL-2 dependent T-cell line proliferation. In fact, when tested at a dilution of 1:20 in the presence of the co-precipitating matrix Igsorb (lyophilized Staphylococcus aureus, The Enzyme Center, Cambridge, MA.), 4E12B2H5 culture supernate completely abrogated T-cell line replication as monitored after twenty-four hour culture in the presence of 3 U/ml IL-2. Three discrete experimental arguments suggested that the inhibitory activity was associatatedwith a monoclonal IgG antibody directed against IL-2 determinants. Passage and subsequent acid elution of cloned hybrid cell culture supernate over a Protein A Sepharose column yielded purified immunoglobulin G fractions which inhibited mouse, rat and human IL-2 activity. Secondly, hybridoma derived IgG in concert with Igsorb was capable of precipitating both "cold" and biosynthetically radio-labeled IL-2 activity. Finally, Sepharose conjugated with purified anti-IL-2 IgG provided an extremely active IL-2 absorption matrix (9). As a further indication of the ability of monoclonal antibody to react with labelled IL-2, we were curious to determine from a biochemical stand point whether precipitable cpm detected in immune precipitation (10) trials were indeed associated with IL-2. To address this question, we conducted anti-IL-2 and Igsorb precipitations in RIPA buffer (0.05 M Tris HCl pH 7.2, 1% Triton X-100, 0.1% SDS, 1% sodium deoxycholate, 0.15 M NaCl, 2 mM PMSF). Following Igsorb incubation, precipitates were washed, suspended in RIPA buffer containing 0.1% SDS, heated to 70°C for 15 minutes and immediately electrophoresed through 2 mm thick, 13½ % acrylamide slab gels. Resultant gels were stained with Coomassie blue, fixed and destained, dried and used as templates for 72 hr autoradiographs.

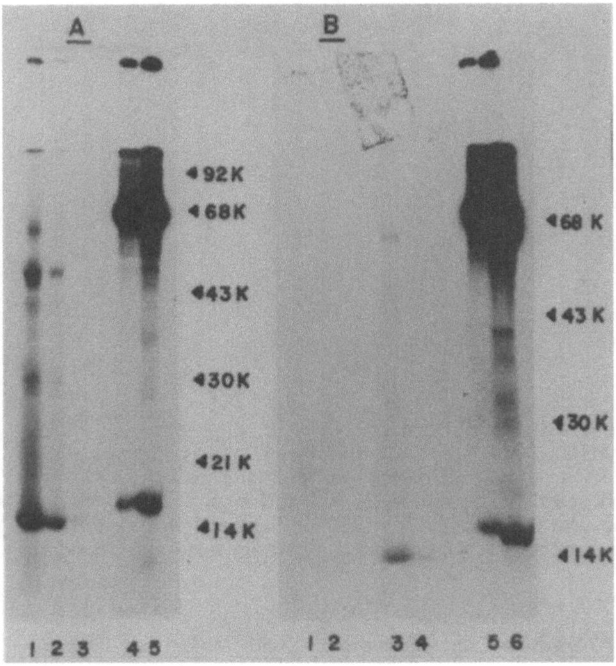

Fig. 3. Seventy-two h autoradiograph of radioimmune precipitation conducted using anti-IL-2 IgG and purified human IL-2 post-labeled with ^{125}I in the presence of carrier protein (BSA). Figure A, Lanes 4 and 4 depict electrophoresis of starting material (2 µl and 8 µl respectively). Lanes 1-3 represent electrophoresis of 5, 2, and 1 µl of immune precipitate formed by the sequential reaction of anti-IL-2 IgG (10 µg/ml) and Igsorb with 500 µl of ^{125}I-IL-2 BSA. Figure B shows a replicate experiment. Lanes 5 and 6 correspond to electrophoretic traces of labeled starting material (2 µl and 8 µl respectively). Autoradiography patterns depicted in Lanes 3 and 4 resulted from electrophoresis of 5 and 1 µl of anti-IL-2 immune precipitate prepared from reaction of anti-IL-2 IgG, ^{125}I-IL-2 BSA and Igsorb. Lanes 1 and 2 (no precipitation visible) resulted from an identical precipitating mixture; however, immune complexes between anti-IL-2 IgG, ^{125}I IL-2 and Igsorb were not allowed to form during a sequential two-step, 2 h incubation.

Results of representative experiments are depicted in Figure 3. Lanes 4 and 5 of the gel depicted in Figure 3 A resulted from electrophoresis of the starting ^{125}-labelled reagent. Lane 4 represents electrophoresis of a 2 µl sample and Lane 5 an 8 µl sample of purified Jurkat-derived IL-2 to which carrier protein (BSA...note the large, heavily stained band at approximately 68,000 daltons) had been added prior to labelling. Immune precipitation of 500 µl of ^{125}I-IL-2, followed by SDS PAGE resulted in protein patterns depicted in Lanes 1-3. A dramatic precipitation of a protein of approximately 15,000 m.w. can be seen in each lane. This labelled band was not present in 2 or 8 µl electrophoresis tracks of the starting material. The ability of the antibody to collect and precipitate such a large amount of labelled 15,000 m.w. material from 500 µl of sample speaks well of the strength and specificity of the monoclonal reagent. Lanes 1-3 correspond to electrophoresis of varying amounts of precipitated material. At present it is unclear whether minor higher m.w. proteins precipitated (visible most notably in Lane 1) represent: (i) aggregation of 15,000 m.w. IL-2; (ii) adherence of free ^{125}I to heavy and light chain immunoglobulin; or (iii) carrier labelled BSA components. Figure 3 B depicts similar immune precipitation experiments. Once again, Lane 5 and 6 represent electrophoresis of 2 or 8 µl samples of ^{125}I labelled BSA-IL-2. Lanes 3 and 4 reveal immune precipitation of 500 µl of human IL-2. A characteristic 15,000 m.w. band is highly visible. Lanes 1 and 2 represent an identical immune precipitation conducted without 2 hr incubation of antibody and labelled IL-2 and Igsorb. In the absence of incubation, one would anticipate formation of few, if any, antigen antibody complexes. As shown in Lanes 1 and 2, radioimmune precipitation conducted without 2 step 2 hr incubation did not result in precipitation of a characteristic 15,000 m.w. human IL-2 radiolabelled polypeptide band.

V. Approaching the Molecular Biology of Tumor Cell Line/Hybridoma Derived IL-2

With the advent of recombinant DNA technology and its successes in the cloning of other immunomodulating agents, most notably interferon, it has become evident that perhaps application of current molecular biology techniques to IL-2, may allow for more efficient production and more precise molecular characterization of this lymphocyte regulatory molecule.

To this end we have begun initial experimentation aimed at the isolation of IL-2 gene messenger RNA, and production of gene expression detection systems that would be essential for cloning of the IL-2 gene. In initial studies, total cellular RNA was extracted (using conventional phenol/chloroform extraction protocols) from mitogen-induced IL-2 producer tumor cells (approximately 5×10^8 Jurkat-FHCRC cells). Total extracted RNA was then passed over a column of oligo-dT cellulose in an attempt to isolate poly A+ messenger RNA. Low salt (0.01 M tris, pH 7.8) elution of oligo dT columns resulted in isolation of 80-100 µg of mRNA.

To our surprise, <u>in vitro</u> translation of 5 µg of Jurkat poly A+ mRNA in a rabbit reticulocyte system (initial volume of translation reaction = 100 µl) yielded a biologically active product. Less than 0.1 µl of translation product stimulated in excess of 4,000 cpm of murine T-cell line tritiated ^{3}H-Tdr incorporation in a standard 24 hr IL-2 assay. These results, in addition to documenting successful isolation and translation of producer cell, human IL-2 message, also lend evidence that post-translational processing (glycosylation not being possible in reticulocyte systems) may not be necessary for synthesis of biologically active IL-2.

Figure 4 further diagrams SDS-PAGE analysis of both ^{35}S-methonine Jurkat-mRNA translation product and the immune precipitate resulting from monoclonal anti-IL-2 antibody precipitation of the identical translation product. Lanes 3 and 4 of the gel represent the electrophoretic profile of the ^{35}S-methionine translation product. By comparison of this profile with that depicted in Lanes 7 and 8 of the gel (translation product in the absence of Jurkat mRNA), one gains an appreciation for the relative quality of the Jurkat mRNA preparation. Several proteins in excess of 50,000 daltons appear to have been translated in the reticulocyte system. In either case the relatively dense staining band(corresponding to the dye front) more than likely represents free ^{35}S-methionine. Lanes 5 and 6 depict the immunoprecitated product resulting from the precipitation of the non-RNA containing translation product (lanes 7, 8) with anti-IL-2 IgG. Little if any protein was precipitated. Lanes 1 and 2 represent SDS-PAGE analysis of anti-IL-2 precipitated Jurkat-mRNA translation product. Distinct protein bands at 16,000; 27,000; and 48,000 m.w. were visualized. Whether these precipitated species represent biologically active IL-2, in either monomeric or multimeric form, remains to be determined.

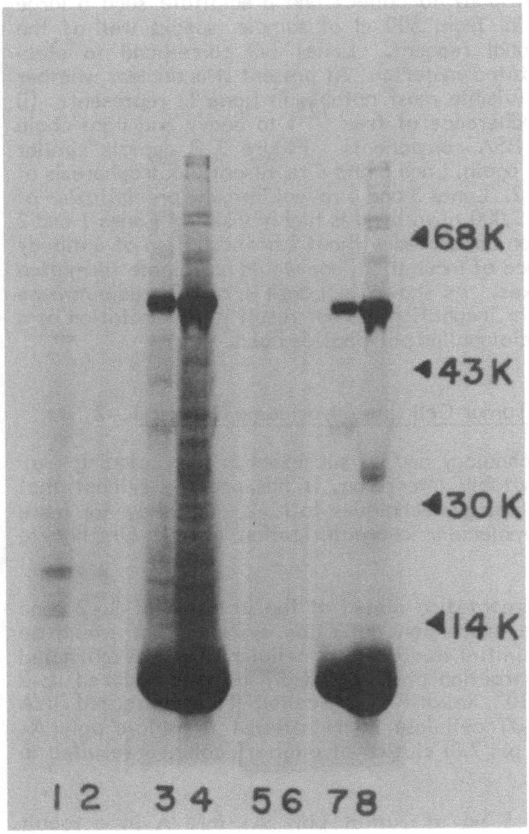

Fig. 4. Seventy-two h autoradiograph detailing SDS-Page analysis of ^{35}S-methionine Jurkat mRNA translation product. Lanes 7 and 8 depict electrophoresis of 5 and 10 µl of ^{35}S-labeled rabbit reticulocyte lysate product following translation conducted in the absence of Jurkat poly A+ RNA. Gel Lanes 3 and 4 detail mobility of labeled protein translated in the presence of IL-2 message. Lanes 5 and 6 resulted from electrophoresis of 5 and 2 µl of anti-IL-2 immune precipitation pellets conducted using 25 µl of the "no RNA" translation depicted in Lane 7 and 8. Lanes 1 and 2 show identical anti-IL-2 precipitation of in vitro translated Jurkat-mRNA, ^{35}S-methionine translation product.

VI. Conclusions

As intimated, the yields of potentially molecularly homogeneous purified IL-2 have been aided immensely by the identification of both murine and human, tumor and hybridoma cells capable of high titer IL-2 production. In addition, adaptation of SDS-gel electrophoresis protocols which allow electroelution of seemingly single polypeptides with retention of biological activity have been of great value in production of what appears to be molecularly homogeneous IL-2. The development of anti-IL-2 monoclonal antibodies has provided another approach to the definitive final purification and molecular manipulation of IL-2 as an immunotherapeutic agent. We hope that by continuing our studies in both biological and molecular arenas we will be able to provide significant data detailing the immunotherapeutic effect of IL-2 and perhaps, via application of recombinant technology, unlock what may be a multi-gene family coding for soluble immune response modifying proteins.

References

1. Ralph P, Broxmeyer M, Moore AS (1978) Induction of myeloid colony-stimulating activity in murine monocyte tumor cell lines by macrophage activators and in a T-cell line by concanavalin A. Cancer Res 38:1414
2. Gillis S, Ferm MM, Ou W, Smith KA (1978) T cell growth factor: parameters of production and a quantitiative microassay for activity. J Immunol 120:2027
3. Gillis S, Scheid M, Watson J (1980) Biochemical and biological characterization of lymhocyte regulatory molecules. III. The isolation and phenotypic characterization of Interleukin 2 producing T cell lymphomas. J Immunol 125:2570
4. Gillis S, Watson J (1980) Biochemical and biological characterization of lymphocyte regulatory molecules. V. Identification of an Interleukin 2 producer human leukemia T cell line. J Exp Med 152:1709
5. Mochizuki DY, Watson J, Gillis S (1980) Biochemical separation of Interleukin-2. J Immunol Methods 39:185
6. Frank MB, Watson JD, Gillis S (1981) Biochemical and biological characterization of lymphocyte regulatory molecules. VIII. Purification of Interleukin 2 from a human T-cell leukemia. J Immunol 127:2361
7. Stull D, Gillis S (1981) Constitutive production of Interleukin 2 activity by a T cell hybridoma. J Immunol 126:1680
8. Wiley HS, Wallace RA (1981) The structure of vitellogenin. Multiple vitellogenins in Xenopus laevis give rise to multiple forms of the yolk proteins. J Biol Chem in press
9. Gillis S, Henney CS (1981) Biochemical and biological characterization of lymphocyte regulatory molecules. VI. Generation of a B cell hybridoma whose antibody product inhibits Interleukin 2 activity. J Immunol 126:1978

The Use of T Cell Hybridomas in the Biochemical and Biological Characterization of Multiple Regulatory Factors Produced by T Cells

J.W. Schrader, I. Clark-Lewis

A. Introduction

The recognition of foreign antigens by T cells is associated with the release of soluble factors affecting not only other lymphocytes but also cells such as hemopoietic progenitor cells, mast cells, macrophages and fibroblasts (reviewed in 1). Conventional techniques for separating various cell populations have proven inadequate for determining unequivocally which of these factors are actually produced by T cells and which are produced by another cell type in response to T cell activation (2). One approach is the generation of lymphokine-producing hybridomas (3). We have demonstrated that the technique of fusing normal, activated T cells to T lymphomas to produce T cell hybridomas is a practical and fruitful approach to this and other questions about lymphokine production and T cell activation. Here we outline some of the possibilities opened by our original observation of the inducible production of the lymphokines by T cell hybridomas (4) and review our initial experiences.

Two lines of evidence indicate that the production of lymphokines by these hybridomas (1,4-6) results from fusion with, and reflects the function of, a T cell. First there is the evidence that the hybrids expressed the allotype of the Thy-1 antigen that was present on the normal T cells present in the fusion reaction. Our hybridomas were produced by fusing activated T cells bearing the Thy-1.2 allotype with T lymphomas from AKR mice bearing the Thy-1.1 allotype and the hybrids expressed both Thy-1.1 and Thy-1.2. Second there is the fact that lymphokine production required induction by a T cell mitogen, concanavalin A (4,5). Thus the spectrum of regulatory factors produced by Con A stimulation of these T cell hybridomas or spleen or lymph node cells (1,4-6) points compellingly to the conclusion that it is the T cells in these populations of normal cells that synthesize these factors after stimulation with Con A.

The hybridoma technique offers a straightforward approach to the analysis of the number and nature of lymphokines produced by different T cell subsets. Observations that different biological activities can be dissociated in different hybridomas, can also provide evidence for the existence of multiple distinct molecular species of regulators. The demonstration that the production of lymphokines by a T cell hybridoma was inducible (4,5), raised the possibility that not only the number and nature of regulatory molecules produced by the T cell, but also the mechanisms governing their release, were potentially amenable to analysis using T cell hybridomas. As noted previously (5,7), it is informative that Con A stimulates these cloned T cell hybridomas directly, in the absence of any accessory cells. Another question is whether a single type of T cell can produce different lymphokines under different circumstances. T cell hybridomas also offer many advantages as a defined source of material for biochemical and genetic analysis of lymphokines. Material for biochemical analysis can be produced in vitro in serum-free conditions (1,4) in the absence of products of non T cells; the ability to grow the T cell hybridoma cells in large numbers and the inducibility of lymphokine-production offer advantages for the preparation and cloning of the relevant genetic material.

In our initial studies, we have concentrated on two related questions, the number of discrete regulatory factors that are produced by T cells and the analysis of their molecular nature.

B. The Production and Maintenance of T Cell Hybridomas

The T cell hybridomas dealt with in this article were produced in two fusions. In the first, performed and reported in detail by Arnold et al. (8), blast cells from lymph nodes removed following immunization with key-hole limpet hemocyanin, were separated by velocity sedimentation and were fused with the thymoma BW5147. The T cell hybridoma 123 resulted from this fusion. In the second fusion (1) spleen cells were stimulated with concanavalin A and were fused to a HAT-sensitive AKR lymphoma TIKAUT. Briefly, CBA spleen cells stimulated by 24 hr incubation in serum-free medium in the presence of Con A (5 μg/ml), were washed and mixed in serum-free medium with TIKAUT at a ratio of spleen:tumour cells of 6:1. Fusion was effected using PEG (50% v/v, MW 4000, Ajax Chemicals, Australia) with dimethyl sulfoxide (15% v/v) present. Cells were plated out in 96 U-bottomed micro-cultures (10^6 cells in 0.1 ml) and 0.1 ml of HAT medium was added at 24 hr. Medium was changed every 3 days and 42 stable cell lines were obtained. Two cloned T cell hybridomas arising from this fusion, T6 and T19.1, will be discussed.

Key factors in obtaining T cell hybridomas producing the described lymphokines are the use of rapid screening tests and early and definitive cloning. Our experience suggests that screening for the production of T cell growth factor (TCGF), the assay for which is rapid and sensitive, is also a useful means of detecting clones producing factors such as colony stimulating factor (CSF), affecting hemopoietic cells, because we have not yet detected clones that make TCGF in the absence of CSF. Also rapid and sensitive is the assay for P cell stimulating factor (12), a factor stimulating the growth of a cell we have operationally termed the persisting (P) cell, because of its prolonged (> 100 days) growth in cultures supplemented with a source of this growth factor (6,7,9,10). The P cell is likely on morphological and biochemical grounds to be a progenitor of mast cells or a subset thereof (9,10). The assays for TCGF and PSF both involve homogeneous populations of target cells, the viability and growth of which is absolutely dependent on the relevant factor. We perform both assays in 10 μl cultures in Terasaki trays, containing 50-100 target cells and score growth visually.

As discussed elsewhere (1), repeated recloning with selection of suitable clones, together with the maintenance of frozen stocks of cells minimize the problem of loss of function due to loss of chromosomes or other causes. We have used the T cell hybridoma 123, for example, for a period now exceeding 3 years.

C. T Cells as a Source of Regulatory Factors

1. Multiple Regulatory Factors Are Produced by Single Monoclonal T Cell Hybridomas

Table 1 lists the factors and their putative targets which we have shown to be produced by the monoclonal T cell hybridoma 123. The range of factors produced by the hybridoma T6 was identical. In the case of 123, the presumptive normal parent was a T cell blast from an antigen-activated lymph node; in the case of T6 the presumptive normal parent was a Con A activated splenic T cell. In both cases, the production of mediators by the hybridomas was dependent on stimulation by Con A. Two points are relevant in evaluating the data in Table 1.

First, in only two of the assays listed, those for TCGF and PSF are the target populations homogeneous. In the other cases there is the formal possibility that the hybridoma-derived factor acts on a second cell type which produces the factor actually affecting the target cell monitored in the assay. T-cell replacing factor, putatively acting on the B cell (11) is one case in point. It has been argued that the TRF assay in many instances at least, measures the effect of TCGF on residual T-helper cells or their precursors. This question will not be

Table 1. Regulatory Factors Produced by Hybridoma 123

Target cell	Factor
Pluripotential hemopoietic stem cell	CFU-s-stimulating activity
Activated T cell; ?T cell precursors	TCGF
Activated B cell	TRF
Granulocyte-macrophage progenitor	GM-CSF
Megakaryocyte progenitor	MK-CSF
P cell (mast cell progenitor)	PSF
Macrophage	"MAF"
Bone marrow cells	TIF

resolved until assays for B cell factors employing homogeneous populations of B cells are available (see below).

A second point, assuming no intermediary cells are involved, relates to whether each of these activities is due to a discrete molecular species or whether individual lymphokines have multiple targets. For example we have observed that supernatants of the hybridoma 123 have activity in an assay for macrophage activating factor (MAF), involving the activation of peritoneal macrophages to kill tumour cells (12) (Table 2). A similar T cell hybridoma produces activities enhancing the production by macrophages of C2 or elastase (13). However it is known that purified granulocyte-macrophage colony-stimulating factor can stimulate macrophages at least in some ways (14) and as 123 produces colony-stimulating factor (4) it is important to examine the biochemical characteristics of this "MAF". A preliminary experiment suggests that the MAF activity produced by 123 has a larger molecular weight than CSF.

Table 2. Activation of Macrophage-mediated Cytotoxicity of Tumour Cells by Supernatants of Hybridoma 123

Added to peritoneal exudate cells	% lysis of ^{111}I-labelled P815
-	0
CAS (1%)	53 ± 3
123-supernatant (1%)	38 ± 5

2. Thy-1 Inducing Factor (TIF)

Recently we have used T cell hybridomas in a series of observations that place limits on the usefulness of the Thy-1 antigen as a marker for the T-lymphocyte lineage among the lymphohemopoietic cells of the mouse. When mouse bone-marrow cells previously depleted of T cells were cultured in the presence of medium conditioned by Con A stimulated spleen cells, within 4 days about 30% of cells expressed the Thy-1 antigen, as detected by immunofluorescence using a monoclonal anti-Thy-1.2 antibody directly conjugated with fluorescein. The specificity of binding was confirmed by the absence of binding to AKR (Thy-1.1) bone-marrow cells cultured the same way; labelling with monoclonal anti Thy-1.1 antibody gave the reciprocal result (Schrader, Battye and Scollay, in preparation).

To determine the nature of these cells, Thy-1 positive cells have been separated out using the fluorescence activated cell sorter (FACS). These experiments have shown that the Thy-1 positive cells are mainly large blasts and include hemopoietic progenitor cells capable of giving rise to colonies of myeloid cells in agar and to P cells in liquid cultures. We have not detected cells capable of growing in the presence of Con A and TCGF in this Thy-1$^+$ population, but cannot at present exclude the presence of T cells or their precursors.

We have given the activity in medium conditioned by Con A activated spleen cells the operational title of "Thy-1 inducing factor" (TIF), and are at present attempting to define its molecular nature and relationship to previously described factors. Table 3 shows results of an experiment demonstrating that the hybridomas 123 and T6 produced TIF, indicating that TIF was a T-cell product. On the other hand another T cell hybridoma T19.1, which produces large amounts of GM-CSF but not TCGF, PSF or CFU-SA, had very little activity in this assay, and TIF is likely to be distinct from GM-CSF. There is also evidence that TIF is distinct from TCGF. Thus medium conditioned by the myelomonocytic tumour WEHI-3B is positive in the assay, although it does not contain TCGF (unpublished). Furthermore biochemical fractionation using hydrophobic chromatography indicates that TIF can be separated from TCGF.

Table 3. Effect of Supernatants of T Cell Hybridomas on the in vitro Appearance of Thy-1$^+$ Cells in 3 Day Cultures of T Cell Depleted Bone Marrow Cells

Supernatants (10% v/v) of medium conditioned by:	% Thy-1$^+$ cells
-	<1
Con A stimulated spleen cells	32
" " T hybridoma 123	64
" " T hybridoma T6	43
" " T hybridoma T19.1	2
Myelomonocytic tumour WEHI-3B	29

These findings emphasize the importance of biochemical characterization of the biological activity in supernatants and the importance of a critical attitude to the use of single antigenic markers to define cell lineages. The data obtained so far clearly rule out both a role for TCGF in this effect and the view that lymphocytoid, Thy-1$^+$ cells are necessarily T cells. They serve as a caveat for the interpretation of experiments where crude supernatants from activated T cells have been claimed to have induced in the generation of T cells from "precursors" solely on the basis of the appearance of Thy-1$^+$ cells. Obviously the presence of Thy-1 on hemopoietic progenitors in the mouse also calls for a careful reassessment of conclusions about T cell differentiation pathways drawn solely from the induction of Thy-1 on lymphohemopoietic cells, or about membership of the T cell lineage, from the expression of Thy-1, for example on NK cells.

D. Significance of the Differential Production of Activities by T Cell Hybridomas

1. Use of a Panel of T Cell Hybridomas

As shown in Table 1 many of the factors present in medium conditioned by Con A stimulated spleen cells (CAS) are produced by a single monoclonal hybridoma 123, but there are other factors which are not. For example we have tested three T cell hybridomas (123, T19.1 and T6) for the production of IFN-γ with negative results (Breschkin, Clark-Lewis and Schrader, unpublished). Such an absence of a particular lymphokine among other products of a T cell hybridoma may prove useful, for example in evaluating the role of that lymphokine in effects mediated by crude spleen-cell conditioned medium. In addition the presence of one biological activity but the absence of another is strong evidence that different molecular species are involved. Elsewhere (1), we have made a comparison of two T cell hybridomas, 123 and T19.1. While 123 produced all the factors listed in Table 1, T19.1 produced only a granulocyte-macrophage colony-stimulating factor (GM-CSF). The absence of other activities in T19.1 conditioned medium clearly indicated that GM-CSF had a different molecular basis to the other activities - a conclusion we later confirmed by biochemical analysis (1,15-17; Table 5).

2. A T Cell-Derived Factor Acting on B Cells

Recently, in collaboration with Vaux, Pike and Nossal, we have examined an activity that is present in medium conditioned by Con A stimulated spleen cells and that enhances the antigen-driven proliferation of purified antigen-binding B cells (18). Because this assay deals with cultures containing single B cells or numbers of highly purified B cells so low (< 200) that the probability of the presence of a non-B cell in individual cultures is very low, the target of the factors added to the assay is unequivocally a B cell. The factor active in this assay is present not only in CAS but also in medium conditioned by Con A-stimulated EL4 cells, and in that EL4 is a thymoma (and produces all the factors listed in Table 1) the factor is likely to be a T cell product. The molecular weight appears to be about 35,000, resembling that of TCGF (unpublished). The use of the T cell hybridoma 123, however has allowed differentiation of this factor from TCGF. Thus while media conditioned by Con A-activated 123, EL4 or spleen cells, all contain readily detectable TCGF, the 123 conditioned medium alone is ineffective in the antigen-stimulated B cell growth assay. Mixing experiments rule out an inhibitor and these experiments thus suggest that a T cell product distinct from TCGF can act directly on B cells in this situation.

3. Regulation of the Expression of Ia Antigens on P Cells

We have recently found evidence for the presence in medium conditioned by Con A-stimulated spleen cells of yet another activity that regulates in this case the expression of Ia antigens on P cells. P cells are a clonable, homogeneous population of cells with a high self-renewal capacity (6,7,9,10) and probably represent a progenitor of mast cells or a subset of mast cells (9,10). The demonstration that this homogeneous population of in vitro cultured cells expresses Ia antigens (7,9) represents the first evidence that hemopoietic progenitor cells directly synthesize Ia antigens and raises the question of the function of Ia antigens on hemopoietic progenitor cells. When cultured in PSF partially purified by hydrophobic chromatography (17) instead of crude medium conditioned by Con A-stimulated spleen cells, P cells do not express Ia antigens. In the experiment shown in Table 4, P cells that had been cultured in conditions not permitting the expression of Ia antigens were washed and then cultured for 24 hours in the presence of the indicated supplements. Expression of Ia antigens was monitored using monoclonal anti-Ia antibodies and immunofluorescence (Table 4). It can be seen that the medium conditioned by the hybridomas 123, T6 and T19.1 was ineffective in inducing the expression of Ia antigens on P cells (Table 4) making it unlikely that any of the factors tested in Table 1 shared this activity. These results correlate with the production of IFN-γ but we have no definitive evidence that the relevant factor is IFN-γ or is produced by a T cell; we are now directly screening T cell hybridomas for the production of this activity. Certainly the regulation of the expression of Ia antigens on P cells raises the interesting possibility that a similar effect may occur with mature mast cells, and highlights the question of the function of Ia antigens on cells of this lineage.

Table 4. Effect of Supernatants of Con A-stimulated Cells on the Induction of Ia Antigens on P Cells

Conditioned medium (10% v/v) from Con A-stimulated:	% Ia positive P cells
-	<1
Spleen cells	96
T hybridoma 123	<1
T hybridoma T6	<1
T hybridoma T19.1	<1

E. Molecular Nature of Lymphokines Produced by T Cell Hybridomas

Table 5 summarizes the results of biochemical studies (15-17,1) on a number of lymphokines produced by the T cell hybridoma 123. It can be seen that by the criteria of relative heat resistance, hydrophobicity and molecular weight, the factors can be divided into two broad groups, TCGF and TRF in the one, and GM-CSF, PSF and CFUs-SA in the other. However GM-CSF can be separated from the other two members of the latter group (PSF and CFUs-SA) by virtue of its more acidic isoelectric point, the separation being clearer after treatment with neuraminidase (16,17,1). Thus at this level of biochemical analysis, there are a minimum of three distinct molecular species involved, although of course there could be many more not resolved by these techniques.

Table 5. Biochemical Characteristics of Lymphokines Produced by Hybridoma 123

Characteristics (Refs. 1,15-17)	TRF TCGF	GM-CSF	PSF CFUs-SA
Apparent MW			
(in saline)	35,000-40,000	29,000	29,000
(in guanidine)	30,000	23,000	23,000
Heat sensitivity	sensitive	resistant	resistant
Hydrophobicity	+++	+	+
pI (after neuraminidase)	4.9	4.7	6.5
Lectin binding	WGA*	WGA*	(WGA, Con A)†

*WGA, wheatgerm agglutinin †Done only for PSF

1. T Hybridoma Versus Spleen Cell-derived Factors

With respect to their apparent molecular weight upon gel filtration in saline or guanidine chloride (6M), and behaviour on hydrophobic chromatography, the hybridoma-derived factors are indistinguishable from their spleen cell-derived counterparts (15,17). The hybridoma-derived activities tend to be slightly more acidic and homogeneous following isoelectric focusing than the spleen cell derived activities and these differences are less evident following treatment with neuraminidase (16,17).

2. Evidence that TCGF Is Glycosylated

Previous investigators have failed to find evidence that TCGF was glycosylated (19,20). However we have found that both hybridoma-derived and spleen cell-derived TCGF binds to the lectin wheatgerm agglutinin and can be eluted with the specific sugar N-acetyl-glucosamine (16). Furthermore, treatment of hybridoma or spleen-cell derived material with neuraminidase results in a shift of the bulk of the TCGF activity to a more basic pI, an effect consistent with that of removal of sialic acid (16). Also material from cultures of 123 hybridoma cells, grown in the presence of tunicamycin, an inhibitor of protein glycosylation (16), was more homogeneous with respect to charge. Thus TCGF and also T cell-derived GM-CSF and PSF appear to be glycoproteins, although the glycosyl residues do not appear to be important for biological activity, at least in vitro.

3. Molecular Weight of T Cell Growth Factor

Recently it has been claimed that murine TCGF is composed of two subunits each with a molecular weight of 16,000 (21). Although we have obtained similar results using gel filtration in SDS, we regard our molecular weight determination of 30,000, using gel filtration in the presence of 6 molar guanidine chloride (15, Table 5) as more likely to be accurate. However as discussed elsewhere (15), recoveries of activity after denaturation with guanidine hydrochloride were low, and the situation may be more complex.

4. Molecular Weight of Hemopoietic Regulators

The hemopoietic regulatory factors we have monitored (GM-CSF, PSF and CFUs-SA) all have apparent molecular weights using gel filtration of about 30,000 in buffered saline or 23,000 in guanidine chloride (6M), whether derived from spleen cells (15,17,1), the T cell hybridomas 123 (15,17,1) or T19.1 (22) or the T lymphoma EL4 (unpublished). In contrast the CSF produced by T cell lines (23) has been reported to have a higher molecular weight (45-50,000). A GM-CSF produced by EL4 has been reported to have a molecular weight equal to that of TCGF (24). It is possible that these discrepancies relate to the presence of large amounts of serum in the samples used by these investigators. We regard the molecular weight of 23,000 determined by gel filtration in the presence of guanidine hydrochloride as the most reliable figure and note that this corresponds with the molecular weight of a number of CSF's from non-T cell sources (25).

F. Concluding Remarks

The use of T cell hybridomas has in a short period added greatly to our knowledge of the breadth of the effects that directly follow from the recognition of antigen by the T cell. A number of lymphokines can now be unequivocally designated as direct T cell products and it is clear that molecules synthesized by activated T cells affect elements of the hemopoietic differentiation pathways from pluripotential stem cells to mature end cells such as macrophages (1). Application of the hybridoma technique should result in rapid progress in delineating the precise number of molecules involved, the sets of such factors produced by different sets of T cells, and the qualitative and quantitative regulation of lymphokine production. Most excitingly, the existence of inducible tumours capable of producing a given lymphokine, should facilitate cloning of the relevant genes and the unequivocal determination of the primary structure of the biologically active molecules.

Acknowledgments

We acknowledge the collaboration of Drs. B. Arnold and P.F. Bartlett in the production of the hybridomas and Dr. J. Breschkin in the assay of interferon. We thank Ms. D. Grail and I. Frangeskakis for excellent technical assistance. This work was supported by the NH&MRC, Canberra, Australia, The Bushell Trust and Windermere Hospital Foundation.

References

1. Schrader JW, Clark-Lewis I, Bartlett PF (1981) T cell hybridoma derived lymphokines: Biology and biochemistry of factors regulating hemopoietic and lymphoid cells. Lymphokines 5:in press
2. Metcalf D (1979) Production of colony-stimulating factors by lymphoid tissues. In: S. Cohen, E. Pick and JJ Oppenheim (eds) Biology of the lymphokines. Academic Press, New York, pp 515-540
3. Howard MC, Burgess AW, McPhee D, Metcalf D (1979) T cell hybridoma secreting hemopoietic regulating molecules: granulocyte-macrophage and eosinophil colony stimulating factors. Cell 18:993-999
4. Schrader JW, Arnold B, Clark-Lewis I (1980) A Con A stimulated T-cell hybridoma releases factors affecting haematopoietic colony-forming cells and B cell antibody responses. Nature 283:197-199
5. Schrader JW, Clark-Lewis I (1981) T Cell Hybridoma-derived Regulatory Factors. I. Production of T Cell Growth Factor Following Stimulation by Concanavalin A. J Immunol 126:1101-1105
6. Schrader JW, Clark-Lewis I, Bartlett PF (1980) Lymphoid stem cells. In: RP Gale and CF Fox (eds) Biology of Bone-Marrow Transplantation. ICN-UCLA Symposia on Molecular and Cellular Biology, Vol XVII. Academic Press, New York, pp 443-460
7. Schrader JW, Nossal GJV (1980) Strategies for the analysis of accessory function: the in vitro cloning and characterization of the P cell. Immunol Rev 53:61-85
8. Arnold B, Battye FL, Miller JFAP (1979) An improved method for selecting T cell hybridomas by the fluorescence activated cell sorter. J Immunol Meth 29:353-360
9. Schrader JW (1981) The In Vitro Production and Cloning of the P Cell, a Bone Marrow-Derived Null Cell that Expresses H-2 and Ia-Antigens, has Mast Cell-Like Granules, and is Regulated by a Factor Released by Activated T Cells. J Immunol 126:452-458
10. Schrader JW, Lewis SJ, Clark-Lewis I, Culvenor JG (1981) The persisting (P) cell: histamine content, regulation by a T cell-derived factor, origin from a bone-marrow precursor and relationship to mast cells. Proc Natl Acad Sci 78:323-327
11. Schimpl A, Wecker E (1975) A third signal in B cell activation given by TRF. Transplant Rev 23:176-188
12. Pace JL, Russell SW (1981) Activation of mouse macrophages for tumor cell killing. I. Quantitative analysis of interactions between lymphokine and lipopolysaccharide. J Immunol 126:1863-1867
13. Herberman RB (1981) Production of both macrophage activating and macrophage inhibiting activities by a murine T-lymphocyte hybridoma. Nature 291:502-503
14. Handman E, Burgess AW (1979) Stimulation by granulocyte macrophage colony-stimulating factor of Leishmania tropica killing by macrophages. J Immunol 122:1134-1137
15. Clark-Lewis I, Schrader JW (1982) Biochemical Characterization of Regulatory factors derived from T cell hybridomas and spleen cells. I. Separation of T cell growth factor and T cell-replacing factor from granulocyte macrophage colony stimulating factor. J Immunol 128:168-174
16. Clark-Lewis I, Schrader JW (1982) Biochemical Characterization of Regulatory factors derived from T cell hybridomas and spleen cells. II. Evidence for glycosylation of T cell growth factor, T cell replacing factor and granulocyte macrophage colony-stimulating factor. J Immunol 128:175-180

17. Clark-Lewis I, Schrader JW (1981) P cell stimulating factor: Biochemical characterization of a new T cell derived factor. J Immunol 127:1941-1947
18. Vaux D, Pike BL, Nossal GJV (1981) Antibody production by single hapten specific B lymphocytes: An antigen derived cloning system free of filler or accessory cells. Proc Natl Acad Sci 78:7702-7706
19. Watson J, Aarden LA, Shaw J, Paetkau V (1979) Molecular and quantitative analysis of helper T cell replacing factor on the induction of antigen sensitive T and B lymphocytes. J Immunol 122:1633-1638
20. Mier JW, Gallo RC (1980) Purification and some characteristics of human T cell growth factor from phytohemagglutinin-stimulated lymphocyte conditioned media. Proc Natl Acad Sci 77:6134-6138
21. Caplan B, Gibbs C, Paetkau, V (1981) Properties of sodium dodecyl sulphate denatured interleukin 2. J Immunol 126:1351-1354
22. Burgess AW, Bartlett PF, Metcalf D, Nicola NA, Clark-Lewis I, Schrader JW (1981) Granulocyte macrophage colony stimulating factor produced by an inducible murine T cell hybridoma molecular properties and cellular specificity. Exp Hematol 9:893-903
23. Nabel G, Greenberger JS, Sakakeeny MA, Cantor H (1981) Multiple biologic activities of a cloned inducer T cell population. Proc Natl Acad Sci 78:1157-1161
24. Hilfiker ML, Moore RN, Farrar JJ (1981) Biologic properties of chromatographically separated murine thymoma derived interleukin 2 and colony stimulating factor. J Immunol 127:1983-1987
25. Burgess AW, Metcalf D, Russell SHM, Nicola NA (1980) Granulocyte/macrophage-megakaryocyte-eosinophil and erythroid colony-stimulating factors produced by mouse spleen cells. Biochem J 185:301-314

Different Factors Active in Lymphoid and Hematopoietic Proliferation Produced by Single Clones of Helper T Cell Hybridomas

C. Corbel

A. Introduction

Single clones of antigen-specific, H-2-restricted helper T cells have been found to produce a variety of lymphokines and hematopoietic factors when stimulated by antigen and adherent cells. These include factors which are active on lymphoid cells, i.e., T cell growth factor (TCGF) (1), B cell growth and maturation factor (BRMF) (2), and T cell replacing factor (TRF) (3). Various colony-stimulating activities (CSF) have been shown to stimulate proliferation of hematopoietic precursor cells. Such activities have also been described previously in supernatants from spleen cells stimulated with either mitogen (4,5) or antigen (6) or from the WEHI-3 macrophage cell line (7,8). The hybridoma technique can immortalize this capacity of normal T cells (9). T cell hybridomas offer advantages as a defined source of material for biochemical and genetic analysis of these different growth factors. It would, however, seem advantageous to have T cell hybridomas which produce only a single factor in order to determine their structure and to distinguish their activity with normal populations of cells from other activities with the same cell populations. I have therefore studied T cell hybridomas with different specificities from different fusion events as well as a series of reclones of a single T cell hybridoma. Of BRMF activities only the LPS blast-replicating activity has been determined, called B cell growth factor (BCGF) in the following. While the relative and absolute amounts of TCGF and BCGF vary even between different reclones of the same cloned T cell hybridoma I have not yet found a hybridoma producing only one activity.

B. Materials and Methods

I. Cells

The T cell hybridomas were generated similar to the way described elsewhere in this volume (10) by fusion of long-term helper T cell lines with either BW5147 (16) or the BIII/4 T hybridoma (17). Both these parental lines were thioguanine-resistant ($HPRT^-$) and ouabain-resistant (OUA^r). The following T cell hybridomas were analyzed:

<u>(BW x V_3)25-7</u> and its three subclones, D_2, D_4 and D_7. The hybridoma was isolated by the fusion of an influenza-specific helper T cell line (V_3) with BW5147 and cloned twice.

<u>(BW x V_2)-10</u> was isolated by the fusion of an influenza-specific helper T cell line (V_2) with BW5147 and cloned once (10).

(III/4 x V_1)-5 was isolated by the fusion of an influenza-specific helper T cell line (V_1) with the BIII/4 T hybridoma and cloned once (10).

(III/4 x 4037)-28 was from a fusion of a horse erythrocyte- (HRC) specific helper T cell line (4037) with the BIII/4 hybridoma and was cloned once.

Cloning was done by limiting dilution. All the mentioned T hybridomas produced TCGF in response to antigen with syngeneic adherent cells and showed helper activity in "bystander" assays with sheep erythrocytes (SRC) added as a "bystander" antigen (11).

T cell hybridomas were continuously grown in Dulbecco's modified Eagle's medium (DMEM) with $5x10^{-5}$ M β-mercaptoethanol and 5% fetal bovine serum.

II. Con A-Activated T Hybridoma Supernatants

T hybridoma cells as well as the parental lines (BW5147 and BIII/4) were grown for 24 hrs at a concentration of $1x10^6$ cells/ml in serum-free Iscove's modified DMEM (IMDM) with supplements as described (12) in the presence of 5 µg/ml Con A (Pharmacia, Uppsala, Sweden). Supernatants were harvested by centrifugation and sterilized through a 0.45-µ Millipore filter and stored at 4^oC for up to one month.

III. TCGF Assay

TCGF activity was assayed using the TCGF-dependent CTLL line, kindly provided by Dr. Kendall Smith, Dartmouth Medical College, essentially as described (13). Proliferation of $2x10^4$ cells/ml was measured after 24 hrs in the presence of Con A supernatants from T hybridomas according to a dose titration. α-methylmannoside was included at 10 mg/ml. ^{3}H-thymidine incorporation was determined after a 2-hr pulse of 2 µCi/ml ^{3}H-thymidine (specific activity 2 Ci/mmol).

IV. BCGF Assay

Supernatant media were serially diluted in culture medium and the cultures set up with 10^4 lipopolysaccharide- (LPS) activated B cell blasts as described above for the TCGF test with CTLL cells. LPS-activated B cell blasts were obtained in a 48-hr stimulation period by incubation of $5x10^5$ C57BL/6J nu/nu spleen cells with 50 µg LPS per ml of culture medium. LPS was a gift from Drs. C. Galanos and O. Lüderitz, Max-Planck-Institut für Immunbiologie, Freiburg, F.R.G. The LPS-activated B blasts were enriched by velocity sedimentation at 1 x g (14) before use in the assay. BCGF-mediated stimulation of replication of the B blasts was assayed at 48 hrs of culture by incorporation of radioactive thymidine as described above for TCGF activity tests. I define as relative unit of activity the concentration of BCGF giving half-maximal values for thymidine uptake with the supernatant of a Con A-activated T hybridoma having the highest BCGF activity. TCGF activity was quantitated in the same way as BCGF.

V. Assay for T Cell-Replacing Activity (TRF)

Supernatant media were tested at 25% in culture medium. Two 0.2-ml cultures were set up in Falcon Microtest II plates with $3\text{-}5 \times 10^6$ nu/nu C57BL/6J spleen cells and 2.5×10^6 sheep erythrocytes (SRC) per ml. The development of SRC-specific direct, IgM-secreting PFC was scored at day 5 or 6 of culture as described (15).

VI. Semisolid Cultures

Methylcellulose cultures were prepared as described (6). Methylcellulose (Methocel MC 4000, Fluka) was used at 0.8%. The culture medium IMDM contained deionized BSA (1.2%), human transferrin saturated with $FeCl_3$ (35 µg/ml), soybean lipids (150 µg/ml), and fetal calf serum (4%). Purified erythropoietin was used at 1 U/ml (kindly given by Dr. N.N. Iscove from our Institute). The inhibitor of Con A, α-methylmannoside, was always used in the presence of conditioned media from Con A-activated cells.

Bone marrow cells ($1\text{-}2 \times 10^5$ cells/ml) and 12-13-day fetal liver cells (2×10^4 cells/ml) from CBA/J, C3H/HeJ, and $(C57BL/6J \times DBA/2)F_1$ mice were cultured in 35-mm plastic Petri dishes containing 1 ml methylcellulose medium with 20% crude conditioned medium. Pokeweed mitogen-(PWM) activated, spleen cell-conditioned medium (SCM) (kindly given to us by Dr. G.R. Johnson, The Walter and Eliza Hall Insititute of Medical Research, Melbourne, Australia) and macrophage line WEHI-3 cell (originally developed by Dr. P. Ralph, Sloan-Kettering Institute, New York) supernatant were used in our study as controls. WEHI-3 cells were grown in serum-free medium (IMDM containing BSA, soybean lipids and transferrin); supernatant was taken when cells started to die.

C. Results and Discussion

Three cloned antigen-specific T cell hybridomas with different antigen specificities from different fusion experiments were tested for Con A-induced production of factors which stimulate lymphoid and hematopoietic cells. Two of these hybridomas arose from fusions with the Con A-inducible T hybridoma BIII/4 and could therefore potentially derive part or all of their factor production from that line. They were hybridomas generated from long-term antigen-specific helper T cell lines, both of which could be induced as normal T cells by Con A to produce factors. The third hybridoma (BW x V_2-10) was derived from a fusion between a Con A-inducible factor-producing normal T cell line, V_2, with the noninducible thymic lymphoma BW5147 (see also Fig. 1) (10). For two factors under study, i.e. TCGF and BCGF, quantitative assays were available (Figs. 1 and 2). TRF, CSF and the various activities of them all can only be measured qualitatively as TRF probably involves more than one factor for activity (15) and as CSF and their various subactivities are scored by less direct quantitative assays of single colonies in semisolid media. All six Con A T hybridoma supernatants contained activities that stimulate growth of pluripotential myeloid/erythroid precursors as well as of more restricted precursor cells (Table 1). Pluripotential myeloid/erythroid precursors give rise to mixed erythroid colonies containing erythrocytes, macrophages, granulocytes, and megakaryocytes. Restricted precursors give rise to either pure erythroid or erythroid/megakaryocyte, or pure macrophage, or pure granulocyte, or macrophage/granulocyte colonies. As shown in Table 1 no differences in these activities were observed.

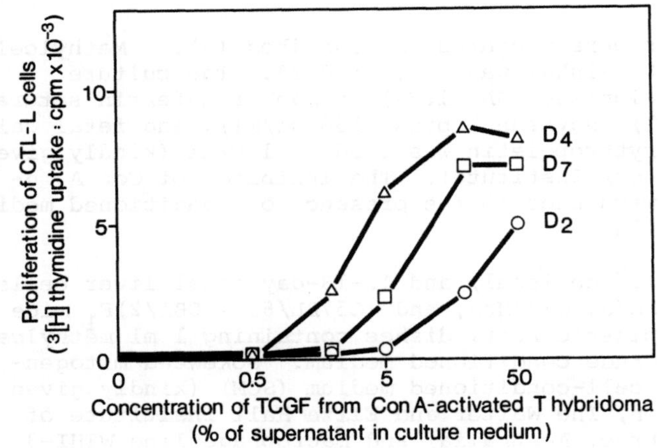

Fig. 1. Titration of TCGF in the supernatant from three different Con A-induced subclones (D_2, D_4 and D_7) of T cell hybridoma (BW x V_3)-25-7. For the assay of TCGF see Materials and Methods.

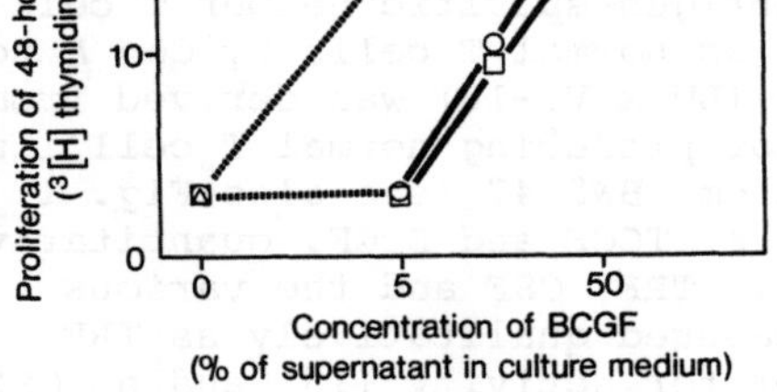

Fig. 2. Titration of B cell growth factor (BCGF) in the supernatant from three different Con A-induced subclones (D_2, D_4 and D_7) of the T cell hybridoma (BW x V_3)-25-7. For the assay for B cell growth factors see Materials and Methods.

Table 1. Colony-stimulating activities (CSF) for hematopoietic precursors* in Con A-induced supernatants from T cell hybridomas

Cell line	Total hematopoietic colonies/ 10^5 bone marrow cells	Percent of colonies		
		macrophage/ granulocyte	erythroid	mixed-erythroid
(BW x V_3)-25-7 D_2	85	87	6	7
(BW x V_3)-25-7 D_7	104	88	6	6
(BW x V_3)-25-7 D_4	92	71	18	11
(BW x V_2)-10**	+	+	+	+
(III/4 x 4037)-28	68	73	8	9
(III/4 x V_1)-5**	+	+	+	+
BW5147	-	-	-	-

*Tested on cultures of bone marrow cells from CBA/J mice; similar results were obtained with cultures from (C57BL/6J x DBA/2)F_1 and C3H/HeJ mice.

**Tested on cultures of 13-day fetal liver from CBA/J mice.

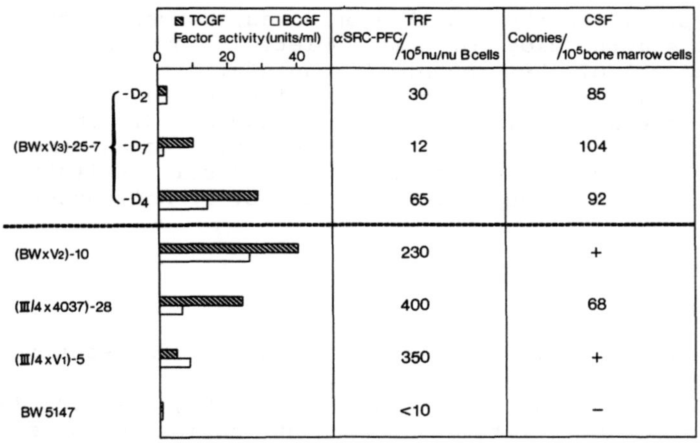

Fig. 3. Multiple activities present in supernatants from different T cell hybridomas stimulated with Con A. The upper part shows TCGF, BCGF, TRF, and CSF activities in supernatants from the three subclones of (BW x V_3)-25-7 (D_2, D_4 and D_7). The lower part shows the same activities for three different T hybridomas and the parental line, BW5147. TCGF and BCGF activities are given as arbitrary units, expressed as 50% of maximum cpm relative to the strongest activity. CSF activity was in most cases tested on bone marrow cultures. In the cases marked "+" CSF activity was assayed with fetal liver cells.

It is evident from the data in Fig. 3 that all T hybridomas produced TRF and CSF of various subactivities (Table 1) upon Con A stimulation.

Absolute amounts of TCGF and BCGF produced by the three hybridomas were different, as were the relative amounts of TCGF and BCGF. The thymic lymphoma line BW5147 could not be induced by Con A to generate supernatant active in any of these assays.

In a fourth experiment the thymic lymphoma line BW5147 was fused to a long-term virus-specific helper T cell line which upon Con A stimulation did not produce TCGF or BCGF (unpublished results). After fusion, however, the resulting T hybridomas were found to produce TCGF and BCGF, indicating that fusion can derepress or greatly enhance factor production which is too low to detect in the parental cells. The T hybridoma was twice recloned and three reclones, D_2, D_4 and D_7, were tested for Con A-inducible factor production. All three clones produced TRF and CSF (Fig. 3), including the subactivities contained in CSF (Table 1), as well as TCGF and BCGF. The absolute and (to each other) relative amounts of TCGF and BCGF were again different. No clone, however, has so far been found which produces TCGF or BCGF alone. The variable expression of factor production in different clones of the same T hybridoma indicates regulation of expression of these different lymphoid and hematopoietic factors. This suggests the possiblity that the expression of all these factors is coordinately controlled.

Acknowledgments. I thank Drs. Jan Andersson, Fritz Melchers and Jesper Zeuthen for providing the T cell hybridomas, Norman Iscove for introducing me to the methodology of hematopoietic colony assay in semisolid media, and all of these people for helpful discussions. The able technical assistance of Denise Richterich, Lotta Haraldson, Annick Peter, and Bente Jensen is gratefully acknowledged. The continuous propagation of T cell lines and T cell hybridomas done by Denise Richterich is especially appreciated. The Basel Institute for Immunology was founded and is supported entirely by F. Hoffmann-La Roche and Co., Ltd.

References

1. Gillis, S, Smith, KA (1977) Long-term culture of tumor-specific cytotoxic T cells. Nature 268: 154-156
2. Melchers, F, Andersson, J, Lernhardt, W, Schreier, M (1980) H-2-unrestricted, polyclonal maturation without replication of small B cells induced by antigen-activated T cell help factors. Eur J Immunol 10: 679-685
3. Schimpl, A, Wecker, E (1975) A third signal in B cell activation given by TRF. Transplant Rev 23: 176-188
4. Iscove, NN, Guilbert, LJ (1978) Erythropoietin-independence of early erythropoiesis and a two-regulator model of proliferative control in the hemopoietic system. In: Murphy Jr, MJ (ed) In Vitro Aspects of Erythropoiesis. Springer Verlag, New York, pp 3-7
5. Johnson, GR, Metcalf, D (1977) Pure and mixed erythroid colony formation in vitro stimulated by spleen-conditioned medium with no detectable erythropoietin. Proc Natl Acad Sci USA 74: 3879-3882
6. Schreier, MH, Iscove, NN (1980) Haematopoietic growth factors are released in cultures of H-2-restricted helper T cells, accessory cells and specific antigen. Nature 287: 228-230

7. Iscove, NN, Schreier, MH (1979) Involvement of T cells and macrophages in generation of burst-promoting activity. Exp Hemat 7: 4.
8. Wagemaker, G (1980) Early erthropoietin-independent stage of in vitro erythropoiesis: Relevance to stem cell differentiation. In: Braun, SJ, Ledney, GD, van Bekkum, DW (eds) Experimental Hematology Today, 1980. Karger, Basel, pp 47-60
9. Schrader, JW, Arnold, B, Clark-Lewis, I (1980) A Con A-stimulated T-cell hybridoma releases factors affecting haematopoietic colony-forming cells and B cell antibody responses. Nature 283: 197-199
10. Melchers, F, Zeuthen, J, Gerhard, W (1982) Influenza virus-specific murine T cell hybridomas which recognize virus hemagglutinin in conjunction with $H-2^d$ and display helper functions for B cells. (This volume.)
11. Hartmann, K-U (1970) Induction of a hemolysin response in vitro. Interaction of cells of bone-marrow origin and thymic origin. J Exp Med 132: 1267-1278
12. Iscove, NN, Melchers, F (1978) Complete replacement of serum by albumin, transferrin and soybean lipid in cultures of lipopolysaccharide-reactive B lymphocytes. J Exp Med 147: 923-933
13. Gillis, S, Ferm, MM, Ou, W, Smith, KA (1978) T cell growth factors: Parameters of production and a quantitative microassay for activity. J Immunol 120: 2027-2032
14. Miller, RG, Phillips, RA (1969) Separation of cells by velocity sedimentation. J Cell Physiol 73: 191-201
15. Andersson, J, Schreier, MH, Melchers, F (1980) T-cell-dependent B-cell stimulation is H-2-restricted and antigen-dependent only at the resting B-cell level. Proc Natl Acad Sci USA 77: 1612-1616
16. Trowbridge, IS, Hyman, R, Mazauskas, C (1978) The synthesis and properties of T_{25} glycoprotein in Thy-1-negative mutant lymphoma cells. Cell 14: 21-32
17. Grönvik, K-O, Andersson, J, Melchers, F, Zeuthen, J (1982) A stable TCGF-producing T cell hybridoma and its thioguanine-resistant variant suitable as a tool for the construction of new functional T hybridomas. (This volume.)

Lymphotoxin and Immune (γ) Interferon Production by T Cell Lines and Hybrids

N.H. Ruddle, B.S. Conta

A. Introduction

The goal of our work is an in-depth analysis of the thymus derived lymphocyte (T cell) and its products. T cell functions, products and molecular biology can now be analyzed through the use of continual T cell lines and hybrids. Primary T cells can be immortalized within the milieu of a proliferating AKR Thy 1.1^+ $HGPRT^-$ lymphoma, BW 5147, through the technique of somatic cell hybridization and selection in HAT medium. Constitutive traits such as isozymes, Thy 1 and H-2 antigens of both parents, and antigen specific activity of the primary parent can be expressed by such hybrids (1, 2, 3), and expression of many such activities is regulated by culture phase (4). We (5) and others (6, 7) have demonstrated constitutive production of antigen specific suppressor factors by such hybrids. However, it is not always possible to obtain a T cell hybrid that maintains expression of all functions of the primary parent, as chromosomes may segregate from T cell hybrids with concomitant loss of traits. This situation also occurs in B cell hybridomas (8) and all other types of somatic cell hybrids (9).

Lymphokines are T cell products that are induced by mitogen or specific antigen whose final activity is neither antigen specific nor H-2 restricted, though in some cases the target is species restricted. Their production is positively correlated with delayed hypersensitivity. Lymphotoxin (LT) is produced predominantly by Ly 1 T cells, though production by Ly 2 cells can also occur (10). Immune interferon (IFN-γ) is another lymphokine that is produced by antigen sensitized T cells after reexposure to specific antigen (11), though the Ly phenotype of such cells has not been reported. Previous workers have studied interleukin-2 (IL-2) induction in T cell hybrids by concanavalin A (Con A) (12) and by specific antigen (13) and constitutive production of low levels of IFN-γ has been reported in T cell growth factor (TCGF) maintained lines (14). There have been no previous reports of LT production by T cell lines or hybrids nor IFN-γ production in antigen specific T cell lines.

The specific goal of the work reported here was the establishment of T cell lines and hybrids that produce LT and/or IFN-γ so that we might compare the regulation, physiology and biochemistry of these two lymphokines. Preparations of each frequently contain activities associated with the other (15, 16, 17), so we wished to be able to clearly distinguish between them and to obtain substantial quantities of homogeneous IFN-γ and LT, which are both of such potential importance as therapeutic agents in cancer.

B. Materials and Methods

I. Sensitization and T Cell Preparation

Mice were injected subcutaneously in the tail with 50 or 100μg of antigen in complete Freund's adjuvant as outlined in a protocol known to induce an intense delayed hypersensitivity (18, 19, 20). Eight days later the draining periaortic and inguinal lymph nodes were removed and T cells were isolated by nylon wool purification (20) or by panning on rabbit or goat anti-mouse immunoglobulin coated plates (21).

II. T Cell Growth Factor Preparations

TCGF production was induced in Lewis rat spleen cells at a concentration of 1 x 10^7 cells/ml in RPMI plus 5% HI FCS, 5 x 10^5M 2-mercaptoethanol and 5μg Con A/ml (22). Supernatants were harvested and membrane-filtered at 24 and 48 hours, and frozen at -80°C. TCGF was also made by Con A (4μg/ml for 24 hours) induction of the T cell hybrid FS6-14.13 (kindly supplied by Dr. P. Marrack). All TCGF preparations were screened for LT and IFN activity prior to use. Individual preparations from Con A induced rat spleen cells frequently showed high levels of LT activity, whereas those from FS6-14.13 were free of contaminating LT and IFN activity. All of the results reported here were obtained with FS6-14.13 or rat spleen TCGF preparations that contained less than 10% LT activity and 4 IU/ml IFN.

III. T Cell Lines

T cell lines were developed according to the procedures of Sredni et al. (23) using cells obtained from C57BL/6 mice sensitized with 100μg 4-hydroxy-3-nitrophenyl chicken gamma globulin (Cappel Laboratory) (NP-CGG). Cells were maintained in microtiter wells (Costar 96-well flat bottom wells #3596) in Click's medium supplemented with 10% heat inactivated fetal calf serum, 250μg/ml CGG, 1 x 10^5/ml 2000r x-irradiated syngeneic spleen cells and 5% (v/v) TCGF. The T cells can be frozen in liquid nitrogen in maintenance medium supplemented with 20% HI FCS and 10% glycerol at a concentration of 2 x 10^6 cells/ml. Viability after thawing is greater than 50% and the cellular proliferation characteristics are unaltered.

IV. T Cell Hybrids

Hybrids were made as described previously (1) by means of polyethylene glycol (Baker 1540 MW, 40% weight/volume) induced fusion of BW 5147, an AKR HGPRT$^-$ Thy 1.1$^+$ lymphoma with primary T cells or T cell lines. For preparation of hybrids of the Hyb 64, 80 and 85 series, the primary T cells were obtained 8 days after sensitization with ovalbumin (OVA), purified as described above and placed in culture (1 x 10^7/ml) with 5 x 10^5/ml 2000r x-irradiated syngeneic spleen cells and 500μg OVA/ml. After 17 hrs, cells were washed and hybridized with BW 5147 at a 3:1 ratio. For preparation of the Hyb 82 and 84 series, T cell lines were maintained in the presence of spleen cells, TCGF and antigen as described above, purified of spleen cells by sedimentation through Lympholyte M prior to hybridization, and hybridized with BW 5147 at a 2:1 ratio. All fusion mixtures were dispensed in HAT medium into microtiter plates in

0.2ml volume, at concentrations of 6 x 10^4 BW 5147 cells/well and fed two or three times weekly. When cells were confluent in small wells, they were expanded to 2 or 10ml cultures. Hybrids can be frozen in liquid nitrogen in medium supplemented with 10% glycerol, 20% serum and recovered. In order to rule out the possibility that some of the lines were revertants of BW 5147 or were T cells that had survived even in the absence of TCGF, hybrids have been tested for constitutive and facultative markers characteristic of each parent. These have included isozymes, Thy 1 and H-2 (1, 4). In some cases, the presence of length polymorphisms in restriction enzyme fragments of the regions of chromosome 12 coding for the constant regions of α and μ immunoglobulin chains have been studied (24).

V. Lymphotoxin Assay

Cells or supernatants were analyzed for LT activity as described previously (10, 20). T cells or their supernatants, after incubation with specific antigen, were added to "innocent-bystander" A9 or L929 cells. Seventy two hours later the surviving fibroblasts were enumerated with a Coulter counter at a setting which excluded the (smaller) lymphocytes and included the (larger) target cells. Cytotoxicity was calculated as:

$$\% \text{ Inhibition} = 100 \times 1.0 - \frac{\text{No. A9 cells in cultures with supernatants from test cells}}{\text{No. A9 cells in cultures with control medium}}$$

Control effects of antigen, TCGF or spleen cells on A9 growth were analyzed where applicable.

VI. Interferon Assay

IFN activity was analyzed as described by Dahl and Degré (25) and measured as the protection of L929 cells from the cytopathic effect (CPE) of vesicular stomatitis virus (VSV). Test supernatants or interferon standards (IFN-β, L929 origin) were added to L929 cells. Eighteen hours later, 4 x 10^7 plaque forming units (PFU) of VSV (Indiana strain, obtained through the generosity of Dr. Peter Lengyel) were added to each well. Twenty four hours later viral cytopathic effects (ranging from 3+ to -) were scored blind. Quantitation of interferon international units (IFN IU/ml) was determined by comparison with an NIH reference standard (NIH #G-002-904-511).

VII. Acid Treatment of Interferon Preparations

Interferon produced by both antigen stimulation and Con A induction of TCGF-maintained cells was tested for acid lability by dialysis for 4 hours vs. 0.1M glycine-HCl at pH 2, followed by dialysis for 24 hrs vs. PBS. As a control, interferon containing samples were dialysed for equivalent times at pH 7. After dialysis, samples were membrane filtered and frozen at -20°C. No loss of activity was detected as a result of membrane filtration or dialysis at pH 7.

C. Results

I. LT and IFN-γ Production by T Cell Lines

The growth factor maintained lines described here are all Thy 1.2 positive, surface immunoglobulin negative, by immunofluorescence. Early after establishment, growth is stimulated by the presence of specific antigen (CGG) (Table 1). At high concentrations of TCGF (>10%) the antigen dependency for growth is not as stringent, though still apparent. The cells of most of the lines studied here have a doubling time of approximately 35 hours in the presence of antigen, 5% TCGF, and x-irradiated syngeneic spleen cells. Cells may be grown up to concentrations of 4×10^5/ml in microtiter wells, but their growth in bulk cultures is less reliable. The cell lines described here have been maintained in culture from 5 to 9 months.

Table 1. Antigen specific stimulation of DNA synthesis in line 19[a]

cpm 6 days after addition of			
TCGF — —	TCGF spleen cells —	TCGF spleen cells Hy	TCGF spleen cells CGG
536[b]	1450	1348	17,901
2157[c]	19,575	14,924	59,195

[a] T cells were isolated from C57BL/6 mice 8 days after sensitization with 100μg NP-CGG subcutaneously in the tail, and maintained with TCGF, CGG, and syngeneic x-irradiated spleen cells.
[b] 2.5% TCGF
[c] 25% TCGF

We have been able to detect LT and IFN in supernatants of certain T cell lines grown in the presence of specific antigen. Supernatants sampled from cultures of line 32 seeded at 1×10^5 cells/ml at 8 weeks after establishment contain both activities within 24 hours of plating and activity increases to maximal levels at approximately 96 hours. The addition of Con A elicits both an earlier peak and a higher titer of IFN (1048 IU/ml at 72 hrs in the experiment depicted in Table 2). The LT titer is also higher in these supernatants. The amount of LT and IFN produced by line 32 at all times is higher than that produced by primary T cells after 3 days in culture (Table 2).

Not all TCGF lines described here make and release both LT and IFN. Cells of line 19 are cytotoxic to innocent bystander cells, but only release LT into the supernatant when the T cells are grown in the presence of Con A. IFN is also released into the supernatants only if the cells are treated with mitogen. The supernatants of some of the sublines of line 32 contain only LT and not IFN, though the reverse, i.e. production of IFN in the absence of LT, has not been detected as yet.

Table 2. LT and IFN production in the presence of antigen and Con A by line 32 T cells at different times after establishment

Time[a]	Culture condition[b]	LT titer[c]	IFN IU/ml
3 days	CGG	1:4	8
8 weeks	TCGF + CGG	1:8	128
	TCGF + CGG + Con A	1:2000	1024
11 weeks	TCGF + CGG	1:4	64
26 weeks	TCGF	>1:16	256
	TCGF + CGG	>1:16	128
	TCGF + CGG + Con A	>1:16	256

[a] Days/weeks after removal and purification of LNC from mice 8 days after sensitization with NP-CGG.
[b] Supernatants were taken 72 hours after addition of x-irradiated spleen cells from CGG or CGG and Con A (5μg/ml) to 1×10^6 T cells (3 days) or 2×10^4 T cells (8,11,or 26 weeks).
[c] Titer is defined as that dilution which results in more than 30% killing of A9 cells.

The IFN produced by the T cell lines studied here fulfills criteria for immune (γ) IFN. It is produced by T cells, is elicited by specific antigen or mitogen, and is acid labile. Dialysis of supernatants against 0.1M glycine-HCl at pH 2 for 4 hours and then against PBS completely eliminates the activity. Control dialysis at pH 7 only marginally affects it. Preliminary evidence indicates that the LT produced by these T cell lines is also acid labile.

Lymphokine production and cellular proliferation are not necessarily coupled in these T cell lines. In many lines the antigen dependency for proliferation is more pronounced than for lymphokine production. Furthermore, though Con A is mitogenic for, and induces lymphokines from normal T cells, it inhibits the growth of the T cell lines described here, while at the same time it stimulates LT and IFN production. Similar Con A induction of a more differentiated phenotype (IL-2 production) and inhibition of cell growth has been observed in the case of a T cell hybrid (12), human T cell lines (25) and EL4, a T cell tumor (25). Con A inhibition of the growth of T cell tumors has even been suggested as one means of determining the cellular origin of a lymphoid tumor (26, 27). In this respect the cell lines described here are more similar to malignant than to normal T cells.

As indicated in Table 2, T cell lines tend to undergo changes after prolonged times in culture, indicating the importance of frequent subcloning. The tendency is towards a less stringent requirement for the presence of antigen during growth, and a higher constitutive level of LT and IFN-γ. This is reflected as a decreased sensitivity to antigen or mitogen induction of IFN-γ and LT, though the cells retain their sensitivity to the growth inhibitory properties of Con A.

II. Lymphotoxin Production by T Cell Hybrids

T cell hybrids which produce LT constitutively were obtained when the primary parental T cells were isolated from mice sensitized in vivo and restimulated in vitro with specific antigen prior to hybridization. LT production by such lines (Hyb 64, 80, 85) has been maintained for at least 15 weeks although loss of this property is frequently observed earlier (Table 3). Subcloning may permit isolation of those individual cells in a population which are continuing to produce LT. Supernatants from cells of the Hyb 85 series were tested for the presence of IFN. All were negative.

The cessation of constitutive LT production is probably due to segregation of chromosomes containing regulatory or structural genes involved in LT production. Assignment of these genes to a particular chromosome requires the use of cloned populations of hybrid cells producing high levels of LT. Using these clones we have been able to eliminate the involvement of chromosomes 7, 9, and 12 in LT production. Hyb 64C11, which stopped producing LT one month after hybridization, continued to express the Thy 1.2 (chromosome 9) and the GPI (chromosome 7) alleles of the primary parent. Three months after hybridization, the cells retained at least one copy of the C57BL/6 primary parent chromosome 12 detected by nucleic acid analysis with cloned immunoglobulin gene probes and made a unique Cμ RNA not made by BW 5147 (24). This suggests that the presence of chromosome 12, which contains structural genes for immunoglobulin heavy chain production, is not sufficient to allow LT production.

Hybrids were prepared between BW 5147 and T cell growth factor lines (Table 3). The parent T cell line of the Hyb 82 series was producing low levels of LT in the presence of CGG at the time of hybridization and was inducible for both LT and IFN production by Con A. Three out of 5 hybrids of this series were positive for constitutive LT production for a period of 8 weeks. None of these hybrids were inducible for LT or IFN production by either antigen or mitogen, and none made IFN constitutively. The Hyb 84 series was made with T cell growth factor line 32 which was producing significant levels of both LT and IFN-γ. Seven out of 140 hybrids were obtained which were constitutive LT producers (>30% LT), and 23/140 hybrids were inducible for LT production by Con A. By 7 weeks, 15 of the lines remain inducible by Con A for LT production. None of the 140 hybrids of the Hyb 84 series were constitutive producers of IFN, nor could they be induced to produce IFN by Con A. Three of the highest inducible LT producers (>60% LT) have been selected for subcloning and induction of LT and IFN by antigen and syngeneic spleen cells.

D. Discussion

We have presented evidence that IL-2 maintained T cell lines can produce LT and IFN-γ when stimulated by specific antigen or mitogen. These levels are higher than those produced by primary T cells immediately after isolation from mice. The IFN produced by these T cell lines after antigen or mitogen induction is acid labile, consistent with its identification as immune or IFN-γ. The IFN-γ levels reported here (2048 IU/3 x 10^5 cells/ml) are significantly higher than those in the only other report of IFN-γ production by mouse T cell lines (14). The highest level obtained in that publication at equivalent Con A levels (5μg/ml) was 800 IU/10^6 cells/ml.

Table 3. Analysis of lymphotoxin and IFN-γ production by BW 5147 T cell hybrids

Cells	Source of 1° T cells	Test time (weeks)	LT production Constitutive[a]	LT production Inducible[b]	IFN-γ production Constitutive	IFN-γ production Inducible
Hyb 64	OVA sens. T	3	4/4	NT	NT	NT
1-4	cells + OVA	4	0/4	NT	NT	NT
Hyb 80	OVA sens. T	4	2/7	NT	NT	NT
1-7	cells + OVA	14	1/7	NT	NT	NT
		15	0/7	NT	NT	NT
Hyb 85	OVA sens. T	>5	4/10	NT	0/9	NT
1-10	cells + OVA					
Hyb 82	Cell line #19	4	3/5	0/5	0/5	0/5
1-5		8	1/1	0/1	NT	NT
		10	0/1	0/1	0/5	0/5
Hyb 84	Cell line #32	5	7/140	23/140	0/140	0/140
1-140		>7	NT	15/30	0/30	0/30

[a] Number clones >30% LT positive/number clones tested.
[b] Supernatants were taken after exposure of 5×10^5 cells/ml for 72 hrs to Con A (5μg/ml) or CGG and x-irradiated spleen cells.

Levels of human IFN-γ comparable to those reported here have been detected in supernatants of a mitogen-treated human T cell tumor (28). In our hands, individual T cell lines exhibit different properties. Line 19 cells, but not supernatants kill A9 cells in the presence of antigen, but release LT and IFN-γ into the medium only upon induction with mitogen. Line 32 originally produced and released both LT and IFN-γ when stimulated by antigen, and these titers were elevated considerably upon addition of mitogen. Certain sublines of line 32 release only LT, and not IFN into the medium after antigen stimulation. It is probable that these lines undergo selection during maintenance in culture, and thus exhibit different properties.

LT can be produced in T cell hybrids and its production elevated by Con A. IFN has not been detected in such supernatants. The abrupt cessation of constitutive LT production in many T cell hybrids indicates that this trait is considerably less stable than others we have studied (29). Though T cells are not usually considered to be targets of LT activity, it may be that production of LT is slightly detrimental to T cell growth. Cells which have lost the ability to produce LT, perhaps through chromosome segregation, may possess a slight growth advantage. This chromosome segregation allows gene mapping of individual lymphokines. Our observation that cells of the Hyb 84 series are inducible with Con A indicates that these cells will be useful for the study of T cell activation.

Our work demonstrates that the technology is available for the preparation of T cell lines and hybrids which produce lymphokines. These lines have already provided important information concerning the regulation of lymphokine production. Our results with many T cell hybrids that are capable of making LT, but not IFN, and sublines of 32 which make LT but not IFN-γ, indicate that the production of LT is not dependent upon the simultaneous synthesis of IFN, and that these two activities are physically separable. There are technical problems inherent in the use of both T cell lines and hybrids. Difficulties in cloning, adaptation to growth in bulk culture, the requirement for the presence of spleen cells, and TCGF preparations which may contain other lymphokines are all disadvantages in the use of T cell lines. The presence of the BW 5147 genome is a complicating factor in the use of T cell hybrids, as is marker instability due to chromosome segregation and epigenetic factors. Nevertheless, our work indicates that it is possible to obtain large quantities of lymphokines for biochemical analysis and to begin studying cellular regulation of lymphokine production in antigen specific populations of cloned cells.

References

1. Ruddle NH (1978) T cell hybrids with specificity for individual antigens. Curr Top Microbiol Immunol 81: 203-211
2. Whitaker RB, Ruddle NH (1980) Antigen specific T cell hybrids I. Ovalbumin binding T cell hybrid. Cellular Immunol 55: 56-65
3. Ruddle NH, Beezley BB, Lewis GK, Goodman JW (1980) Antigen specific T cell hybrids II. T cell hybrids which bind azobenzenearsonate. Molec Immunol 17: 925-931
4. Ruddle NH, Beezley BB, Eardley DD (1980) Regulation of self-recognition by T cell hybrids. Cellular Immunol 55: 42-55
5. Goodman JW, Lewis GK, Primi D, Hornbeck P, Ruddle NH (1980) Antigen specific molecules from murine T lymphocytes and T

lymphocytes and T cell hybridomas. Molecular Immunol 17: 933-945

6. Kapp JA, Aranco BA, Clevinger BL (1980) Suppression of antibody and T cell proliferative responses to L-glutamic acid[60]-alanine[30]-L-tyrosine[10] (GAT) by a specific monoclonal T cell factor. J Exp Med 152: 235-240
7. Taniguchi M, Saito T, Tada T (1979) Antigen-specific suppressive factor produced by a transplantable I-J bearing T-cell hybridoma. Nature 278: 5704 555-558
8. Kohler G, Pearson T, Milstein C (1977) Fusion of T and B cells. Somatic Cell Genet 3: 303-312
9. Giles RE, Ruddle FH (1973) Production and characterization of proliferating somatic cell hybrids. in: Tissue Culture, Academic Press, New York, 475-500
10. Eardley D, Shen FW, Gershon RK, Ruddle NH (1980) Lymphotoxin production by subsets of T cells. J Immunol 124: 1199-1202
11. Epstein LB (1981) Interferon as a model lymphokine. Fed Proc 40: 56-61
12. Harwell L, Skidmore B, Marrack T, Kappler J (1980) Concanavalin A-inducible, interleukin-2-producing T cell hybridomas. J Exp Med 152: 893-904
13. Kappler JW, Skidmore BJ, White J, Marrack P (1981) Antigen-inducible, H-2 restricted, IL-2 producing hybridomas: lack of independent antigen and H-2 recognition. J Exp Med 153: 1198-1214
14. Marcucci F, Waller M, Kirchner H, Krammer P (1981) Production of immune interferon by murine T cell clones from long-term cultures. Nature 291: 79-81
15. Ware CF, Granger GA (1979) A physicochemical and immunologic comparison of the cell growth inhibitory activity of human lymphotoxins and interferons in vitro. J Immunol 122(5): 1763-1770
16. Blalock JE, Georgiades JA, Langford MP, Johnson HM (1980) Purified human immune interferon has more potent anticellular activity than fibroblast or leukocyte interferon. Cell Immunol 49 (2): 390-394
17. Ito M, Buffett RF (1981) Cytocidal effect of purified human fibroblast interferon on tumor cells _in vitro_. J Nat Canc Inst 66: 819-825
18. Ruddle NH (1974) Cytotoxicity reactions mediated by antigen-activated rat and mouse lymphocytes. in: Mechanisms of Cell-Mediated Immunity (McCluskey and Cohen), Wiley, Chichester, pp. 401-407
19. Ruddle NH (1978) Delayed hypersensitivity to soluble antigens in mice I. Analysis _in vivo_. Int Archs Allergy Appl Immun 57: 560-566
20. Ruddle NH (1979) Delayed hypersensitivity to soluble antigens in mice II. Analysis _in vitro_. Int Archs Allergy Appl Immun 58: 44-52
21. Wysocki LJ, Sato VL (1978) "Panning" for lymphocytes: A method for cell selection. Proc Natl Acad Sci 75: 2844-2848
22. Gillis S, Ferm MM, Ou W, Smith KA (1978) T cell growth factor: Parameters of production and quantitative microassay for activity. J Immunol 120: 2027-2032
23. Sredni B, Tse H, Schwartz R (1980) Direct cloning and extended culture of antigen-specific MHC-restricted proliferating T lymphocytes. Nature 283: 581-583
24. Zuniga MC, D'Eustachio P, Ruddle NH (1982) Immunoglobulin heavy chain gene rearrangement and transcription in murine T cell hybrids and T lymphomas. PNAS in press

25. Stadler BM, Dougherty SF, Farrar JJ, Oppenheim JJ (1981) Relationships of cellcycle to recovery of IL-2 activity from human mononuclear cells, human and mouse T cell lines. J Immunol 127: 1936-1940
26. Ralph P (1973) Retention of lymphocytic characteristics of myelomas and θ+ lymphomas: sensitivity to cortisol and phytohemagglutinin. J Immunol 110: 1470-1475
27. Ralph B, Nakoinz I (1973) Inhibitory effects of lectins and lymphocyte mitogens on murine lymphomas and myelomas. J Nat Canc Inst 51: 883-890
28. Nathan I, Groopman JE, Quan SG, Bersch N, Golde DW (1981) Immune (γ) interferon produced by a human lymphoblast cell line. Nature 292: 842-844
29. Ruddle NH (1982) T cell hybrids: Their functions, products, and regulation. Lymphokine Reports 5: 49-76

Antigen Specificity of Continuous T Cell Lines

J.F.A.P. Miller, P.L. Mottram, W.R. Thomas, J. McKimm-Breschkin, I.L. Walker, J. Gamble

Introduction

T cells are an extremely heterogeneous population composed of various subsets which interact in a complex manner during an immune response. Furthermore each T cell expresses an antigen-specific receptor unique in its discriminating capacity and in its restriction by gene products of the major histocompatibility complex (MHC). The availability of defined T cell clones should help elucidate the nature of T cell specificity and the function of particular T cells in immune circuits. We have produced T cell lines either by growing T cells from immune mice with antigen, antigen-presenting cells (APC) and T cell growth factors (interleukin-2, IL-2), or by immortalizing T cells after fusing them to T lymphoma cell lines. Some of the results obtained in this work are summarized here.

Hapten-specific T Cell Lines

T cells from mice sensitized to 4-ethoxymethylene-2-phenyl oxazolone (OX), picryl chloride (PCl) and azobenzenearsonate (ABA) were cultured as detailed elsewhere (1). They were continuously stimulated with antigen presented by syngeneic, irradiated APC and supplemented with a source of IL-2. Three independent assays of antigen-specificity and MHC restriction were performed: in vivo local delayed-type hypersensitivity (DTH), in vitro antigen-dependent proliferation and interferon (IFNγ) production. These tests have been described elsewhere in detail (1,2).

The surface markers of 4 continuous cell lines examined are given in Table 1. All lines were Thy-1$^+$ and expressed Lyt-1 or 2 to different degrees.

The antigen specificity and MHC restriction of these lines were tested in vivo and in vitro and the results are summarized in Tables 2 through 5. Antigen specificity was evident in all tests. The lines produced swellings in local DTH assays at cell doses from 10 to 100-fold less than those required by uncultured immune cells. They could be activated either by hapten-self moieties or hapten-foreign protein conjugates (Table 2). This might not have been expected from previous studies in guinea pigs which responded poorly, after contact sensitization, to hapten presented on foreign carriers (3). The results do however agree with those on hapten-specific DTH to NP (4) and ABA (5) in mice.

Table 1. Continuous T cell lines from mice immunized with contact sensitizers

Line			% of cells fluorescent after staining with		
number	strain	antigen specificity	anti-Thy-1.2	anti-Lyt-1	anti-Lyt-2
05	BALB/c	OX	>97	50	<5
010	CBA	OX	>97	<5	<5
P2	BALB/c	PCl	>97	90	<5
A50	A/J	ABA	>97	<5	<5

Table 2. In vivo hapten specificity of T cell lines

Exp.	Immune cells injected into syngeneic recipients	Number of cells	Challenge	Ear swelling*(10^{-2}mm) in response to Immune cells	Normal lymph node cells
1	BALB/c OX-immune lymph node cells+	$2.5x10^6$	0.5% OX	6.2 (0.6)	0.7 (0.7)
		$1.2x10^6$	0.5% OX	5.4 (1.0)	2.0 (0.8)
2	BALB/c OX 05	$4x10^5$	0.5% OX	8.0 (0.6)	2.7 (0.4)
		$6x10^4$	0.5% OX	7.3 (0.5)	-
		10^4	0.5% OX	4.9 (1.2)	-
		$3x10^5$	0.1% PCl	1.4 (0.6)	1.6 (0.5)
3	BALB/c OX 05	$3x10^5$	OX KLH	11.1 (2.2)	1.6 (0.4)
		$3x10^5$	TNP KLH	2.4 (1.7)	1.8 (2.0)
4	CBA OX 010	$5x10^5$	0.5% OX	18.1 (2.5)	3.3 (1.3)
		$5x10^5$	0.5% PCl	4.7 (1.3)	-
5	BALB/c PCl P2	$5x10^5$	0.5% PCl	14.6 (2.0)	4.9 (1.6)
		$5x10^5$	0.5% OX	2.4 (1.0)	0.8 (0.8)
6	A/J ABA A50	$5x10^5$	0.2% ABA	9.9 (1.8)	2.9 (0.8)
		$5x10^5$	0.5% OX	3.7 (0.4)	2.3 (0.5)

* Mean (± SE) 5 mice per group

\+ Draining lymph node cells 3 days after painting 3% OX containing about 75% T cells.

In all tests, the lines reacted when challenged with the correct antigen presented by APC of the correct MHC (Tables 3-5). This MHC restriction enabled us to determine whether irradiated APC might have contributed to the reactions: when the CBA line "010" was cultured for 1 week with irradiated (CBA x BALB/c)F_1 APC and filler cells and tested for DTH in CBA or BALB/c mice, a reaction occurred only in CBA mice (Table 3).

Table 3. In vivo MHC restriction of T cell lines

Cells injected		Ear swelling (10^{-2}mm)*	
Line	Recipient	Cell line	Normal lymph node cells
CBA OX 010 cultured on CBA APC	CBA (H-2^k)	15.1 (3.1)	1.4 (1.0)
	BALB/c (H-2^d)	2.7 (2.3)	1.2 (0.6)
	BALB/k (H-2^k)	7.5 (0.2)	1.3 (0.6)
	$(CBAxBALB/c)F_1$	13.9 (0.9)	1.6 (0.6)
CBA OX 010 cultured on $(CBAxBALB/c)F_1$ APC	CBA	23.4 (1.8)	2.0 (0.5)
	BALB/c	0.4 (0.2)	1.5 (0.3)
BALB/c OX 05	BALB/c	16.2 (1.6)	2.6 (0.5)
	CBA	3.0 (0.8)	3.0 (0.7)

* Mean (±SE) 5 mice per group

Table 4. Antigen specificity and MHC restriction of T cell lines proliferating in vitro

Exp.	Line	Antigen	APC strain ($5x10^4$ cells)	cpm (± SE)*
1	BALB/c OX 05	OX	BALB/c	25,286 ± 1,290
		PC1	BALB/c	4,126 ± 324
		OX + PC1⁺	BALB/c	31,528 ± 366
2	BALB/c OX 05	OX	BALB/c	106,002 ± 9,514
		OX	CBA	1,510 ± 300
		OX	$(CBAxBALB/c)F_1$	88,028 ± 8,360
3	A/J ABA A50	ABA¶	A/J	37,427 ± 1,978
		PC1¶	A/J	4,709 ± 597

* ^{3}H-TdR incorporation by $5x10^4$ cells measured after 3 days
\+ $5x10^4$ PC1 + $5x10^4$ OX APC
¶ coupled in vitro

We also investigated the ability of the cell lines to produce IFNγ after antigen stimulation. As shown in Table 5, IFNγ production required not only the correct antigenic stimulation but also the APC of the appropriate MHC: it was restricted by the I region. A minimum of 2 x 10^4 APC was necessary for the induction of detectable IFNγ and optimal induction occurred with a ratio of APC to line lymphocytes >10:1. IFNγ was first detected at 6 to 12 hours and reached a maximum at 30 hours. Thereafter the level remained constant for at least another 3 days (2).

Some of the above cell lines were fused to T lymphomas such as BW 5147 or TIKAUT and some antigen-specific T cell hybridomas were obtained. They will be described elsewhere.

Table 5. Antigen specificity and MHC restriction of IFNγ induction by T cell lines

Source of APC			IFNγ titre (PRD_{50})	
Number	Antigen	Strain	BALB/c OX 05*	CBA OX 010
10^5	PCl	CBA (H-2^k)	-	<5
$3x10^4$	OX	CBA	-	5
10^5	OX	CBA	-	20
$3x10^5$	OX	CBA	<5	100
10^6	OX	CBA	-	200
$3x10^5$	OX	BALB/c (H-2^d)	20	<5
$3x10^5$	OX	BALB/c (H-2^k)	<5	200
$3x10^5$	OX	$(CBAxBALB/c)F_1$	50	100
$3x10^5$	OX	A.TL	<5	200

* 10^5 antigen reactive cells; supernatants harvested after 48 hours for IFN assay

Pitfalls Encountered in Screening for Antigen or Idiotype-specific T Cell Hybridomas

We have encountered certain pitfalls in screening for antigen-specific T cell hybridomas. One was the presence, in some T cell hybridoma and lymphoma supernatants, of material nonspecifically inhibitory in radioimmunoassays: an apparent specificity resulted from difference in the susceptibility of the assays to nonspecific interference. This has been described in detail elsewhere (6). Another pitfall is described here.

T cell hybridomas were constructed by fusing ABA-immune splenic T cells to the T lymphoma EL-4. Two individual hybridoma clones, 9 and 14, were identified which bound a purified monoclonal Ig, "7.1.3". This is specific for ABA, is an IgG_{2a} molecule and bears the entire cross reactive idiotype characteristic of 30 to 70% of the anti-ABA antibodies of A/J mice (7). The 7.1.3 molecules were purified by affinity chromatography on an ABA-BSA-Sepharose column and were eluted with 0.1M glycine, pH 2.8 (HCl). The column was eluted with 6M guanidine HCl, 0.1M Tris pH 7.5 (HCl) immediately after and before use to remove avidly bound Ig molecules which were not eluted with low pH glycine buffer.

The two hybridomas, 9 and 14, appeared initially to have the same receptor specificity as a well defined category of suppressor T cells (Ts_2) induced by administration of ABA conjugates to A/J mice (8). The specificity of the binding reaction was exemplified by the ability of purified 7.1.3 or purified serum anti-ABA antibodies to act as inhibitors and by the failure of MOPC 173 and other class and allotype matched monoclonal Ig or normal mouse serum to do so (Fig. 1).

The parental T lymphoma and a T cell hybridoma derived from T cells immune to oxazolone did not exhibit enhanced binding of 7.1.3. Taken together these data suggested that hybridomas 9 and 14 had clonally expressed receptors specific for a major idiotype rather than receptors for Fc. A relationship between specific binding ability and Fc receptors was however demonstrated by the inability of the hybridoma clones to bind either the Fab or $F(ab)_2$ fragments of the CRI-bearing monoclonal Ig. Furthermore, removal of a small fraction (<5%) of ^{125}I-7.1.3 by absorption with clone 9 cells resulted in the loss of most or all of the molecules which could "specifically" bind to these cells (Fig. 2). Since over 60% of ^{125}I-7.1.3 molecules in the same preparation were idiotypically intact, at least 55% of the ^{125}I-7.1.3 remaining after absorption would have had binding affinity for a genuine idiotype-specific receptor. Thus, a small fraction of 7.1.3 molecules were responsible for the high binding values observed. It is not clear whether these molecules were aggregated as a result of affinity chromatography and elution or whether they formed immune complexes with ABA-BSA molecules after their covalent detachment from the affinity matrix. The Ig, 23B3 and 21A5, although of the same class and allotype, failed to bind to clone 9 cells (Fig. 1) even though both had been purified by acid elution from protein A Sepharose. Thus acid elution alone did not seem to be responsible for the formation of aggregates of 7.1.3 Ig.

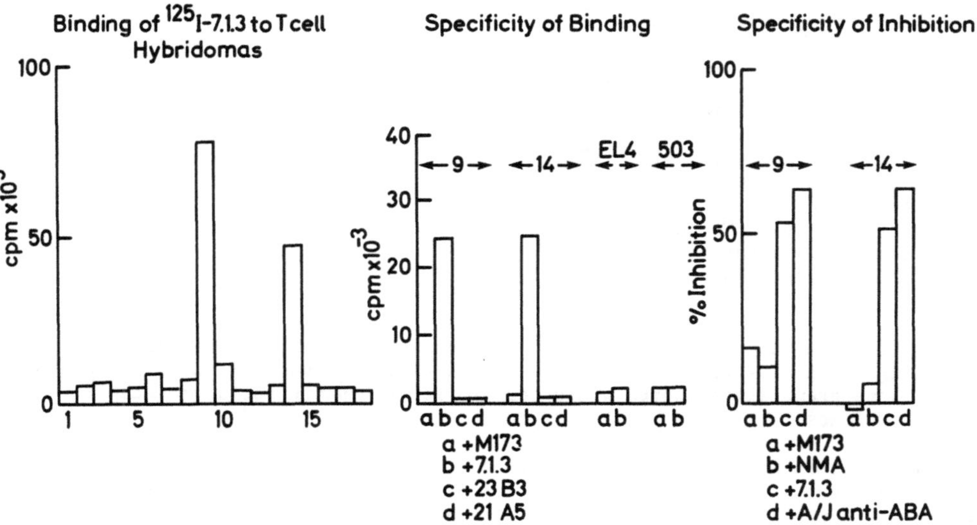

Fig. 1 (1) Left: 10^7 cells from 18 cloned T cell hybridomas derived from an A/J mouse immune to ABA were incubated with 10^6 cpm of ^{125}I-7.1.3 in a total volume of 100 µl for 1 hour at room temperature. Cells were then washed and bound radioactivity determined in a Packard gamma counter. (2) Centre: binding of several class-matched Ig molecules to T cell hybridoma clones 9 and 14 (from (1)), 503 (a hybridoma resulting from a fusion of oxazolone-primed cells) and the T lymphoma EL4, was carried out as in (1). MOPC 173 is a BALB/c IgG_{2a} myeloma protein. 7.1.3, 23B3 and 21A5 are all A/J IgG_{2a} hybridoma proteins directed towards ABA, TEPC 15 and TEPC 15, respectively. (3) Right: binding of ^{125}I-7.1.3 to T hybridomas 9 and 14 was carried out in the presence of 5 µg/ml M173, 5 µg/ml 7.1.3, 5 µg/ml affinity purified A/J anti-ABA Ig, or 5% v/v normal mouse ascites (NMA).

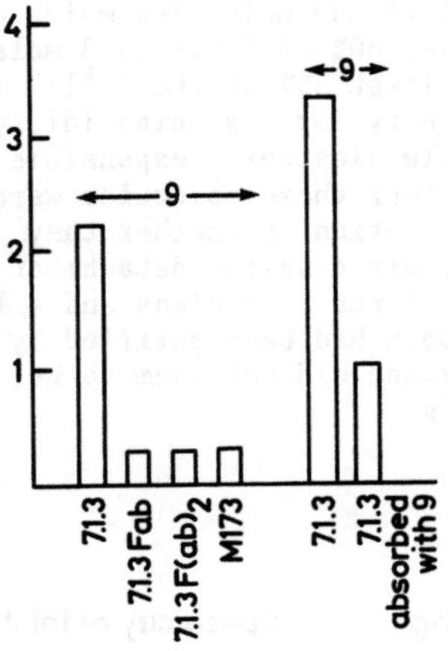

Fig. 2 Binding to hybridoma 9 of ^{125}I-7.1.3, 7.1.3 Fab, 7.1.3 $F(ab)_2$ and M173. The preparation of ^{125}I-7.1.3 used in the right panel was absorbed for 1 hour with clone 9 cells (10^7cpm, 10^8 cells). The unbound radioactive material was then rebound to fresh cells (10^6cpm, 10^7 cells).

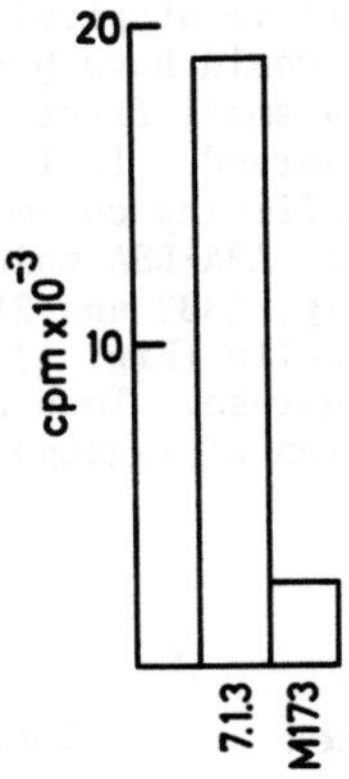

Fig. 3 Binding of ^{125}I-M173 and 7.1.3 to line 18.81, an Abelson pre-B lymphoma, was assessed as described in Fig. 1 (1).

The ability of both purified 7.1.3 and serum anti-ABA to act as unlabeled inhibitors of binding correlates with the likely presence of low amounts of immune complexes within these preparations which would compete for receptors for aggregated Ig or immune complexes.

Finally 10.81, an Abelson virus-transformed pre-B cell line (9), was also shown to exhibit apparent specificity in binding 7.1.3 Ig (Fig. 3). This line is known to possess Fc receptors but no functional Ig receptors since it produces μ chains without κ or λ light chains.

The results of this study underline the necessity to distinguish rigorously Fc receptors from immunospecific T cell receptors in binding tests using specific antibodies.

Acknowledgments. The work in the authors' laboratory was supported by a grant from the National Health and Medical Research Council of Australia.

References

1. Thomas WR, Mottram P, Miller JFAP (1982) Hapten-specific T cell lines mediating delayed hypersensitivity to contact sensitizing agents. J Exp Med in press.
2. McKimm-Breschkin JL, Mottram P, Thomas WR, Miller JFAP (1982) Antigen specific production of immune interferon by T cell lines. J Exp Med in press.
3. Benacerraf B, Gell PGH (1959) Studies on hypersensitivity. III. The relation between delayed reactivity to the picryl group of conjugates and contact sensitivity. Immunol 2: 219-229.
4. Weinberger JZ, Benacerraf B, Dorf ME (1979) Ir gene controlled carrier effects in the induction and elicitation of hapten-specific delayed-type hypersensitivity response. J Exp Med 150: 1255-1259.
5. Bach BA, Sherman L, Benacerraf B, Green MI (1978) Mechanisms of regulation of cell-mediated immunity. II. Induction and suppression of delayed-type hypersensitivity to azobenzene arsonate-coupled syngeneic cells. J Immunol 121: 1460-1468.
6. Miller JFAP, Walker IL. Arnold B, Gamble J (1982) Products of T cell lines. In "Lymphokines: A Forum for Non-antibody Lymphocyte Products", vol. 2, Pick E, ed Acad Press, New York: 1-23.
7. Kuettner MG, Wang AL, Nisonoff A (1972) Quantitative investigations of idiotypic antibodies. VI. Idiotypic specificity as a potential genetic marker for the variable regions of mouse immunoglobulin polypeptide chains. J Exp Med 135: 579-595.
8. Sy M-S, Dietz MH, Germain RN, Benacerraf B, Greene MI (1980) Antigen and receptor driven regulatory mechanisms. IV. Idiotype-bearing I-J$^+$ suppressor T cell factors induce second order suppressor cells which express anti-idiotypic receptors. J Exp Med 151: 1183-1195.
9. Siden EJ, Baltimore D, Clark D, Rosenberg NE (1979) Immunoglobulin synthesis by lymphoid cells transformed in vitro by Abelson murine leukemia virus. Cell 16: 389-396.

List of Contributors

L. Adorini, CNEN-Euratom Immunogentics Group, Laboratory of Radiopathology, CNS Casaccia, Rome, Italy

F. Albert, Centre d'Immunologie, INSERM-CNRS de Marseille-Luminy, Case 906, F-13288 Marseille Cedex 9, France

A. Altman, Department of Molecular Immunology, Scripps Clinic and Research Foundation, La Jolla, CA 92037

J. Anderson, Basel Institute for Immunology, 487 Grenzacherstrasse, Postfach, CH-4005 Basel, Switzerland

B.A. Araneo, Department of Pathology and Laboratory Medicine, The Jewish Hospital of St. Louis and Department of Pathology and of Microbiology and Immunology, Washington University School of Medicine, St. Louis, Missouri 63110

E. Arman, Department of Chemical Immunology, The Weizman Institute of Science, Rehovot 76100, Israel

G.L. Asherson, Division of Immunological Medicine, Clinical Research Centre, Watford Road, Harrow, UK

E. Barbanti, CNR Center of Cytopharmacology, University of Milan, Via Vanvitelli, 32, 20129 Milan, Italy

G. Berke, Department of Cell Biology, The Weizman Institute of Science, Rehovot 76100, Israel

S. Bitton-Großfeld, Department of Chemical Immunology, The Weizman Institute of Science, Rehovot 76100, Israel

C. Bron, University of Lausanne, Institute of Biochemistry, CH-1066 Epalinges, Switzerland

M. Buferne, Centre d'Immunologie, INSERM-CNRS de Marseille-Luminy, Case 906, F-13288 Marseille Cedex 9, France

S. Carel, University of Lausanne, Institute of Biochemistry, CH-1066 Epalinges, Switzerland

I. Clark-Lewis, The Walter and Eliza Hall Institute of Medical Research, Post Office, Royal Melbourne Hospital, Melbourne, Victoria 3050, Australia

B.S. Conta, Department of Epidemiology and Public Health, Yale University School of Medicine, P.O. Box 3333, 60 College Street, New Haven, CT 06510, USA

M.A. Cooley, Centre d'Immunology, INSERM-CNRS de Marseille-Luminy, Case 906, F-13288 Marseille Cedex 9, France

C. Corbel, Basel Institute for Immunology, 487 Grenzacherstrasse, Postfach, CH-4005 Basel, Switzerland

G. Corradin, University of Lausanne, Institute of Biochemistry, CH-1066 Epalinges, Switzerland

M. Cramer, Institute for Genetics, University of Cologne, Weyertal 121, D-5000 Cologne, FRG

C.M. Croce, The Wistar Institute, 36th Street at Spruce, Philadelphia, PA 19104, USA

E.J. Culbert, ICRF Tumour Immunology Unit, University College, Gower Street, London WC1E 6BT, UK

E.C. DeFreitas, The Wistar Institute, 36th Street at Spruce, Philadelphia, PA 19104, USA

H. Deguchi, The Third Department of Internal Medicine, Osaka University Hospital, Fukushima-ku, Osaka 553, Japan

C. De Préval, Centre d'Immunologie, INSERM-CNRS de Marseille-Luminy, Case 906, F-13288 Marseille Cedex 9, France

M. Dorf, Department of Pathology, Harvard Medical School, Building D-2, Room 467, 25 Shattuck Street, Boston, MA 02115, USA

G. Doria, CNEN-Euratom, Immunogenetics Group, Laboratory of Radiopathology, CSN Casaccia, Rome, Italy

L.M. Douglas, ICRF Tumour Immunology Unit, University College, Gower Street, London WC1E 6BT, UK

Z. Eshhar, Department of Chemical Immunology, The Weizman Institute of Science, Rehovot 76100, Israel

M. Feldmann, ICRF Tumour Immunology Unit, University College, Gower Street, London WC1E 6BT, UK

K. Fischer Lindahl, Basel Institute for Immunology, 487 Grenzacherstrasse, Postfach, 4005 Basel, Switzerland

J. Gamble, The Walter and Eliza Hall Institute of Medical Research, Post Office, Royal Melbourne Hospital, Melbourne, Victoria 3050, Australia

W. Gerhard, The Wistar Institute, 36th Street at Spruce, Philadelphia, PA 19104, USA

S. Gillis, Basic Immunology Program, Hutchinson Cancer Center, 1124 Columbia Street, Seattle, WA 98104, USA

K.-O. Grönvik, Department of Immunology, B.M.C., Box 582, S-751 23 Uppsala, Sweden

J. Grooten, Department of Chemical Immunology, The Weizman Institute of Science, Rehovot 76100, Israel

A. Guimezanes, Centre d'Immunologie, INSERM-CNRS de Marseille-Luminy, Case 906, F-13288 Marseille Cedex 9, France

G. Hämmerling, Institute for Immunology and Genetics, German Cancer Center, Postfach 101949, D-6900 Heidelberg, FRG

D. Hansburg, National Institute of Allergy and Infectious Diseases, National Institutes of Health, Building 10, Room 11D14, Bethesda, MD 20205, USA

C. Healy, Roche Institute of Molecular Biology, Nutley, NJ 07110, USA

E. Heber-Katz, National Institute of Allergy and Infectious Diseases, National Institutes of Health, Building 10, Room 11D14, Bethesda, MA 20205, USA

A. Holliman, Department of Immunolgoy, ARC Institute of Animal Physiology, Babraham, Cambridge, UK

T. Honjo, Osaka University Medical School, Department of Genetics, 3-57 Nakanoshima, 4-Chome Kita-ku, Osaka 530, Japan

A. Ishizaka, The Third Department of Internal Medicine, Osaka University Hospital, Fukushima-ku, Osaka 553, Japan

M. Kanno, Department of Immunolgoy, School of Medicine, Chiba University, 1-8-1 Inohana, Chiba 280, Japan

J.A. Kapp, Department of Pathology and Laboratory Medicine, The Jewish Hospital of St. Louis and Department of Patholgoy and of Microbiology and Immunology, Washington University School of Medicine, St. Louis, Missouri 63110, USA

D.H. Katz, Department of Molecular Immunology, Scripps Clinic and Research Foundation, La Jolla, CA 92037, USA

Y. Kaufmann, Department of Cell Biology, The Weizman Institute of Science, Rehovot 76100, Israel

S. Kishimoto, The Third Department of Internal Medicine, Osaka University Hospital, Fukushima-ku, Osaka 553, Japan

T. Kishimoto, The Third Department of Internal Medicine, Osaka University Hospital, Fukushima-ku, Osaka 553, Japan

S. Kobatake, The Third Department of Internal Medicine, Osaka University Hospital, Fukushima-ku, Osaka 553, Japan

S. Kontiainen, Department of Bacteriology and Immunology, University of Helsinki, Haartmaninkatu 3, SF-00290 Helsinki 29, Finland

H. Koprowski, The Wistar Institute, 36th Street at Spruce, Philadelphia, PA 19104, USA

K. Krupen, Roche Institute of Molecular Biology, Nutley, NJ 07110, USA

A. Linnenbach, The Wistar Institute, 36th Street at Spruce, Philadelphia, PA 19104, USA

P. Lonai, Department of Chemical Immunology, The Weizman Institute of Science, Rehovot 76100, Israel

J. McKimm-Breschkin, The Walter and Eliza Hall Institute of Medical Research, Post Office, Royal Melbourne Hospital, Melbourne, Victoria 3050, Australia

F. Melchers, Basel Institute for Immunology, 487 Grenzacherstrasse, Postfach, CH-4005 Basel, Switzerland

R. Mierau, Institute for Genetics, University of Cologne, Weyertal 121, D-5000 Cologne, FRG

J.F.A.P. Miller, The Walter and Eliza Hall Institute of Medical Research, Post Office, Royal Melbourne Hospital, Melbourne, Victoria 3050, Australia

M. Minami, Department of Pathology, Harvard Medical School, Building D-2, Room 467, 25 Shattuck Street, Boston, MA 02115, USA

D. Mochizuki, Basic Immunology Program, Hutchinson Cancer Center, 1124 Columbia Street, Seattle, WA 98104, USA

P.L. Mottram, The Walter and Eliza Hall Institute of Medical Research, Post Office, Royal Melbourne Hospital, Melbourne, Victoria 3050, Australia

E. Mozes, Department of Chemical Immunology, The Weizman Institute of Science, Rehovot 76100, Israel

A.A. Nordin, Gerontology Research Center, Baltimore City Hospital, Baltimore, MD 21224, USA

L. Nordin, Gerontology Research Center, Baltimore City Hospital, Baltimore, MD21224, USA

I. Nowowiejski, Roche Institute of Molecular Biology, Nutley, NJ 07110, USA

K. Okuda, Department of Pathology, Harvard Medical School, Building D-2, Room 467, 25 Shattuck Street, Boston, MA 02115, USA

T. Oren, Department of Cell Biology, The Weizman Institute of Science, Rehovot 76100, Israel

K. Pfizenmaier, Institute for Medical Microbiology, University of Mainz, Hochhaus am Augustplatz, D-6500 Mainz, FRG

P. Ricciardi-Castagnoli, CNR Center of Cytopharmacology, University of Milan, Via Vanvitelli, 32, I-20129 Milan, Italy

N. Richardson, Department of Immunology, ARC Institute of Animal Physiology, Babraham, Cambridge, UK

F. Robbiati, CNR Center of Cytopharmacology, University of Milan, Via Vanvitelli, 32, I-20129 Milan, Italy

M. Röllinghoff, Institute for Medical Microbiology, University of Mainz, Hochhaus am Augustplatz, D-6500 Mainz, FRG

N.H. Ruddle, Department of Epidemiology and Public Health, Yale University School of Medicine, P.O. Box 3333, 60 College Street, New Haven, CT 06510, USA

A.-M. Schmitt-Verhulst, Centre d'Immunologie, INSERM-CNRS de Marseille-Luminy, Case 906, F-13288 Marseille Cedex 9, France

J.W. Schrader, The Walter and Eliza Hall Institute of Medical Research, Post Office, Royal Melbourne Hospital, Melbourne, Victoria 3050, Australia

R.D. Schreiber, Department of Molecular Immunolgoy, Scripps Clinic and Research Foundation, La Jolla, CA 92037, USA

M.H. Schreier, Basel Institute for Immunology, 487 Grenzacherstrasse, Postfach, CH-4005 Basel, Switzerland

R.H. Schwartz, National Institute of Allergy and Infectious Diseases, National Institutes of Health, Building 10, Room 11D14, Bethesda, MD 20205, USA

O. Shiho, The Third Department of Internal Medicine, Osaka University Hospital, Fukushima-ku, Osaka 553, Japan

G. Skibinski, The Ludwik Hirszfeld Institute of Immunology and Experimental Therapy, Wroclaw, Poland

C.M. Sorensen, Roche Institute of Molecular Biology, Nutley, NJ 07110; Department of Patholgoy and Laboratory Medicine, The Jewish Hospital of St. Louis and Department of Pathology and of Microbiology and Immunology, Washington University School of Medicine, St. Louis, MO 63110, USA

S. Stein, Department of Molecular Biology, Hoffmann-LaRoche Inc, Nutley, NJ 07110

M. Suemura, The Third Department of Internal Medicine, Osaka University Hospital, Fukushima-ku, Osaka 553, Japan

K. Sugimura, The Third Department of Internal Medicine, Osaka University Hospital, Fukushima-ku, Osaka 553, Japan

G. Suzuki, Department of Immunology, Faculty of Medicine, University of Tokyo, 7-3-1 Hongo, Bunkyo-ku, Tokyo 113, Japan

T. Tada, Department of Immunology, Faculty of Medicine, University of Tokyo, 7-3-1 Hongo, Bunkyo-ku, Tokyo 113, Japan

T. Takemori, Institute for Genetics, University of Cologne, Weyertal 121, D-5000 Cologne, FRG

M. Taniguchi, Department of Immunology, School of Medicine, Chiba University, 1-8-1 Inohana, Chiba 280, Japan

M.J. Taussig, Department of Immunology, ARC Institute of Animal Physiology, Babraham, Cambridge, UK

R. Tees, Basel Institute for Immunology, 487 Grenzacherstrasse, Postfach, CH-4005 Basel, Switzerland

W.R. Thomas, The Walter and Eliza Hall Institute of Medical Research, Post Office, Royal Melbourne Hospital, Melbourne, Victoria 3050, Australia

T. Tokuhisha, Department of Immunology, School of Medicine, Chiba University, 1-8-1 Inohana, Chiba 280, Japan

S. Vella, The Wistar Institute, 36th Street at Spruce, Philadelphia. PA 19104. USA

H. Wagner. Institute for Medical Microbiology. University of Mainz. Hochhaus am Augustplatz, D-6500 Mainz, FRG

T. Waks, Department of Chemical Immunology, The Weizman Institute of Science, Rehovot 76100, Israel

J.L. Walker, The Walter and Eliza Hall Institute of Medical Research, Post Office, Royal Melbourne Hospital, Melbourne, Victoria 3050, Australia

D.R. Webb, Roche Institute of Molecular Biology, Nutley, NJ 07110, USA

K.J. Wieder, Department of Molecular Genetics, Hoffmann-LaRoche Inc, Nutley, MJ 07110, USA

Y. Yamamura, The Third Department of Internal Medicine, Osaka University Hospital, Fukushima-ku, Osaka 553, Japan

J. Zeuthen, Basel Institute for Immunology, 487 Grenzacherstrasse, Postfach, CH-4005 Basel, Switzerland

H. Zinger, Department of Chemical Immunology, The Weizman Institute of Science, Rehovot 76100, Israel

MIX
Papier aus verantwortungsvollen Quellen
Paper from responsible sources
FSC® C105338

If you have any concerns about our products,
you can contact us on
ProductSafety@springernature.com

In case Publisher is established outside the EU,
the EU authorized representative is:
Springer Nature Customer Service Center GmbH
Europaplatz 3, 69115 Heidelberg, Germany

Printed by Libri Plureos GmbH
in Hamburg, Germany